Water Conservation

Water Conservation

Edited by **Keith Wheatley**

Callisto Reference
New York

Published by Callisto Reference,
106 Park Avenue, Suite 200,
New York, NY 10016, USA
www.callistoreference.com

Water Conservation
Edited by Keith Wheatley

International Standard Book Number: 978-1-63239-648-8 (Hardback)

Printed in China.

Contents

Preface

I am honored to present to you this unique book which encompasses the most up-to-date data in the field. I was extremely pleased to get this opportunity of editing the work of experts from across the globe. I have also written papers in this field and researched the various aspects revolving around the progress of the discipline. I have tried to unify my knowledge along with that of stalwarts from every corner of the world, to produce a text which not only benefits the readers but also facilitates the growth of the field.

This book unfolds the innovative aspects of water conservation which will be crucial for the progress of this field in the future. Water conservation refers to the study and practice of optimum utilization of water. Although, the population is rapidly growing but the water as a resource is not. Thus, scientists and ecologists argue that soon our society will face a severe water crisis. Therefore, water conservation has become an essential part of our political and social reforms. This book will unravel the recent studies in the field of water conservation. It will discuss the most vital concepts and emerging trends in this field. This book will serve as a reference to a broad spectrum of readers who are interested in the field of water conservation and want to further their knowledge. It includes contributions of experts and scientists which will provide innovative insights into this field.

Finally, I would like to thank all the contributing authors for their valuable time and contributions. This book would not have been possible without their efforts. I would also like to thank my friends and family for their constant support.

Editor

1

Restoration of Damaged Ecosystem Services on Wetlands in Northern Part of the Volga-Akhtuba Floodplain (Russia)

A. V. Lukonina[1], G. Y. Klinkova[2], N. B. Lopantzeva[3]

[1]Botanic Garden of "Volgograd State Socio-Pedagogical University", Volgograd, Russia
[2]Chair of Park-and-Garden and Landscape Construction of "Volgograd State Socio-Pedagogical University", Volgograd, Russia
[3]Lower Volga and Northern Caspian Region of UNDP/GEF-MNR RF Project "Mainstreaming Biodiversity Conservation into Russia's Energy Sector Policies and Operations", Volgograd, Russia
Email: alukonina@rambler.ru, gklinkova@mail.ru, Natalya.Lopantzeva@volgawetlands.ru

Abstract

Due to the regulation of the river Volga flow, expanding of the territories intended for construction and development of agriculture wetlands Volga-Akhtuba floodplain are being significantly transformed and are losing many of their qualities. Non-regulated loads lead to the break of natural physical and chemical characteristics of water bodies, loss of biodiversity, degradation of structure and functions of biological communities and, as a result, to decreasing of potential ecosystem services. Restoration of the lost qualities of water ecosystems is a key factor in increasing sustainability of reclaimed landscapes and creating comfortable conditions for living of locals and their economic activity. On the territory of the biosphere reserve "Nature park" "Volga-Akhtuba floodplain" a number of model sites were defined in order for research projects and practical work on restoration of degraded landscapes to be executed. The widest range of the lost ecosystem qualities is being restored during work on Lake Zapornoye. Restoration of ecosystem functions is being implemented according to the main economic activities carried out in this area. Ecosystem services of the area are specified in accordance with its functional purpose. The existing landscape and ecology conditions have been studied and estimated to see if the planned ecosystem services can be provided. The project suggests the following activities: restoration of hydro-morphological characteristics, reconstruction of habitats for biodiversity and their ecosystem services, improvement to restore and preserve recreational qualities of landscape. The project includes the monitoring the effectiveness of restoration. Monitoring program is based on indices of environmental assessment.

Keywords

Restoration of Wetlands, Ecosystem Services, the Volga-Akhtuba Floodplain

1. Introduction

The Volga river basin is one of the mostly developed areas in Russia. More than 60 million people live here and 45% of industry is located in this area [1].

It is well known that intensive and non-regulated anthropogenic loads affect natural physical and chemical qualities of water bodies, lead to biodiversity loss, degradation of structure and functions of biological communities and, as a result, to the reduction of potential ecosystem services from which people can benefit [2].

In the 20th century chain of power plants was built on the Volga river that drastically changed natural regime of water flow including its lower part—the Volga-Akhtuba floodplain. Being impacted by the hydrological regime change, by expanding of areas intended for building and developing agriculture in the Lower Volga valley, environmental parameters of water bodies are being transformed and many ecosystem qualities significant for people are being lost. Many shallow water bodies and eriks in the floodplain are in need of rehabilitation or full reclamation.

Restoration of water ecosystem qualities that have been lost is a key factor in improving sustainability of developed landscapes and creating comfortable conditions for living and working. Existing and potential economic activities of the area (for example, plant growing, recreation) as well as the regime of its use should be taken into consideration in restoration of degraded water bodies. It helps define range of potential ecosystem services. Estimate of the existing landscape and environmental conditions shows the possibilities of implementation of ecosystem services planned.

As world practice shows, success of this work does not only depend on state politics but also on local authorities, landowners, land users and locals who are concerned about the future of their home land and want to preserve it for their children.

Within the frames of United Nations Development Programme/Global Environment Facility and Ministry of Natural Resources of the Russian Federation (hereinafter referred to as UNDP/GEF-MNR RF) Project "Conservation of wetland biodiversity in the Lower Volga region" (2009-2013) and UNDP/GEF-MNR RF Project "Mainstreaming biodiversity conservation into Russia's energy sector policies and operations" (2014-2015) and grant program "Every drop matters" of UNDP Bureau (in Europe and CIS) and Coca-Cola Russia (2013-2015), employees of Chair of park-and-garden and landscape construction and Botany garden of Volgograd State Socio-Pedagogical University jointly with their partners managed to get research experience and practical skills in restoration of degraded wetlands and their ecosystem services. The projects were implemented by specialist from various spheres: ecology, hydrology, botanic science, zoology, ichthyology, architecture. Proposals suggested within the projects were agreed with representatives of nature protection authorities, municipal entities, land-owners, public.

2. Condition of the Object Today

On the territory of UNESCO biosphere reserve "Nature park" "Volga-Akhtuba floodplain" (Volgograd region) in different functional areas a number of model sites were defined for research and practical work to restore affected landscapes. Lake Zapornoye is among them.

Condition of the object today: Lake Zapornoye is located in the northern part of the Volga-Akhtuba floodplain (**Figure 1**). It is a typical example of degradation because of a high anthropogenic load: the water body is located near a busy road, which is surrounded by settlements and dams are built to secure the settlements in flood periods; the lands adjacent to the water body are ploughed up and are being used by locals for agriculture.

In the past during flood periods water was coming to the lake from north and south which made the lake a flowing water body and kept its sustainability. After the area used for building expanded and a highway was built the northern inflow was totally blocked and the lake no longer had streaming water that result in increasing sediments, bogging and invasion of air and aquatic plants. Based on the space image of 2002 the lake water bed in flood period had the square area of surface 30.3 ha, and in 2014 the square was 7.8 ha. Thus the lakebed area inhabited by helophytes increased from 12.1 ha to 34.3 ha (**Figure 2**). The water body is in critical condition because of a severe eutrophication.

Because the northern part of the lake bottom has been deepened many times, there are almost no submerged macrophytes, fauna of hydrobionts is very poor, blue-green algae is extensively developing. Fish population are rather poor in species, fish capacity is dropping.

Land vegetation is gone through a serious transformation due to a high anthropogenic load. Species tolerant to

Figure 1. Lake Zapornoye.

Figure 2. Area being inhabited by air and aquatic plants. Left-2002, Right-2014.

a heavy anthropogenic impact prevail. Most part of the local flora consists of adventive species which act as subdominants in some areas. Animal species of the model territory comply with those inhabiting the landscape anthropogenically affected.

Thus, Lake Zapornoye has almost lost most part of its ecosystem functions that need to be restored.

3. Activities Planned for Landscape Reclamation

Before you begin to format your paper, first write and save the content as a separate text file. Keep your text and graphic files separate until after the text has been formatted and styled. Do not use hard tabs, and limit use of hard returns to only one return at the end of a paragraph. Do not add any kind of pagination anywhere in the paper. Do not number text heads—the template will do that for you.

Finally, complete content and organizational editing before formatting. Please take note of the following items when proofreading spelling and grammar:

3.1. Restoration of Hydro-Morphological Parameters

Hydrological situation would be cardinally improved provided the lake stream capacity was restored; however, the existing constructions prevent it. Thus, there is a possibility to increase the flood water inflow into the urochische as well as increase the amount of water that is left in the lake basin after flood periods by means of restoring hydro-morphological parameters of the lake by eliminating sediments and alluvium from the lakebed.

Southern part of the lake that is almost totally filled with sediments and covered with air-and-aquatic plants needs to be deepened and, if not an open water surface, than a number of channels providing flood water inflow to the main lakebed during the flood period should be made. As an experiment special machinery partially takes out alluvium and builds "islands" with "channels" between them. Formation of this landscape will lead to increasing littoral habitats and creating a more complicated coastline, filling potential environmental niches.

3.2. Restoration of Degraded Habitats and Biodiversity

A number of bottom dredging works in the northern part of the lake resulted in a simpler morphology of the lakebed: the coastline is straight; the area of shallow water is minimized, deep water areas are growing fast. These factors lead to a fast reduction of diversity of habitats for all biological groups and because of that the ecosystem condition is estimated as very low.

The functioning of the ecosystem is improved and the lost functions are restored by increasing the angularity of the coastline and creating shallow waters of various types by means of land work. This will improve biodiversity indicators and influence the water quality. In order to restore the habitats species typical for the Volga-Akhtuba floodplain are introduced there (the plants were cultivated in Wild flora nursery of VSSTTU Botanic garden).

In accordance with environmental conditions of the landscape to be restored, flood water level restore trees and bushes, floodplain meadow communities, littoral submerged vegetation and helophyte communities.

Trees and bushes support specific groups of dendrophиl biota. In the process of restoration by planting local species dendroflora species are increased that will help attract various birds and insects.

Meadows support a significant group of biodiversity. They determine the typical image and esthetics of the valley, having been formed in specific conditions of the Volga-Akhtuba floodplain. Meadows are very diverse in species and inhabited by plants from different ecological groups. That is why, besides introducing plants on specially prepared sites [3], restoration of vegetation communities includes also replenishing soil with seeds that will provide a complete pool of species by means of distributing hay and seed dust [4]. This technology includes preparation of hay with genesic organs on donor sites in several stages and distributing it in the areas to be restored.

Coastal (air-and-aquatic) plants are habitats for a large group of invertebrate animals and serve as food for phytophilous fish. Because the lake is invaded by air-and-aquatic vegetation (mainly by P*hragmites australis* (Cav.) Trin.ex Steud. (*P. communis* Trin.)) spawning areas for phytophilous fish are significantly decreasing as well as the fish growing area, this also affects development of plankton and bottom living organisms which are the main food for fish, and it also affects oxygen regime. In this connection work is being executed to reduce the lake bed area covered by high helophytes in compliance with protective regime of the biosphere reserve: such as biological reclamation that is introducing herbivorous fish into the lake; mechanical elimination of the vegetation by means of mowing; chemical cultivation using herbicides allowed in Russia to be used in water conservation area of fish farming water bodies. Other helophyte species that are more attractive for hydrobionts are introduced in the habitats to substitute for the eliminated species.

More functional connections of invertebrate hydrobionts being the main food for fish can be traced in associations of submerged soft vegetation. In Lake Zapornoye this phytocenosis is located in small areas. Restoration is done by introducing seeds and primordiums of submerged water plants typical for the Volga-Akhtuba floodplain.

3.3. Territory Improvement for Better Recreational Characteristics of the Landscape

After the bridge over the Volga river was put into operation, recreational load on the floodplain ecosystems started growing very fast. Non-regulated recreational activities result in degradation of the northern part of the floodplain, including Lake Zapornoye which has always been significantly affected by the main highway and close location to Volgograd. One of the ways out from this serious situation is creating specially equipped and recreationally attractive recreation areas with supervised access. Lake Zapornoye can serve as a model area.

Recreational improvement of the territory is making the landscape of this area attractive by means of creating esthetically attractive scenery with the help of vegetation and creating recreational infrastructure. Ecological landscape project of the territory is elaborated to carry out the work within Lake Zapornoye water conservation zone in accordance with recommendations on restoration of biodiversity of aquatic and semi-aquatic plants, hydrobionts and ichthyofauna with the existing anthropogenic load and economic use. To make the landscape more attractive landscape groups were created, structure of the area was made more complicated, the number of ornamental wild plants was increased, habitats for song birds and insects were created. Recreational infrastructure elements were designed (bridges, ramps) as well as viewing spots, beach areas.

Possible positive impact of the activities planned on ecosystem services on Lake Zapornoye is given in **Table 1**.

Table 1. Plan for restoring ecosystem services.

Ecosystem services under restoration/activities planned	Improving hydromorphological parameters and water supply	Creating habitats	Restoration of plant communities				
			Communities of trees and bushes	Elevated meadows	Medium level meadows	Low level meadows	Submerged littoral vegetation
Providing							
Food							
For people			++		+	++	++
For domestic animals				+	++		
Fresh water (capacity)	+++	+					
Wood (firewood)		+	+				
Genetic resources			+	+	+	+	+
Regulating							
Microclimate regulation	+++	+++	+++			++	
Water regulation							
Water purification					+	+	+++
Pollination			+++	+++	++	+	
Supporting							
Soil-formation			++	++	++		
Substance circulation			++	++	++	++	++
Cultural							
Recreation	+++		+++	+	++	++	++
Tourism	+	+++			++	++	++
Educational			++	++	++	++	++
Comfort and esthetics	+	++	+++		+	+	

Pre-project, current and post-project monitoring activities are planned. Monitoring programme is based on environmental indicators elaborated in the region. It is aimed at estimating the effectiveness of work on restoration and further management of the area.

The results of this research and practice work have a great significance for creating practice of restoring degraded ecosystems of the Lower Volga.

References

[1] Klinkova, G.Yu., Selischeva, V.A. and Planeta, M. (2011) Nature Complexes and Biodiversity of the Lower Volga: Workbook, Elaborated, Published and Distributed Free of Charge within the Frames of the UNDP/GEF-MNR RF Project "Conservation of Wetland Biodiversity in the Lower Volga Region" with Financial Support of Global Environmental Facility, 256.

[2] Millennium Ecosystem Assessment (2005) Ecosystems and Human Well-Being. Synthesis Report, Island Press, Washington DC, 160.

[3] Klinkova, G.Yu., Lukonina, A.V., Planeta, M., *et al.* (2014) Restoration of Floodplain Forest Ecosystems: Study Guide Published within the Frames of the UNDP/Coca-Cola Grant Project "Every Drop Matters". 128.

[4] Dzybov, D.S. (2001) Agrosteppe Method: Fast Restoration of Natural Vegetation. Study Guide. Publishing House Nauchnaya kniga, Saratov, 40.

2

Strategy & Techniques for Restoration of Healthy Aquaecosystem from Toxic Super Eutrophic Water Body

Peimin Pu[1], Jiangping Pu[2]

[1]Nanjing Institute of Geography & Limnology, Academia Sinica, Nanjing, China
[2]PLA University of Science and Technology, Nanjing, China
Email: pupm2@qq.com, pupm1@yahoo.com

Abstract

There are complex river-lake systems in the Taihu Lake catchment with total water surface area of 6174.7 km^2, and population density of 1079/km^2, including Taihu Lake water surface area of 2338 km^2. The water systems in this catchment have healthy aquaecosystems during long history. However, in some riverlets in this catchment the water quality was estimated as "acute toxicity for higher organisms" and over standards for many heavy metal elements content; there were no any living plants and macro organisms in the water body, because there were developed a series of industry with abundant release of heavy metals and difficult decomposition organic chemical components along the riverlets during last decades. The even more serious situation was observed in sediments of the riverlets. How to restore such riverlet into a healthy aquaeosystem with abound plants and higher organisms? The main strategy and techniques are described in this paper as summarizing a report of engineering in a riverlet in Wuxi New District during last years, which leads to restore the aquaecosystem into a healthy one with abundant surface plant cultured on floating islands and observed living fish, lobster, frog, toad, mollusk and others in the riverlet. The main techniques are: 1) softwall buffer technic; 2) floating eco-island technic by using which can culture any plant which can be cultured in solution; 3) immobilized nitrogen cycle bacteria (INCB) technic; 4) tattering esters and other big-molecule organic chemicals by using electronic pulse technic and photosensitization technic; 5) mist spray facility technic for improving dissolved oxygen in deep water layers; 6) technic for buffering and suppressing H_2S release from water; 7) the appropriate portion of surface with cultured plant to the total water surface area is about 1/3; 8) Cress [*Oenanthe Ljavanica* (Bl.) DC.] and *Myriophyllum verticilatum* L. may be cultured in Taihu Lake catchment during the whole year as main plants with mosaic combination of other supplement plants in different seasons.

Keywords

Restoration of Healthy Aquaecosystem, Toxic Water Body, Electronic Pulse Technique for Decomposition Esters, Photosensitization for Decomposition of Organic Matters, Floating Eco-Islands, Immobilized Nitrogen Cycle Bacteria

1. Introduction

Taihu Lake has water surface of 2338 km^2. There are complex river-lake systems in its catchment with total water surface area of 6174.7 km^2, and population density of 1079/km^2. The water systems in this catchment have healthy aquaecosystems during long history. The river-lake systems were used as the easiest way for storing all the industrial, agricultural, and domestic wastewater. The pollution loading was exceeded usually the purification capacity of the water body. Therefore, the water quality in the riverlet becomes worse and worse at the densely populated and economic developed zone, spot by spot. The aquaecosystems in the riverlet deteriorate and the purification capacity decreases gradually. The vicious cycle of aquaecosystem deterioration happened during last decades. The water plants disappear gradually. Many pollutants store at the bottom and usually release detrimental solid, dissolved liquid, and gaseous (such as toxic H_2S) substances. As an example, there is a riverlet in Wuxi New District, in which there were no any living plants and macro organisms in the water body, because there were densely populated and developed a series of industry with abundant release of heavy metals and difficult decomposition organic chemical components along the riverlets during last decades. The even more serious situation was observed in sediments of the riverlets.

How to restore such riverlet into a healthy aquaeosystem with abound plants and higher organisms? The usual approaches for restoring eutrophic/hyper-eutrophic water system could not be used in such case, because of macrophyte could not live in such a water body. The new strategy and techniques should be developed. For this purpose, a demonstrative project has been performing since 2011. The main results are summarized in this paper.

2. Background Information

2.1. Geographic and Hydrological Situation

The riverlet segment of the project is located in Wuxi New District in East-West direction in Taihu Lake catchment, belonging to the Wangyuhe river system. It has width of 12 - 20 m and length of 1000 m. The across section has U shape with steep coasts and 3.5 m deepest bottom under normal water level. The hydrological conditions are related with the surroundings and rainfall. The maximun amplitude of water level reaches round by 1 m during heavy rainfall (it was 1.5 m during last rainfall 16 - 17 June 2015, while all the facilities of project were safety). The current direction is changeable with maximum speed of over 15 cm/s. The bottom was dredged several years ago.

This riverlet had a healthy aquaecosystem with very good water quality and dense macrophyte population a couple decades ago. But the water quality becomes worse and worse in relation with rapid development of a series of industry with abundant release of heavy metals and difficult decomposition organic chemical components and rapid population growth along the riverlet during last decades. Macrophyte in the riverlet disappeared gradually in recent years.

2.2. Water Quality State in the Beginning of the Project

The water colour was dark black with frequent release of bad smell of H_2S, should work there with gas mask. The water surface was covered with oil and esters frequently, especially when the bubbles appeared on the surface. The COD_{Cr}, TP, TN was measured as 372 - 8770 mg/L, 4.74 - 13.20 mg/L, and 34.0 - 55.6 mg/L separately at 6 sites along the reverlet on 1 November 2012. In the dense area of esters The COD_{Cr} reached 14,911 - 28,448 mg/L on August 2011.

The detected data by ICP-MS show the concentration of Al, Mn, Fe, Ni, Pb is exceed the national standards, especially the Al concentration is 5 - 19 times more than national standard.

The biological acute toxicity of the water samples was determined by standard method of GB/T15441-1995 and was estimated according to the American State Standard as "acute toxicity for higher organisms".

3. Technical Approaches

3.1. Scheme of the Technical Approaches

The high polluted water has non-uniform pattern usually, as water masses with different characteristics. We should take different approaches for different water masses, which are listed in the scheme as follows:

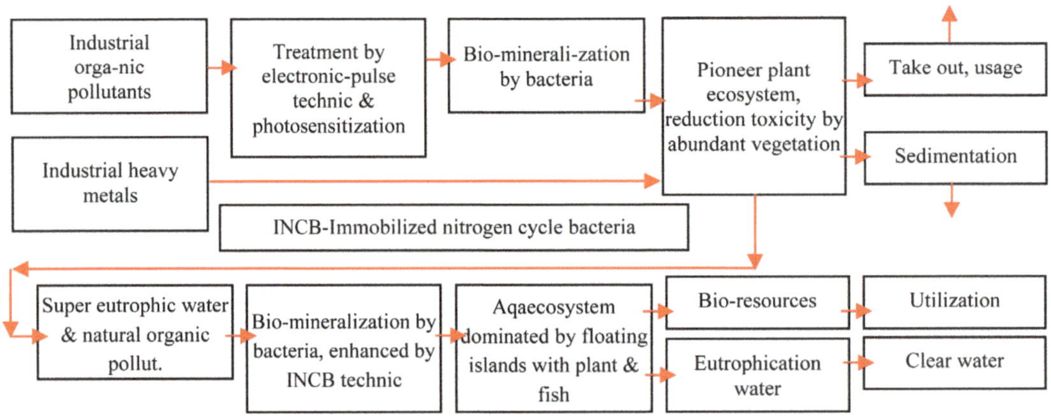

3.2. Main strategy and technics

The main strategy and technics are:

1) Limit the high concentration pollutants at the sites of their release in as small area as possible, and protec the better ecosystems, decreasing the negative influences of worse water mass, controlling the exchange rate between the masses, by using softwall technic [1]-[3];

2) Take different treatment technics for different water masses;

3) Take complex enhanced measures, so the threshold would be exceeded for restoring healthy ecosystem and improving water quality in a limited area;

4) The horizontal non-uniform distribution method could be used for improving vertical exchange of water characteristics. For example, the water temperature in the open place is higher than that under the floating eco-islands during sun shine day, the non-uniform horizontal water temperature will generate the vertical exchange currents, and oppositely, the water temperature in the open place is lower than that under the floating eco-islands during night, the vertical exchange movement will occur in opposite direction;

5) The mosaic distribution of plants benefits for preventing plant diseases and insect pests and absorbing nutrients in balance for ecosystem stable during season's exchange;

6) Because of the industrial organic matters could not be easy destroyed by bacteria, so, the industrial organic pollutants should be removed by management and destroyed into organic matters easy be biomineralized by bacteria, by using electronic-pulse technic and photosensitization, then; the nutrients be absorbed by plants partly, cultured on the floating islands [4], reducing the toxicity of water quality; and in this case the water quality would be the similar one of the super eutrophic water;

7) The super eutrophic water would be bio-mineralized by bacteria, enhanced by immobilized nitrogen cycle bacteria (INCB) technic [5] [6];

8) Construction of aquaecosystem dominated by floating islands technic with plant and total planting area of about 1/3 to the total water area as the optimized ratio;

9) The oxygenation technic and technic for removing hydrogen sulfide (H_2S) in water should be used in case, if necessary.

3.3. Details for Some Technics

1) Electronic-pulse technic for tattering esters and other big-molecule organic chemicals. The electric voltage of 24 VDC in the riverlet was used for safety in the riverlet. The special frequency, intensity of the pulse, ratio of pulse width/interval and electrode substances (especially the substance for anode) should be well tested and chose. The esters become more diluted and easier to be further decomposed after a few hours treatment.

2) Photosensitization technic for absorb esters and oil by using plant, which could died after attaching oil and transformed into humus. The humus may intensively absorb the photo energy with wave length less than 500 nm, part absorbed energy of that would be transported to the artificial synthesis substances and acting as the decomposition factors. The esters and oil round the dead plant disappear gradually. The organic matters decomposed by bacteria and should be absorbed by other living plant gradually.

3) There are no submerged plants, so the dissolved oxygen is transported from the air-water interface. The

oxygen consumption process in water, especially in lower layer of the riverlet leads to oxygen deficit in lower water. Oxygenation by air pump in deep layer needs to overcome the water pressure and to create as small as possible air bubbles. A fact had been revealed that the dissolved oxygen in the water collected from water-drops after mist spray of water from very low saturation level of dissolved oxygen (such as ≤0.2 mg/L) has oversaturation of 108%. This kind of oversaturated water can be collected in the bucket floating on the water surface with a water level in it higher than that of the water level in the river/reservoir. Then the water in the bucket with high dissolved oxygen can be lead to the deep layers in the river/reservoir by a tube, using the effect of syphonage.

4) Controlling of hydrogen sulfide (H_2S): There are so many esters and other organic pollutants stored in bottom of the riverlet, which would release toxic H_2S during anaerobic decomposition process. If we use $Fe_2O_3{\cdot}H_2O$ during oxygenation, the H_2S may be transformed into Fe_2S_3 and FeS as follows:

$Fe_2O_3{\cdot}H_2O + 3H_2S \rightarrow Fe_2S_3{\cdot}H_2O + 3H_2O$ $\Delta H = 63$ KJ/mol; $Fe_2O_3{\cdot}H_2O + 3H_2S \rightarrow 2FeS + S + 4H_2O$ $\Delta H = 103$ KJ/mol.

We may use the industrial desulfurizer to absorb the H_2S, if it appears as follows:

$2Fe(OH)_3{\cdot}xH_2O + 3H_2S \rightarrow Fe_2S_3{\cdot}xH_2O + 6H_2O$ (desulfurization)

$Fe_2O_3{\cdot}xH_2O + 3H_2S = Fe_2S_3{\cdot}xH_2O + 3H_2O$ (desulfurization); $Fe_2S_3 = 2FeS + S$ (decompozation), and to regenerate it when the dissolved oxygen is enough, or take it into air condition:

$Fe_2O_3{\cdot}xH_2O + O_2 = Fe_2S_3{\cdot}xH_2O + 3S$ (regeneration); $2FeS + O_2 + xH_2O = Fe_2O_3{\cdot}xH_2O + 2S$ (regeneration)

For this purpose, we may set the industrial desulfurizer in the net tubelets, which are fixed on the industrial textile for setting on the submerged islands (**Figure 1(a)**). All these facilities function as buffering and suppressing H_2S release from water. The desulfurizer becomes black colour after absorbing H_2S (**Figure 1(b)**), and returns to brown colour after regeneration in air/or in water with high dissolved oxygen concentration (**Figure 1(c)**).

5) Distribution of islands for culturing plants: A synthetic fibre rope net can be constructed according to the riverlet shape at site for fixing the islands. The floating eco-islands (2 m*4 m) [4] [6] occupy the surface area with appropriate ratio about 1/3 to the total water surface. The other 2/3 surface may be occupied by submerged islands in case necessary for restoration submerged macrophyte (coloured by blue rectangles). The distribution scheme of the islands is shown in **Figure 2**.

6) Mosaic culture of plants: Cress [*Oenanthe Ljavanica* (Bl.) DC.] and *Myriophyllum verticilatum* L. may be cultured in Taihu Lake catchment during the whole year as main plants with mosaic combination of other supplement plants in different seasons. The toxic super eutrophic riverlet had been restored to a healthy aquaecosystem with abundant surface plant cultured on floating islands and observed living fish, lobster, frog, toad, mollusk and others in it-a healthy aquaecosystem (**Figure 3**).

4. Purification Effects of the Aquaecosystem Dominant by Surface Plant

The purification effects of the aquaecosystem dominant by surface plant may be summarized in **Figure 4**.

The density of heavy metal decreased above a couple of times to maximum of 25 - 43 times (**Figure 4(a)**). For example, the cress may increase DO with mean value of 0.63 mg/L, or 74.7% during 20150402-20150519, and occasionally up to 2.5 mg/L. The TDS (total dissolved solid) can be used as an indicator of density of nutritive salts-pollutants. In certain period the high density pollution was loaded (**Figure 4(c)**, red ellipsoid) with maximum

(a) (b) (c)

Figure 1. Setiing the industrial desulfurizer in the net tubelets, which are fixed on the industrial textile for setting on the submerged islands (a), desulfurization in water with H_2S (b), and regenerization after 2^h 27^m in air (c).

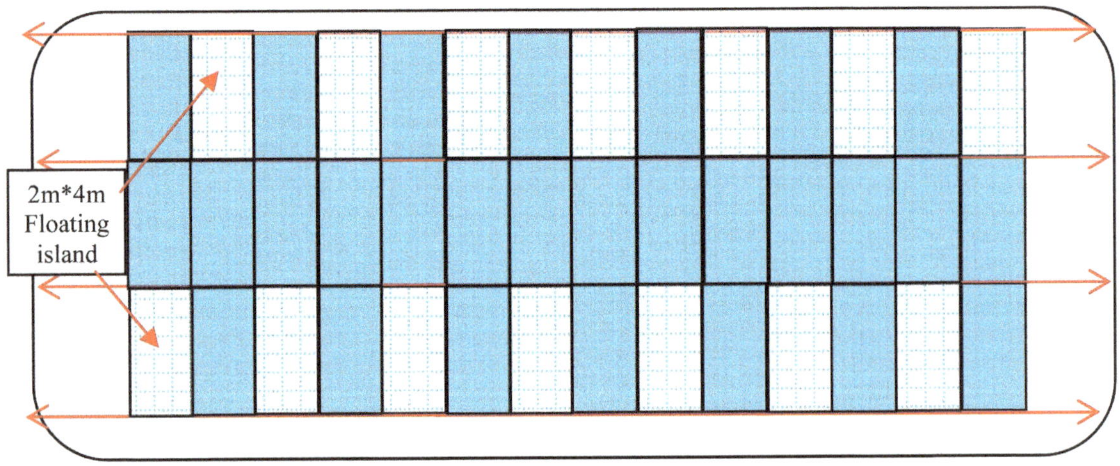

Figure 2. Scheme for distribution of floating eco-islands (2 m * 4 m), occupying the surface area with appropriate ratio about 1/3 to the total water surface. The other 2/3 surface may be occupied by submerged islands in case necessary for restoration submerged macro-phyte (coloured by blue rectangles).

Figure 3. Abundant surface plant cultured on floating islands and observed living fish, lobster, frog, toad, mollusk and others in it—a healthy aquaecosystem.

of TDS = 897 mg/L (average along the riverlet on 20141227). It was 252 mg/L (average along the riverlet on 20150516), decreased on 3.6 times. The square of correlation coefficient between TDS under plant and in adjacent riverlet was 0.939 (**Figure 4(d)**). There are reports showing that the cress may remove TP 74.8% and its root may release oxygen and other wetland plants have similar function: releasing oxygen and absorbing nutrients and heavy metals [7]-[9].

The monitoring data on COD_{Cr}, COD_{Mn}, TP, TN, NH_4^+, etc. showed that they were high/very high at the pollution loading entrances into the riverlet and were as low as standards for landscape reuse water. It seems we have had constructed a sewage plant. It needs about ¥10^7 for construction and ¥1650/d for daily management, as the population in this area is about 10,000, according to the data for adjacent Meicun sewage plant [10]. This is much more expensive than that for performing this restoration aquaecosystem project and its maintenance. As there are abundant cress and other vegetation on floating eco-island, which purify the water and contain nutrition and less concentration density of heavy metals, leading to decrement of toxity, and therefor, can be used as the natural resources. There are about 3837 km^2 riverlets—small lakes in Taihu Lake catchment. If we can use a part of these water surface, we may organize and develop an agricultural-environmental industry.

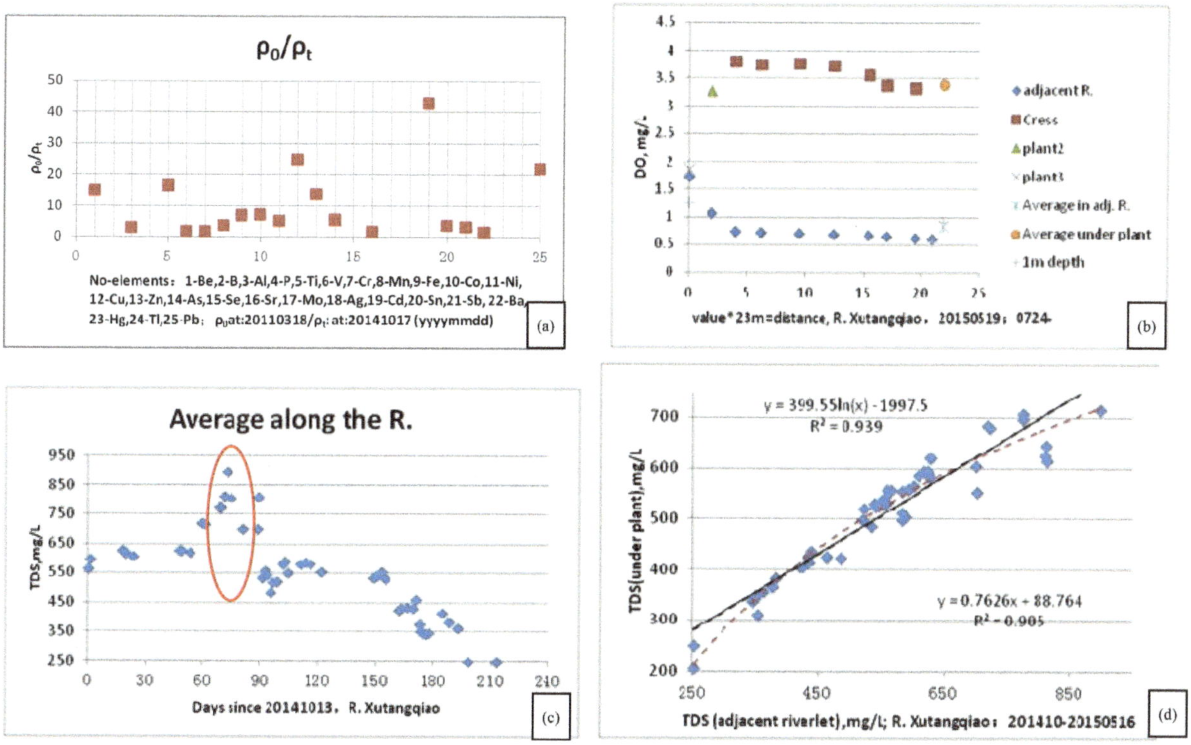

Figure 4. Improvement of water quality in ratio of original density of heavy metal ρ_0 to recent ρ_t (a); Dissolved oxygen under plants and in adjacent R. (b); Variation of average TDS along the R. (c); Relationship between TDS under plant and in adjacent R. (d).

5. Summary

Treatment of industrial pollution is essential for improving environment and should be execute strictly. The domestic and agricultural pollution is difficult to collect, especially during heavy rainfall and flood. The riverlet—small lake net is the natural resource with various functions in history, but it becomes pollutants reservoir at present in many regions. The results reported in this paper may serve as one of the possible approaches with high cost performance for restoring the historical landscape.

Acknowledgements

The authors acknowledge the leadership of Hongshan district office for supporting this project and Dr. Xiaoyin Cheng, Jiangnan University, for her and her lab's help in chemical analysis during performance of this project.

References

[1] Pu, P.M., Hu, W.P., Pang, Y., Wei, Y.C., *et al.* (1993) A Physico-Ecological Enginering Experiment for Purifying Raw Water Quality in Lake. *J. Lake Sciences*, **5**, 171-180.

[2] Pu, P.M., *et al.* (1993) Physic-Ecological Engineering for IMPROVING Taihu Lake water Quality in Intake Area of Mashan Drinking Water Plant. *Proceedings of* 5*th International Conference on the Conservation and Management of Lakes*, Stresa, 480-483.

[3] Hu, W.P. and Pu, P.M. (1995) Study on the Ecological Efficiency of Soft Enclosure in Lake Water. 6*th International Conference on the Conservation and Management of Lakes-Kasmigaura*'95, 1886-1889.

[4] Pu, P.M. (2011) The Role of Multifunctional Eco-Island Technic in Remediation for Healthy Aquaecosystem. Learning and Gathering of China. China Federation of Press, Beijing, 408-414.

[5] Li, Z.K. and Pu, P.M. China National Invitation Patent: ZL 2005 1 0038831.9 and ZL 2005 1 0038832.3.

[6] Pu, P.M. Application for China National Patent: No. 201110035542.9 and 201120048592.6.

[7] Liu, Z.K. (2010) Research on Oxydation of Root System of Wetland Plants and Its Impact on Removal of N, P Nu-

trient in Dianchi. Thesis for Master Degree, Taiyuan Polytechnic University.

[8] Jaynes, M.L. and Carpenters, R. (1986) Effects of Vascular and Nonvascular Macrophytes on Sediment Redox and Solute Dynamics. *Ecology*, **67**, 875-882. http://dx.doi.org/10.2307/1939810

[9] Li, G.H. and Liu, J.G. (2010) Root Oxygen Release and Heavy Metal Uptake of Various Wetland Plants. *Water Resources Protection*, **26**.

[10] Shen, X.L., Hang, W.L., Hu, B., Zhou, M.W. and Sun S.R. (2009) Engineering Design for Second Face of Meicun Sewage Plant, Wuxi City, Jiangsu, China. *Symposium on New Technics and New Technologies for Upgrade of Waterworks and Sewage Plant and Energy Conservation and Carbon Emission Reduction*, 52-56.

3

Application of Google Earth in Modern River Sedimentology Research*

Xuewen Zhou, Hongliang Wang#

School of Energy Resources, China University of Geosciences, Beijing, China
Email: zhouxw_zxw@qq.com, *whliang@cugb.edu.cn

Abstract

Google Earth, a software based on satellite's images and database of navigation, gains features of high resolution, integrated images, quick update, convenience, simpleness and free cost. Practice of this software shows that it plays an important role in mapping, construction and river research. Based on the fundamental features of Google Earth, this passage makes an introduction to its application in studying modern river sedimentology through the case study of the Yellow River. The results show that Google Earth satellite's high-resolution images and image overlay functions make it easy for users to quickly navigate river, determine the type of river, track the river flows and measure the terrain slope; the software's "Ruler" and "Add Path" functions make it easy for users to measure channel width, curvature, amplitude, wavelength, size, morphology of point-bar and other river parameters; the software's "Historical Image" function has important significance in the study of modern river migration, sedimentary evolution and river geomorphological shape during different seasons.

Keywords

Google Earth, River Type, Terrain Slope, River Parameters, Modern River Evaluation

1. Introduction

Google Earth is a set of software based on satellite and GPS data, which allows users to view satellite image, maps, terrain and 3D buildings on a 3D interactive sphere [1], and to measure the length and height of target objects as well. As one of the most popular digital earth platforms at present, it has characteristics of high resolution, integrated images, quick update, convenience, simpleness and free cost. Practice of the software found that

*Fund project: National Undergraduates Innovating Experimentation Project (A): China University of Geosciences Innovation and Entrepreneurship Training Project, project number: 201411415024.
#Corresponding author.

it can play an important role in many previous studies, including application in construction [1]-[4], surveying and mapping [5]-[7], setting up the database [8] and visualization model [9]. In the study of rivers, Alice Goudie [10] used Google Earth to measure the stream distribution and geomorphological characteristics of salt marsh in England and Wales. Result shows that the density and the bending of a stream are closely related with tidal range. Shi Shuyuan [11] put forward the method to build geological knowledge database of meandering river based on the Google Earth software. Li Yupeng [12] used Google Earth software to measure modern rivers and concluded that the river width had a positive correlation with the length of point-bar. He pointed out that it had a function of guiding the quantitative analysis of reservoir configuration under the point-bar of meandering river. Zhang bin [13] measured and defined the parameters of meandering river based on the Google Earth software. He pointed out that the deep meander of Jialing River was one of the world's most complex, irregular and curved channels. Zhang Changmin [14] used Google Earth software to study modern deposition of Dongting Lake and Poyang Lake. He found two types of delta. The application of Google Earth to river geomorphology and sedimentology shows that it does have a lot of uses, especially in the study of sedimentary and geomorphic features of rivers in macro. However, when the previous uses Google Earth to study the modern river sedimentology, they often only focus on a single aspect. So on the basis of characteristics of Google Earth software, this paper uses the Yellow River as an example, introducing the application of Google Earth software to the measurement of river types, terrain slope, river parameters and analysis about evolution of modern rivers.

2. Google Earth Overview

2.1. Software Introduction

Google Earth is virtual globe software which was first launched by Google Company in 2005. It decorates the satellite images, aerial photography and GIS on a 3D model of the earth [15]. Users can get free high-resolution satellite images around the world through the Google Earth software.

2.2. Image Accuracy

Google Earth integrates satellite images with aerial data. Parts of satellite images are from the Digital Globe and Earth Sat companies. Aerial images are derived from the Blue Sky, Sanborn, IKONOS and SPOT5 companies. The SPOT5 Company can provide images with the resolution of 2.5 m. The IKONOS Company can provide image with the resolution of 1 m, and the Quick Bird Company can offer 0.61 m high-precision image which is the highest level of global business [3] [16]. High-precision images are advantageous to the meticulous research of river types and sedimentary characteristics. Although there is a certain error when using the Google Earth software to extract related parameters of the river, the influence to macro geomorphology research can be ignored.

2.3. Platform Interaction

Google Earth is not only a satellite image browsing software, but also an interactive platform which combines satellite images, maps with related information and communication. The interaction of this platform shows in two aspects: "Layer" and "Landmark" function. The "Layer" function is similar to the layer management function of AutoCAD software. It can classify and manage information through different layers. Each layer can be displayed on and off in need to maintain a clear and clean interface. The "Landmark" is the address label, it can be inserted to the map in order to make annotations, reminders. The "Landmark" can be saved and exported as a separate file [5] [17].

3 Application of Google Earth in the Study of Modern River Sedimentology

3.1. River Classification

In the study of modern river sedimentology, different types of river have different features of depositional model and geomorphologic shape. With the launch of Google Earth software, users can quickly determine the river types by observing sedimentary and geomorphological features of rivers through high-resolution satellite images. In the field of sedimentology, rivers are divided into braided river, meandering river, anastomosing river and Straight River. There is great difference in their macro geomorphological and sedimentary characteristics (**Table 1**).

Braided River (**Figure 1(a)**) has features of flat valley, low curvature, and high terrain slope. It has many channel bars in the whole valley and its channel branches off around diara. It is like intertwined "braid" in the valley and the channels and bars are very unstable, constantly migrating its path in the deposition process.

Meandering River (**Figure 1(b)**) has features of single channel, high curvature, and low terrain slope. Its riverbank can resist strong corrosion. The entire deposition process is that the concave bank constantly being eroded while the convex bank keeps accumulating deposition. Point bars can see clearly on the Google Earth satellite images.

Anastomosing River (**Figure 1(c)**) has features of multi-channel, straight or curved, low terrain slope. It has stable riverbank composed of fine-grained sediments (silt or clay) and vegetation. These channels are separated by river flood plain composed of natural levee, Vegetation Island and wetland. The difference between the braided river and the anastomosing river is that the anastomosing channels are stable while the braided channels are unstable.

Straight River (**Figure 1(d)**) has features of single channel, low curvature. If the channel's curved index < 1.5, it is Straight River. It can be used as criteria to distinguish Straight River and braided river.

In the actual study, steps of using Google Earth to divide river types are as follows:

1) Open client software, type the name of the river or place that river flowing through in the upper left corner

Table 1. Comparison table of geomorphological and sedimentary characteristics of different rivers.

River Type	Curvature	Channel Morphology	Channel Stability	Channel Sand	Abandoned Channel Characteristics
Braided River	Low	Two or More Channels; Sand Islands and Sandbanks	Very Unstable, Rapid Migration	Channel Bar	No Oxbow Lake
Meandering River	High	Single Channel	Gradually Lateral Migration	Point Bar	Oxbow Lake
Anastomosing River	Low-Medium	Two or More Channels; Large Stable Sand Islands	Stable	No Channel Sandbanks; Little Point Bar	No Oxbow Lake; Abandoned Channel
Straight River	Very Low-Low	Single Channel	Unstable	Unstable Point Bar; Alternate Distribution	No Oxbow Lake

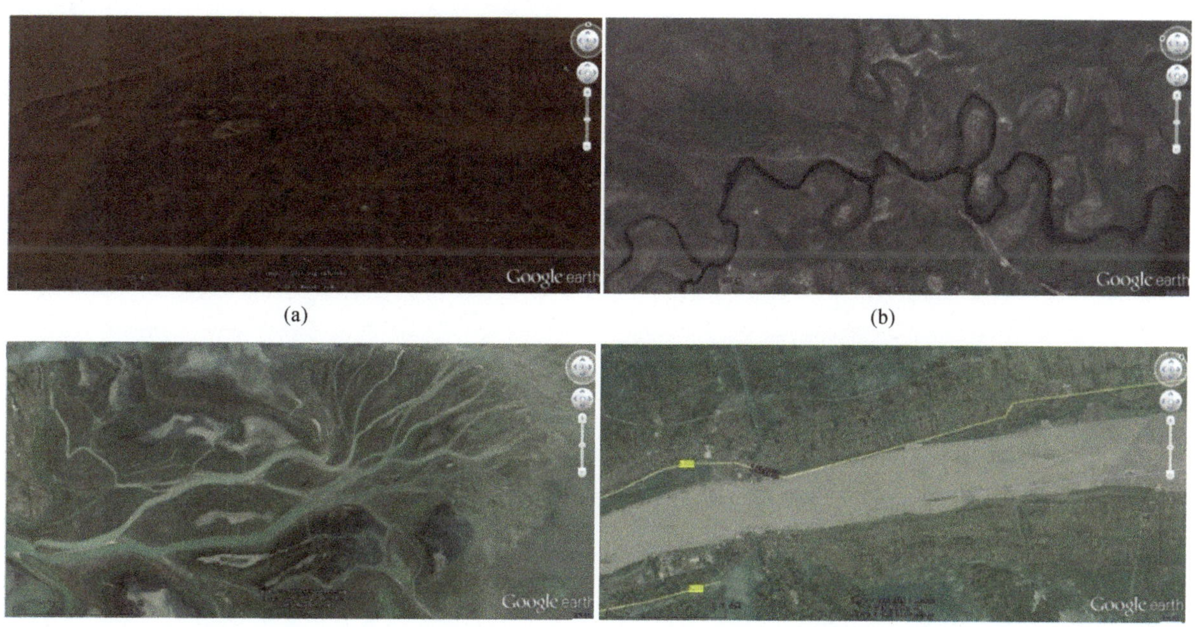

(a) (b) (c) (d)

Figure 1. Satellite's image of different types of river. (a) Braided river has many channel bars in the whole valley and its channel branches off around diara; (b) Meandering River has features of single channel, high curvature, low terrain slope; (c) Anastomosing River has features of multi-channel, straight or curved, and has stable riverbank composed of fine-grained sediments (silt or clay) and vegetation; (d) Straight River has features of single channel, low curvature.

of the search field to locate river (**Figure 2**). At the same time, users can also create corresponding landmarks in order to find it again quickly if necessary;

2) Find the research section of the river on Google Earth satellite images, combine geomorphological and sedimentary characteristics of different types of river (**Figure 1** and **Table 1**) to determine the its type.

According to the methods above, the river type of the Yellow River from Baihe Town (34°51'35.74"N, 112°33'41.26"E) to Zhengzhou City, Garden Town (34°54'20.40"N, 113°42'4.78"E) is braided river. The bifurcation index of the river is high, and its channel bar is significant (**Figure 3**).

3.2. Terrain Slope Measurement

In the study of fluvial sedimentology and geomorphology, the terrain slope is one of the significant factors controlling river types. In general, the terrain slope of braided river is big, while that of meandering river and anastomosing river is small. Measuring the terrain slope can be used to explore the relationship between different types of river and terrain slope in different river basins. The terrain slope, one of the basic parameters of the river, can provide data for the development of water conservancy projects directly when it converted to basin slope. However, the lack of data of early aerial photo limits some scholars' research and measuring work. With the launch of the Google Earth software, users can quickly locate rivers, track the flow direction and batch obtaining altitude data via Google Earth online images. It provides great convenience for the measurement of terrain slope.

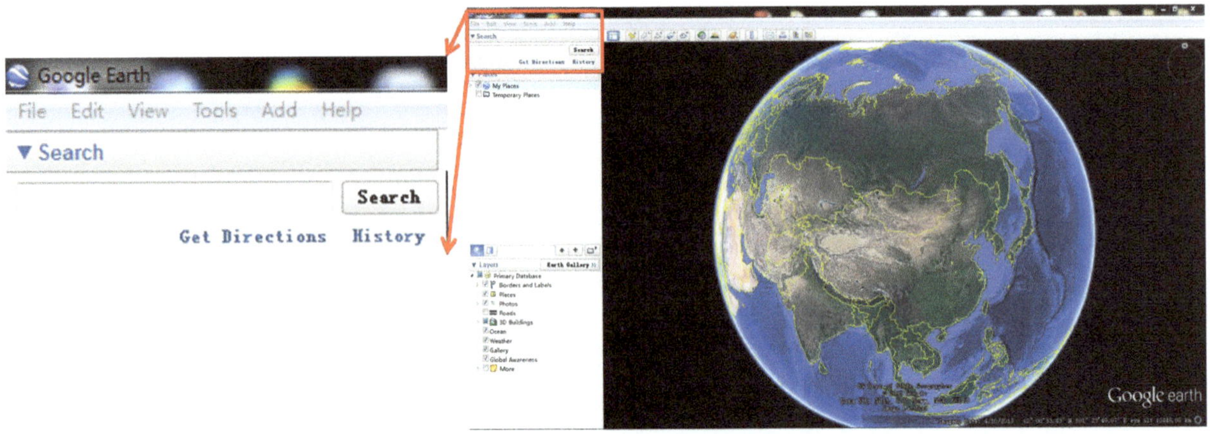

Figure 2. Schematic diagram of searching and navigating river. The red box of the image shows the window of searching for places.

Figure 3. Analysis chart of river type of Yellow River from Baihe town to Zhengzhou city, garden town. the channel bars of the braided river can be seen clearly in the red box of the image.

In the actual study, steps of using Google Earth to measure terrain slope are as follows:

1) Use the search tool to find the selected rivers, and then observe the flow direction and characteristics of the river basin before selecting target parts of river for measurement. The Google Earth software does not display the rivers' name, which brings great inconvenience to the researchers when tracking the rivers. So users can use layer overlay function to superpose third-party street map directly on the Google Earth image to improve efficiency;

2) Create the starting "Landmark a" and ending "Landmark b" on the Google Earth image of the selected river;

3) Read latitude, longitude and altitude Ha, Hb of the two landmarks respectively and measure the "Distance L" between the two landmarks through the "Ruler" function in tool bar (**Figure 4**);

4) Calculate the corresponding terrain slope θ based on the relationship of trigonometric functions:

$$\theta = \arcsin\left(\frac{Ha - Hb}{L}\right).$$

Firstly, use one braided reach of the Yellow River from Zhengzhou City, Sun Village to Garden Town as an example (**Figure 4**, **Table 2**) and measure the terrain slope of this braided river. The altitude of starting "Landmark a" (113°34'32.69"E, 34°56'57.39"N) is 94.368 m and that of ending "Landmark b" (113°41'53.64"E, 34°55'1.34"N) is 91.102 m. Secondly, use the "Ruler" tool to measure the "Distance L" between two landmarks and the result is 11.73 km. Lastly, put the measured data into topographic gradient calculation formula in the step 4) to calculate the terrain slope and the result is 0.016°.

3.3. River Parameters Measurement

Google Earth satellite images gain features of high resolution, high precision. So in modern river sedimentology research, using Google Earth software "Add Path", "Ruler" tool, can directly measure the river's width, curva-

Table 2. Calculation table of terrain slope of braided river in the yellow river from Zhengzhou city, sun village to garden town.

River Length L (km)	Landmark a			Landmark b			Height difference Ha-Hb (m)	Slope (°)
	Longitude (E)	Latitude (N)	Altitude (m)	Longitude (E)	Latitude (N)	Altitude (m)		
11.73	113°34'32.69"E	34°56'57.39"N	94.368	113°41'53.64"E	34°55'1.34"N	91.102	3.266	0.016

Figure 4. Schematic diagram of measuring terrain slope, the red line connecting two yellow landmarks is the "Distance L" that needs to be measured.

ture, amplitude, wavelength, point bar size and shape and other river parameters.

In the actual study, the main measurement steps of measuring the meandering river's point bar parameters are as follows:

1) Find the meandering river's fluvial deposits point which is need to be measured on the Google Earth satellite image and select the appropriate resolution;

2) Use the "Add Path" tool to delineate river unit which is under measurement on Google Earth satellite images;

3) Use "Distance Measurement" function in "Ruler" tool under the model of measurement tool, add an anchor point-by-point along the curve with the mouse and finally measure the length of the curve.

According to the methods above, use a point bar (37°45'28.04"N, 119°8'57.74"E) near the Yellow River estuary as an example (**Figure 5**). Then use Google Earth software to select the appropriate set of measurement points. The result of the measurement is that the channel width is 0.29 km, the length of point bar's arc is 2.19 km and the diameter of it is 0.97 km.

3.4. Modern River Evolution Analysis

Google Earth has the function of checking the historical satellite images. This function is of great importance to the study of modern rivers changes, sedimentary evolution and the river geomorphology in different seasons. At present, the Google Earth can check the earliest historical images in 2000. However, there will be a slight difference in the time of year when checking the historical images in different regions.

In the actual study, the steps of checking historical images are as follows:

1) Find the river on the Google Earth satellite image, and keep the interface stable;

2) Click the "View" column, find the "Historical Image", and check the item;

3) The time axis will appear in the upper left corner, and the two sides of the axis are the earliest and the latest year's images of Google Earth respectively. Move the axis pointer to view the corresponding year's historical images.

Use one meandering reach of the Yellow River near Fengqiu County, Yangzhai Village (34°56'35.32"N, 114°27'14.83"E) as an example. The red lines in the figure are the stable dams fixed on the both sides of the Yellow River, so they are used as the reference in the study of modern river evolution. According to the Google Earth's historical images of this meandering reach in different years from 2000, it can be found that the meander near the dam which is built on the upper left side migrated towards east from 2002 (**Figure 6(a)**) to 2006 (**Figure 6(b)**).There was little changes to the dam on the upper left side while the two meanders near the lower

Figure 5. Schematic diagram of measuring point bar's parameters of the Yellow River. The red line is the arc length. The yellow landmark is the position of the point bar. The green arrow is the channel width. The blue line is the diameter of the point bar.

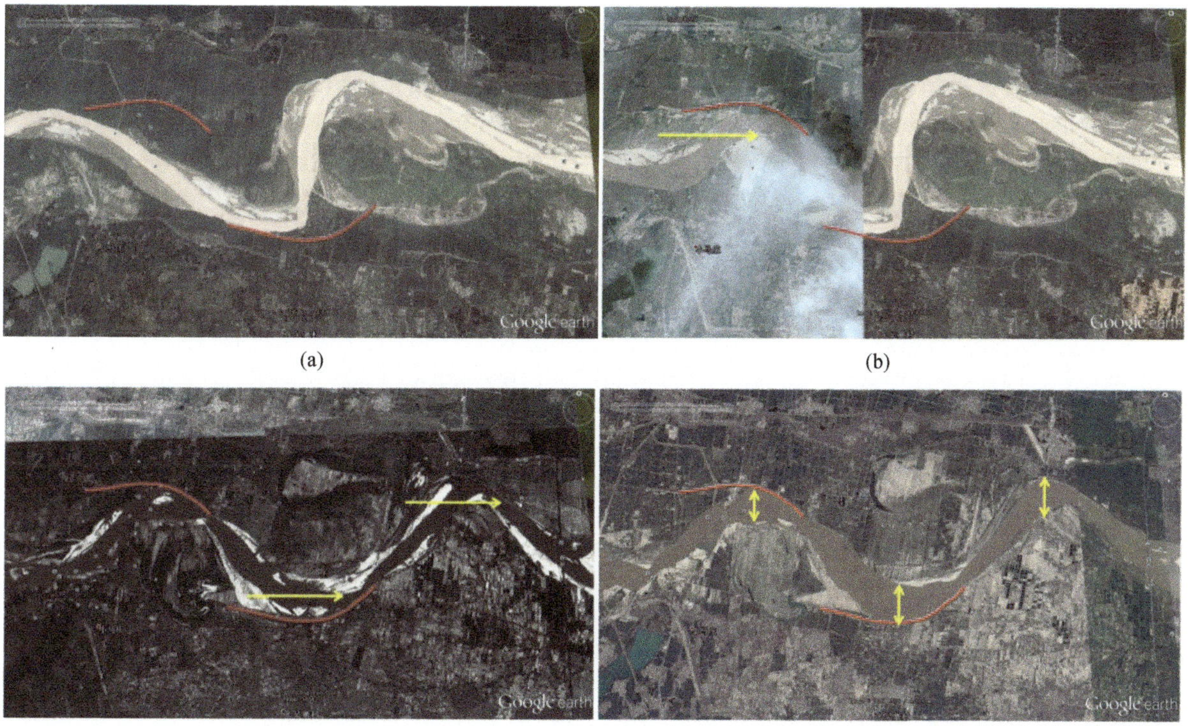

(a) (b)

(c) (d)

Figure 6. Schematic diagram of modern river evolutionary analysis. (a)The image was taken in August 2002 (the two red lines were the dams built on the two river banks; (b) The image was taken in May 2006 (the yellow arrow means the flow direction and the left meander apparently migrated towards east during 2002-2006); (c) The image was taken in December 2012 (the two yellow arrows mean the water flow direction and the meander apparently migrated towards east during 2006-2012); (d)The image was taken in March 2015 (the three yellow arrows mean the extension of the river width during 2012-2015).

right side both migrated towards east from 2006 (**Figure 6(b)**) to 2012 (**Figure 6(c)**). There was no apparent migration occurred to the position of the meanders, however, the channel width became significantly wider from 2012 (**Figure 6(c)**) to 2015 (**Figure 6(d)**). In summary, the whole meandering river migrated towards east and its channel gradually widened from 2002 to 2015.

4. Conclusions

1) The resolution of satellite images from Google Earth software is high; users can study the sedimentary and geomorphological features of the rivers and quickly determine their types by observing the high-resolution satellite images;

2) Users can superimpose appropriate layers on the Google Earth's satellite images to quickly locate the river, track river flow, obtain batch elevation data and easily measure the terrain slope;

3) The Google Earth software's "Ruler", "Add Path" and other functions make it easy for users to directly measure the channel width, radian, amplitude, wavelength and some dam parameters such as size and shape;

4) Google Earth software's "Historical Image" function has important significance in studying modern river migration, sedimentary evolution and topography of river in different seasons.

Therefore, the use of Google Earth can greatly improve the efficiency of modern river sedimentology research.

References

[1] Shi, X.L. (2014) The Application of Google Earth to Coalfield of Panjiayao Block with 3D Seismic Exploration Technology. *Journal of Engineering Geophysics*, **4**, 457-461.

[2] Wang, H.G., Ma, C.L., Ye, Y.L., Jia, W.J. and Zhou, J.P. (2007) Application of Google Earth to Oil Exploration.

Journal of Geophysical Prospecting Equipment, **4**, 306-308.

[3] Ye, F.M. and Han, Z.M. (2009) Application of Google Earth in the Railway Survey Design. *Railway Survey*, **6**, 43-46.

[4] Lin, Y., Li, X.J. and Pei, Z.Q. (2007) Use Google Earth Satellite Images to Assist the Design in the Construction of Geophysical Exploration. *Journal of Geophysical Prospecting Equipment*, **17**, 48-52.

[5] Li, P.C. (2012) Application of Google Earth to Railway Surveying and Mapping Work. *Railway Survey*, **1**, 15-18.

[6] Ren, H.Q., Chen, H. and Zhang, H.Q. (2011) Application of Google Earth Software to the Surveying and Mapping of Ribbon Topographic Map. *Journal of Surveying and Mapping Technology and Equipment*, **3**, 36-37.

[7] Dou, C.H. (2011) Application of Google Earth in Water Conservancy and Hydropower Survey and Mapping. *Journal of Geospatial Information*, **5**, 53-55.

[8] Duan, H.J. and Bian, S.F. (2008) Research on the Visualization of Ocean Physical Field Data Based on the Google Earth. *Journal of Marine Surveying and Mapping*, **28**, 36-39.

[9] Chien, N.Q. and Keat Tan, S. (2011) Google Earth as a Tool in 2-D Hydrodynamic Modeling. *Computers & Geosciences*, **37**, 38-46. http://dx.doi.org/10.1016/j.cageo.2010.03.006

[10] Goudie, A. (2013) Characterising the Distribution and Morphology of Creeks and Pans on Salt Marshes in England and Wales Using Google Earth. *Estuarine, Coastal and Shelf Science*, **129**, 112-123. http://dx.doi.org/10.1016/j.ecss.2013.05.015

[11] Shi, S.Y., Hu, S.Y., Fen, W.J. and Liu, W. (2012) Building Geological Knowledge Database Based on Google Earth Software. *Journal of Sedimentary*, **30**, 869-878.

[12] Li, Y.P., Wu, S.H. and Yue, D.L. (2008) The Quantitative Relationship between Channel Width and Point-Bar Length of Modern Meandering River. *Journal of Daqing Petroleum Geology and Development*, **27**, 19-22.

[13] Zhang, B. (2007) The Form and Origin of Jialing River Meander in China. *Chinese Science Bulletin*, **52**, 2671-1682.

[14] Zhang, C.M., Yin, T.J. and Zhu, Y.J. (2010) The Depositional Model of Shallow Water Delta. *Journal of Sedimentary*, **28**, 933-944.

[15] Guo, X.B. (2012) Application of Google Earth to the Measurement in the Geology Exploration. *Journal of Mining Census*, **3**, 523-526.

[16] Meng, F.X. (2008) Application of Google Earth in Geological Exploration. *Journal of Coal Technology*, **7**, 130-131.

[17] Wang, D. and Zhao, Z.X. (2006) Explanation of Using Google Earth. *Journal of Engineering Geology and Computer Applications*, **9**, 23-31.

4

Instruments for Conservation and Sustainable Use of Water Resources in the Baikal Natural Area

Taisiya Bardakhanova, Svetlana Ayusheeva, Zinaida Goryunova

Laboratory of Nature Management Economics, Baikal Institute for Nature Management, Siberian Branch of the Russian Academy of Sciences, Ulan-Ude, Russia

Email: tbard@binm.bscnet.ru, ayuscheeva2010@yandex.ru, zina--90@mail.ru

Abstract

The present paper contains the results of a review of economic instruments for water resources management in the Republic of Buryatia prepared under the OECD project "Improving the use of economic instruments for water management in the Republic of Buryatia (Lake Baikal basin)" and aimed at supporting the policy dialogue on water in Russia, in cooperation with the EU Water Initiative. Based on the analysis, some recommendations were developed for improving the use of instruments for water resources management in the Republic of Buryatia as a part of the Baikal Natural Area.

Keywords

Economic Instruments, Water Resources, Tariffs, Paying for Water as a Resource, Pollution Charges, Budget Financing, the Baikal Natural Area, the Republic of Buryatia

1. Introduction

The Organization for Economic Cooperation and Development (OECD) project "Improving the use of economic instruments for water management in the Republic of Buryatia (Lake Baikal basin)" is part of a series of projects implemented under the aegis of the OECD within the EU Water Initiative, applied to the Eastern Europe, Caucasus and Central Asia region (EECCA). The OECD engaged with a group of consultants to implement the project on a contractual basis during the period from December 2013 through March 2015. Mr. A. Martoussevitch (OECD) managed the Report preparation. The Report was prepared by a group of consultants led by the Deputy Director of the Municipal Economy department (Institute for Urban Economics), Mr. L. V. Pertsov. The Report was co-authored by a group of consultants (researchers) (Baikal Institute for Nature Management of the Siberian Branch of the Russian Academy of Sciences, the authors of this paper).

The following approach was applied in the analysis of economic instruments for water resources management (WRM): after implementation of a summary review of the general status of water resources and water infrastructure, as well as the institutional and legal frameworks of the water sector, the following points are outlined:

- Key goals and objectives of the socio-economic development of the relevant territory and major problems and challenges facing the water sector (water resources and water infrastructure management) on this territory;
- Key economic instruments for water resources management on this territory (taxes, tariffs, etc.), which are assessed with the help of the OECD-EU methodology, including environmental, financial, economic and social criteria [1] [2], with focus on the practices of application of the reviewed instruments and their conformity with the goals and objectives of the socio-economic development, problems and challenges faced by the water sector on this territory;
- Recommendations for improvement of economic instruments for WRM on the territory in question, based on the assessment of the effectiveness of WRM instruments and international best practices. This approach was earlier successfully applied in Kyrgyzstan [1] and the Russian Federation at the federal level [2].

2. Water Resources of the Republic of Buryatia and Their Use

In terms of available freshwater resources, the Republic of Buryatia (RB) is a truly unique place. The total volume of freshwater resources in Buryatia is estimated at 15.7 thousand cubic kilometers. The major part of Lake Baikal is located on the territory of the Republic. The lake is the deepest freshwater lake in the world and holds 1/5 of global freshwater resources of the highest quality. The region's rivers belong to three larger river basins: the Lake Baikal river basin, and the basins of river Lena and river Enisey. 52% of Buryatia's territory (*i.e.* 182.7 thousand sq·km) lies in the Lake Baikal basin [3].

In 2013, the total volume of freshwater used in Buryatia amounted to 517.8 million cubic meters. Most freshwater is used for the needs of manufacturing industry (83.2%), followed by drinking water supply for domestic needs –7.7% (mainly from groundwater sources). Irrigation and other agricultural needs amount to 6% of the total volume of fresh water abstracted.

3. The Baikal Natural Area

The BNA comprises the territories of three Russian regions (subjects of Russian Federation): the Republic of Buryatia, Irkutsk Oblast and Zabaikalski Krai. The total area of the BNA amounts to 386.2 thousand sq·km and can be categorized as follows: Buryatia, with 220.4 thousand sq.km (57.1% of the BNA), Irkutsk Oblast with 110.1 thousand sq·km (28.5%), and Zabaikalski Krai with 55.6 thousand sq·km (14.4%). The BNA's area is comparable with the total area of all national parks and reserves in Russia (which amount to 464 thousand sq·km in total) [3]. The BNA is a large-scale natural, social and economic system with a complex structure, performing various functions, the most important of which is the conservation of Lake Baikal. The unique environmental system of the lake is a Russian national asset and a UNESCO World Heritage Site (1996). The BNA has a number of unique natural specificities, which have no parallel on our planet, and its diversity and unique nature necessitate a multi-faceted, integrated and systemic approach to the research, conservation and use of the BNA.

The regulation of the conservation and use of the natural resources in the Lake Baikal basin differs from the legal regulations established for other water bodies on the territory of Russia. The specific status of the unique ecosystem of Lake Baikal is reflected in a number of federal legal acts: first in the Federal Law "On Lake Baikal Protection", and the targeted federal program "The conservation of Lake Baikal and the socio-economic development of the Baikal Natural Area for 2012-2020". The Federal Law "On Lake Baikal Protection" established requirements for the conservation of the lake and the sustainable use of natural resources of its basin. The Federal Law defines the regime of conservation and the use of natural resources in the Lake Baikal basin. Within the Baikal Natural Area (BNA), three environmental zones have been established: a central zone, a buffer zone and an atmospheric impact zone. For each zone, specific conditions and restrictions on economic activities and specific standards for environmental impacts have been defined. The most stringent restrictions and bans were set for the territories within the central environmental zone, including Lake Baikal, its water protection zone and the protected natural areas adjacent to Lake Baikal.

4. Water Sector Management Institutions

The federal regulatory framework distributes rather well the core functions for water sector management among several institutions. The federal executive branch plays a key role in water resources management. Institutional responsibilities assigned to regional and municipal authorities are very limited. The mandated implementation activities are regulated in great detail by federal acts.

An Inter-Agency Commission for the conservation of Lake Baikal was established in 2007 to coordinate the activities of the public authorities with regard to the BNA and address regional socio-economic development issues based on the principles of sustainable development. The Commission contributes to enhancing the efficiency of water sector management in the region, although the most important issues to be addressed in the water sector in Buryatia are beyond its mandate.

5. Water Infrastructure

The infrastructure of the water sector of the region is facing a number of urgent problems, including:

- The unsatisfactory condition of the hydro-technical structures in the Republic, including almost 60 crumbling infrastructure schemes which urgently require rehabilitation;
- In 2012, the region ranked 79th among 83 RF constituent subjects for the accessibility of centralized WSS for households; more than a quarter of WSS networks needed to be replaced, but the actual rate of replacement is dozens of times lower than the standard rate;
- More than 50% of wastewater is discharged into water bodies without any treatment;
- The irrigated area of the RB shrank from 213,200 hectares in the late 1980s to 149,300 hectares in 2012. The financing of irrigation systems is not sufficient even to simply rehabilitate existing systems.

6. Goals for Water Resources Management in Buryatia and the Baikal Natural Area

Federal programme on the management of resources and the conservation of Lake Baikal pursue the goal of protecting Lake Baikal and the BNA against adverse effects of human and economic activities as well as natural factors. At the same time, the Republic of Buryatia, when identifying priorities for the conservation and sustainable use of water resources, has to ensure a balance between two goals: the protection of Lake Baikal and the BNA against adverse effects, and the sustainable economic development of the Republic. The task of protecting the population against the negative impact of water-related hazards is shared by both federal and regional programmes. A specificity of water resources in Buryatia is the absence of water scarcity for the majority of the populated areas. Therefore, public authorities managing water resources do not face the task of improving the efficiency of allocation of water between competing uses and users. While this approach seems fully justified due to the abundance of water resources in the Republic, it does not provide incentives for innovation. However, innovations can not only improve the efficiency of water use, but also result in economic benefits (for example, in the growth of production of goods for similar levels of effort and for the same amount of used raw materials, energy, water and other natural resources).

7. Paying for Water as a Resource

The system of paying for water as a resource is fully established in the federal legislation in Russia. Fees for water are made either under contracts for water use (for water abstraction from surface water bodies) or in the form of a water tax (mostly for water abstraction from groundwater bodies, and only to a minor extent for water abstraction from surface water). The water abstraction fee and the water tax are collected on a volumetric basis in the Russian Federation. The system of fees and taxes for water is used as a source of replenishment of the budget and does not perform the function of regulating water abstraction. The recovery of expenditures for the water sector through the aforementioned taxes and fees is not viewed as an objective at the state level in Russia.

The revenue generated by these fees and taxes is fully centralized in the federal budget. Aggregate revenues collected in RB in recent years amounted to RUB 200 - 250 million annually, which is by one or two orders of magnitude smaller than the total expenditure of the water sector in the region. A reduced rate (RUB 70 per thousand cubic meters of water) is established for water abstraction drinking water supply and domestic purposes (depending on the water basin, 10.3% to 28.5% of total payments are made at the full rate, without any discount). There is some duplication within the system of instruments for paying for water as a resource, as fees for water use coexist with the water tax. However, the bodies of state power of the Republic of Buryatia do not believe that such duplication may cut down budget revenue or unreasonably complicates the overall administration of the two instruments.

The specific status of the BNA is accounted for, through a coefficient applied to the water abstraction rate for water reservoirs of Lake Baikal basin. The fee for water abstraction from the lake's basin is the highest in the RF and is the same of all environmental zones of the BNA.

8. Tariff Policy in the Water Supply and Sanitation and Irrigation Sectors

The low level of development of the water supply and sanitation (WSS) sector in RB is, to a great extent, a result of low levels of financing: the proportion of investment expenditure in the prime cost structure over the last five years was almost two times smaller than the average figure for Russia.

The revenue currently generated by the tariffs of vast majority of WSS utilities in Buryatia does not even cover the costs of a basic rehabilitation of fixed assets. Meanwhile, the scope of the budget support to the WSS sector in the Republic has been minimal, compared with other regions of the RF.

The potential for increasing WSS investments is currently limited, as the Federal legislation imposes limitations on increasing WSS tariffs for residential customers—the main users of WSS services—regardless of the obvious need to increase investments in modernizing and developing the WSS sector (primarily through increasing the amount spent on investment in capital expenditure within WSS tariffs), and in spite of the high affordability of WSS services to consumers.

Irrigation systems in the RB are mostly financed through federal budget funds. Both in the Republic and in the rest of Russia, a fee for the transportation of irrigation water is set, although the amount of this fee is very small (on average RUB 2000 per hectare and per season), and the payments collection rate is low. Irrigation systems do not pay for the costs of surface water as a resource.

9. The Water Sector Budget

In the 2000s, the overall financing available from the public budget for water infrastructure and water protection activities was highly unstable. There were acute fluctuations in the scope of financing, the sources and the structure of budget financing: in 2007, budget financing amounted to RUB 231.3 million; in 2009—RUB 915.7 million; and in 2011—RUB 275.1 million. At the end of 2012-beginning of 2013, the scope of budget financing for the water sector was increased, as a number of federal and regional programmes were approved and their implementation started. These programmes envisage a major growth of financing from public funds in the coming years.

The analysis of the difference between actual budget financing and planned financing needs, as well as a review of the water sector target indicators for the period up to 2020, shows that a drastic increase in direct public financing is crucial to halt further degradation of the infrastructure. Nevertheless, the condition of water infrastructure in the region will remain highly problematic, if additional sources of financing (outside of the budget) are not mobilized through economic instruments for water resources management. Moreover, decisions on financing from the federal budget are made at the federal level and, accordingly, such financing is less sustainable and predictable than financing from regional and local sources, which are directly related to the RB and BNA.

10. Cross-Subsidizing the Energy Sector and Neighboring Regions at the Expense of the Republic of Buryatia

In the 1960s, a cascade of hydroelectric power stations (HES) was constructed on the Angara River in Irkutsk Oblast. As a result, the level of the water table of Lake Baikal is higher than before. The major negative resulting environmental and economic effects are borne by the RB, whereas the income generated by the hydroelectric complex is perceived by Irkutsk Oblast.

11. Other Economic Instruments for Water Resources Management

Apart from the fees and tax levied on the use of water as a resource, tariffs charged by WSS utilities, tariffs for water supply to irrigation systems, and budget financing, the major economic instruments for water resources management include charges for a negative impacts on water resources (pollution charges), payments for ecosystem services, and external financing of the water sector and water protection measures (outside of the budget). These instruments may have a certain positive impact on water infrastructure and environmental management, but in general their role has so far been quite limited.

Charges for negative impacts on water or the environment is, obviously, a useful economic mechanism, but respective revenues in RB have so far been insignificant (from RUB 3 to 11 million per year over the last couple of years). Moreover, the mechanism for computation of the amount charged for these negative impacts is currently being reformed, and its role in fostering environmental actions is rather low. The provision of ecosystem services in Buryatia is limited and is mostly related to the use of the specially protected natural areas. According to the authors, not all private investments are accounted for in statistics. However, even with this in mind, it can be said that the role of private investments is not significant.

12. Administrative Restrictions and Their Implications

Administrative instruments for water resources management are based on bans and restrictions imposed on certain economic activities for environmental purposes, on the use of the resources of the BNA and on the drainage and discharge of harmful substances that generate a hazard for the ecosystem of Lake Baikal. The central environmental zone, where the strictest bans are applied, is located within 20 km of the lake's borderline on average. The bans on this territory are imposed on: extraction of mineral resources, logging and wood harvesting, construction of roads and other transport structures (with some exceptions), and construction of new industrial facilities. Serious restrictions are also imposed on wastewater removal and disposal of municipal waste [4]. The role of administrative instruments in the Republic of Buryatia is somewhat controversial:

On the one hand, administrative bans and restrictions produce an enormously positive effect on the ecosystem of Lake Baikal. Once the system of bans was established in the 2000s, the discharges of polluting substances into surface water bodies in Buryatia have been reduced from 46 to 25 thousand tons per year, despite the fact that this happened during a period of economic recovery following a deep transformational recession. At present, the Republic produces a "cleaner" GRP (Gross Regional Product) than average in Russia and in neighbouring regions.

At the same time, the system of bans and restrictions does not reach a balance between protection of the BNA and the need to ensure the economic development of the Republic. The existing restrictions increase production costs and result in lost economic opportunities for the Republic: as estimated by the regional public authorities, every year the Republic loses between 5% and 7% of the GRP due to these restrictions [5]. In terms of per capita GRP, the RB dropped from 44th place in 2007 to 59th place in 2011 (among all constituent subjects of the RF). The ratio of the GRP per capita in the RB to the average for the RF decreased from 56.9% (in 2007) to 51.5% (in 2012).

Environmental restrictions are not the only factor impeding the economic development of the region. However, they obviously produce a negative impact on the standards of living.

13. Major Issues Facing the Water Sector

We concluded that the most acute problems facing the water sector in the RB are as follows:

- Inadequate alignment between target setting and inefficient coordination of public authorities activities in the process of management of the BNA, including its water resources management;
- The slowdown of the socio-economic development of the RB as the result of administrative restrictions and bans established for the BNA (but not always justified);
- A major funding gap for the rehabilitation of water infrastructure of all types;
- Insufficient incentives to discourage polluters from discharging pollutants into water bodies and to implement environmental interventions;
- Deterioration of WSS systems and irrigation systems;
- Littering near water bodies and near the shores or river banks, diffuse pollution and pollution with liquid domestic wastes;
- Inadequate risk management for the negative impacts of water-related hazards (floods, under flooding, mudflows) in the RB territory.

14. Key Areas of Improvement for Economic Instruments for Water Resources Management

The recommendations are based on the need to ensure a balance between two goals: protecting the ecosystem of

Lake Baikal and promoting a sustainable socio-economic development of the Republic of Buryatia. They take into account the current condition of the water sector in the Republic and the specificities of water sector management in the RF in general.

Specifically, the key areas of improvement of economic instruments for water resources management in the RB are as follows:

- Optimizing water abstraction fees, taking into account the fact that there is no shortage of water resources in the RB;
- Improving the collection rate of tariffs for irrigation of agricultural land, and introducing a payment for water as a resource;
- Introducing a large-scale system of payments for ecosystem services, with subsequent ringfencing of the revenue generated in a dedicated fund for targeted use. The introduction of a "resort fee" could become a first step toward payments for ecosystem services;
- Assessing the environmental and economic damage incurred by the RB because of the Angara cascade of HES, and implementation of a system of compensatory payments and re-distribution of hydroelectric revenue inclusive of the RB;
- Imposition of a special tax on products sold in the BNA which may potentially contaminate the environment and could be replaced by environmentally safe equivalents;
- Introduction in the Baikal Natural Area of compulsory environmental risk insurance and relevant liability insurance for elimination and reimbursement of damages to the environment and third parties.

An additional measure deserving attention is:

- Implementation of a set of projects for rehabilitating and upgrading water infrastructure, which will be financed through the establishment of higher rates for WSS tariffs and/or through additional federal financing for the Republic of Buryatia.

The main recommended measures, complementary to the development of the above economic instruments, are as follows:

- Relaxing the complete ban currently in effect on a range of activities necessary for the development of settlements in the central environmental zone (based on thorough research aimed at identifying the limits of such reforms, given the current and expected human influence on the ecosystem of the region);
- Designing a regulatory framework for compulsory disclosure of information on the environment performance indicators of industrial and infrastructure facilities located in the BNA with their subsequent rating and banchmarking, as well as the assignment of eco-status and eco-labelling;
- Implementation of measures that are aimed at ensuring that the fluctuations of Lake Baikal's water table level are as close to natural fluctuations as possible (by regulating operations of the hydroelectric power stations (HES) of the Angara cascade, and through the monitoring of the fluctuations of the water table level).

We would like to highlight that the most important task of the RB Government, as outlined in many strategies and programmes, is the creation of proper conditions for changing the structure of the economy of the Republic in order to ensure the development of environmentally-oriented activities. To address this, further research is needed in order to help evaluate the feasibility of revising current bans on activities, which are necessary to the development of settlements in the central environmental zone.

References

[1] OECD (2013) Economic Instruments for Water Resources Management in the Russian Federation. OECD Publishing, Paris.

[2] (2014) State Report: On the Condition of the Lake Baikal and Its Protection Measures in 2013. Siberian branch FGUNPP "Rosgeolfond", Irkutsk.

[3] Bardakhanova, T.B. (2013) About Measures on Environmental Management in the Baikal Natural Area. *Modern Economy: Problems and Solutions*, **9**, 30-39.

[4] Bardakhanova, T.B., Mikheeva, A.S. and Ayusheeva, S.N. (2014) Institutional Environmental Management and the Formation of Environmental Costs. *Management of Economic Systems: An Electronic Scientific Journal*, **10**. http://www.uecs.ru

[5] OECD (2012) Improving the Use of Economic Instruments for Water Resource Management in Kyrgyzstan: The Case of Lake Issyk-Kul Basin (Final). (ENV/EPOC/EAP(2012)5).

5

A Cost-Effective Wind Power-Driven RO Plant for Treatment of Brackish Water

Shivendra Verma, Ramsingh Meena, Anurag Mudgal*

Department of Mechanical Engineering, School of Technology, Pandit Deendayal Petroleum University, Gandhinagar, India

Email: *anurag.mudgal@sot.pdpu.ac.in

Abstract

Desalination has often been perceived as a high cost means of producing water. Reason being that some of the water treatment methods use electricity to run a pump as in Reverse Osmosis system and some uses thermal energy to produce steam as in Multi Effect Distillation. High cost and less availability of conventional fuels has attracted the focus of researchers to explore other unconventional means and methods to purify water. This paper explores the possibility of using wind power for water treatment. In costal countries, tremendous wind power is available at the coastline which may be used to drive a RO plant. In India particularly, the average wind speed calculated along the coastlines of Gujarat state gives a fair idea about potential of energy that can be generated. This energy can be harnessed to run a rotating pump, or a piston-cylinder arrangement coupled with RO module to provide the necessary pressure required to push saline water through the membrane. Along the western coast of India there is an abundance of wind energy to capitulate high amount of energy in terms of electrical Power. If the rotor diameter is 4 feet (1.3 m) and located at a height of 8 feet (2.5 m), each wind turbine may produce power of around 2 KW. This power may be used to run the reciprocating pump which provides pressure to the RO Apparatus. The paper describes recent advancements in the wind industries technology to produce electrical energy at low cost but it concentrates more on coupling wind energy directly with RO module to save energy conversion cost which is a seldom or different approach than converting electrical energy to mechanical energy for running of RO system.

Keywords

Wind Power, Reverse Osmosis, Desalination

1. Introduction

Water scarcity and population in the present world is increasing day by day. Fresh water resources are limited and modern lifestyle is leading to increase per capacity water consumption desalination can provide a solution to meet required surplus amount. Desalination is mostly achieved by either RO (Reverse Osmosis) or thermal desalination. Both the processes have drawbacks as RO is efficient but uses costly electrical energy whereas ther-

*Corresponding author.

mal desalination is inefficient but uses less expensive thermal energy. This paper mostly focuses on running the RO plant with the use of unconventional energy source. The RO plant would be utilised to purify mostly the brackish water available in rural areas of India, where availability of electricity is less or intermittent which is required to run a plant is a challenge.

The idea of using unconventional energy to operate desalination plant aims on meeting future demand. One of the major factors affecting the total cost of water production by any type of desalination process is the energy cost. Typically in a reverse osmosis plant, 3 to 10 kWh of electric energy is required to produce one cubic meter of freshwater from seawater [1]. Since cost of renewable energy is generally independent of fossil fuel prices, the cost fluctuations due to energy cost instabilities can be avoided if the system is operated with renewable energy technologies. Furthermore the energy required for pressuring feed water can be significantly reduced if high pressure or desired pressure is generated with the help of unconventional energy source.

2. Objectives

For rural applications where decentralized supply of water is needed, the plant should have the following characteristics [2]:

1) It should be small, capable of producing a few cubic meters of treated water per day to satisfy local demands.

2) It should be energy efficient and should run on local energy sources, wind energy preferably (as per this paper).

3) It should be cost effective and affordable by rural communities.

4) It should be rugged, robust and easy to maintain. In case of a fault, a rural mechanic should be able to repair it easily.

3. Problems Faced with Generating Pressure from Motor (Electricity Provided from Wind Turbine)

Installation of underground laid cables which transmits the electricity generated from individual wind mill to the main grid of the plant requires a large capital.

At some places the temperature may drastically change which may lead to destruction of cables (**Figure 1**) if not protected. Underground rodents which may live in nearby area may destroy the cables. In this paper it provides the remedies to solve the problems described above. As we are not using electricity so there is no need of underground cables. For future perspectives, if applied for large scale applications like wind farm this technology could be used with sundry changes. Different wind mills will have their own reservoir tanks (treated purified water). The water source would be brackish water, brought to individual wind mills with the help of pump (booster pump). The further processing is done by the procedure as described below.

Extreme temperatures – both hot and cold – can take their toll on cables.
Photo courtesy of Helukabel USA

Figure 1. Destruction of cable.

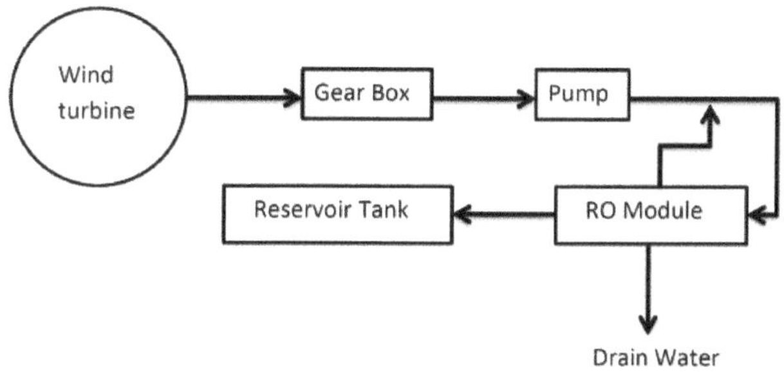

Figure 2. Flow diagram of the process.

4. Methodology

The main components of the RO system may be summarised as follows:

1) RO module.
2) Wind turbine.
3) Pump (run by wind turbine).
4) Wind turbine gearbox.

As shown in above flow diagram (**Figure 2**), we couple the shaft of the wind mill to pump with the help of gear box attached to it. The shaft of the pump rotates with the help of wind mill and the pumps work without generation of electricity. The wind will is installed at a height of about 8 feet about the ground; distance of the centre of hub in wind mill is 8 feet above the ground. Shaft of the wind mill is horizontal, connected to bevel gear which converts the horizontal rotational movement to vertical rotational movement, which is further connected to another gear box. The shaft of this gear box is coupled with the shaft of motor which provides the pressure required by the RO module. Below shows armature of the pump (**Figure 3**) is removed which runs with the help of electricity. Here instead we are not converting electrical energy generated from wind turbine to run the pump. Instead we couple the shaft of the pump to the shaft of wind turbine via a gear box which provides sufficient rotation to generate enough pressure in RO module.

5. Calculations

5.1. Calculating Wind Power Potential

A German physicist Albert Betz concluded in that no wind turbine can convert more than 59.3% of the kinetic energy of the wind into mechanical energy turning a rotor. To this day, this is known as the **Betz Limit or Betz law** [3]. The theoretical maximum power efficiency of any design of wind turbine is 0.59 (*i.e.* no more than 59% of the energy carried by the wind can be extracted by a wind turbine) [4] [5].

$$P_{avail} = \frac{1}{2}\rho A v^3 C_p \left[\text{Power Available}\right] \quad (1)$$

5.2. Calculations with Given Data

The following data:

1) Blade length, $l = 1.2192$ m (4 ft),
2) Wind speed, $v = 12$ m/sec,
3) Air density, $\rho = 1.23$ kg/m^3,

Power coefficient, $C_p = 0.4$.

Inserting the value for blade length as the radius, calculate swept area (**Figure 4**).

$$\begin{aligned}\text{Area} &= \pi \times r^2 \\ &= \pi \times 1.2192^2 \\ &= 4.6674\ \text{m}^2\end{aligned} \quad (2)$$

We can then calculate the power converted from the wind into rotational energy in the turbine. By using Equation (1), we get.

$$= 1/2 \times 1.23 \times 4.6674 \times 12^3 \times 0.4$$
$$= 1.984 \text{ KW}$$
$$= 2.6605 \text{ Hp}$$

The power coefficient is not a static value as defined in the main question; it varies with the tip speed ratio of the turbine. Tip speed ratio is defined as [3] [6]:

$$\lambda = \frac{\text{blade tip speed}}{\text{wind speed}}$$

λ (tip speed ratio) changes as wind speed is variable quantity, if it changes so does C_p as represented by the graph (**Figure 5**).

The blade tip speed can be calculated from the rotational speed of the turbine and the length of the blades used in the turbine using the following equation:

$$v = 2\pi r \times RPM \times (60/1000) \text{ km/hr} \quad (3)$$

5.3. Aggregating the Costs

Machine Equipments	Costs (Rs.)
Wind turbine gearbox, structure, pump set	65,000
Pump	2150
RO module	2200
TDS meter	700
Ph meter	850
Auxiliary investments	1000
Total	71,900 (1105.33$)

Figure 3. The armature used in pump.

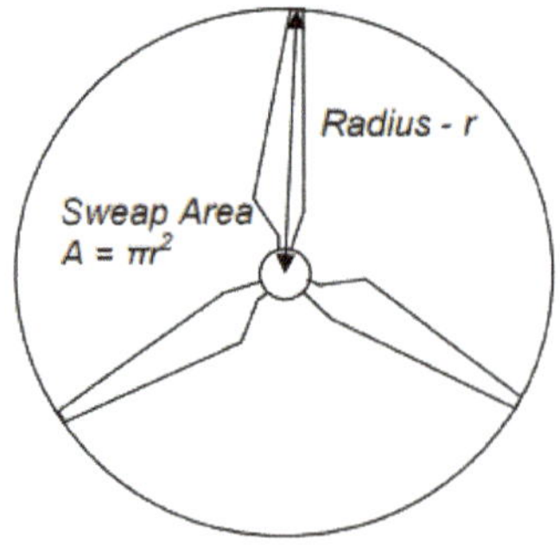

Figure 4. Schematic diagram of wind turbine.

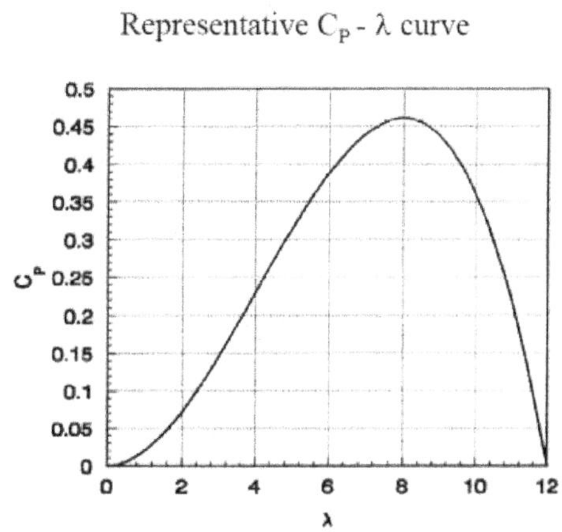

Figure 5. Relationship C_p vs λ.

6. RO-Module Specifications

Membrane Type: Thin film Composite Membranes, Size (approximate): 2.75" Diameter × 15.74".

7. Pump Specifications

Design: Positive displacement 3 chamber Diaphragm Pump (**Figure 6**), Motor: Permanent motors, Ball bearing, water flow: 14 LPH, Inlet Pressure Max: 40 psi, Outlet Pressure: 125 psi, Rotation Speed rate: 800 ± 50 rpm.

8. Gear Box

To calculate the type and number of teeth in the gears used, we take the help of rpm and use it in the following equation:

$$S1 \times T1 = S2 \times T2 \tag{4}$$

$S1$ = Speed of driver gear (rpm) → 94 – 100,
$S2$ = Speed of driven gear (rpm) → 800,
$T1$ = Number of teeth in driver gear,
$T2$ = Number of teeth in driven gear.

$$T1 / T2 = 800 / 100 \tag{5}$$

The gear ratio required is 8:1.

To get an optimum desired rpm we use helical gears which operate more smoothly and quietly compared to spur gears due to the way the teeth interact. The thrust load varies directly with the magnitude of tangent of helix angle. Helical gears can be used to adjust the rotation angle by 90 deg. When mounted on perpendicular shafts. Its normal gear ratio range is 3:2 to 10:1. The gear box could be considered as hub of the mechanical mechanism as it provides required rpm to the shaft of motor which generates pressure to inlet water to pass through RO Module.

9. Break-Even Analysis

With the help of the setup arranged we generate 7 to 8 litres per hour of mineralised water if we cost the customer Rs.1/lt (0.015$/lt).

7 lt × 24 hours = Rs.168/day

168 × 30 days = Rs.5, 040/-

When we increase the rate which includes maintenance of the setup by 3% annually and increase the cost by Rs.1/lt/year. We get the graph as given below (**Figure 7**).

10. Conclusions

Above calculated Break-Even analysis provides you with brief of the result as well as conclusion. The treated

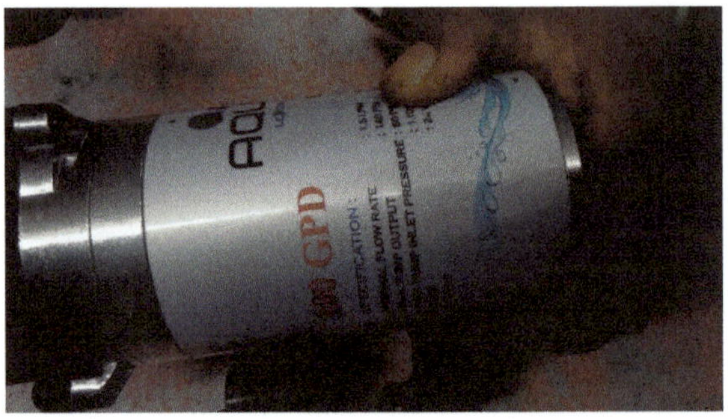

Figure 6. Pump used to generate pressure.

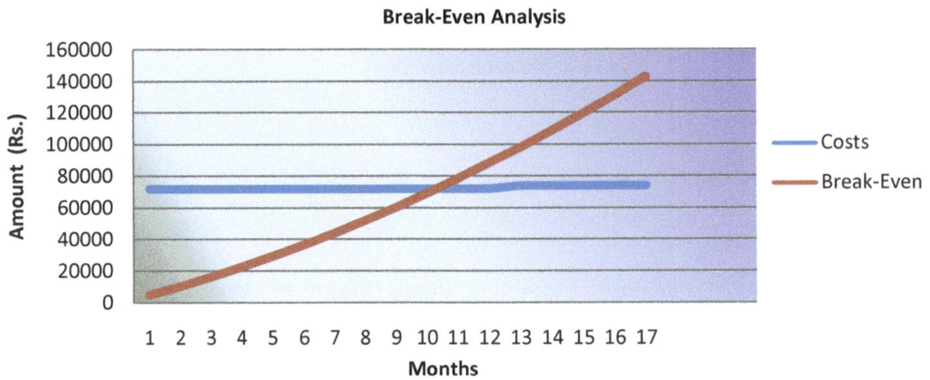

Figure 7. Graph showing break-even analysis.

water obtained has fewer amounts of dissolved salts and the cost invested in installation and running of the setup is recovered in less than a year, which can be seen in graph (**Figure 7**).

The power generated is 2 KW with the help of single setup. This power can be used to run the reciprocating pump which provides pressure to the RO Apparatus. If the experiment is applicable for mass production then the money generated in terms of profit can be further used, more equipment can be planted in many parts of rural areas. Money invested could be recovered in 10 months (**Figure 7**).

11. Future Work

The money collected as profit would be further invested in instalments of the machine in various other places which are deprived of clean water or cannot have enough capital to have water treatment method. The places could be villages where there is dearth of electricity, which is required by common water treatment methods. A prototype is proposed to be fabricated at PDPU (Pandit Deendayal Petroleum University), India. The project proposal is being submitted as ORSP (Office of Research & Sponsored Programs) cell to finance the fabrication work.

Acknowledgements

We acknowledge the support from PDPU (Pandit Deendayal Petroleum University) in terms of Laboratory Facilities and financial support to carry out this research and presenting the paper.

References

[1] Dashtpour, R. and Al-Zubaidy, S.N. (2012) Energy Efficient Reverse Osmosis Desalination Process. International *Journal of Environmental Science and Development*, **3**.

[2] Sen, P.K., Sen, P.V., Mudgal, A., Singh, S.N., Vyas, S.K. and Davies, P. (2011) A Small Scale Multi-Effect Distillation (MED) Unit for Rural Micro Enterprises: Part 1—Design and Fabrication. Desalination 27915-26. http://dx.doi.org/10.1016/j.desal.2010.11.003

[3] Bird, J. (2007) Basic Engineering Mathematics. Elsevier Ltd.

[4] Royal Academy of Engineering (2010) Wind Turbine Power Calculations. RWE Npower Renewables, Mechanical and Electrical Engineering Power Industry.

[5] Ragheb, M. and Ragheb, A.M. (2011) Wind Turbines Theory—The Betz Equation and Optimal Rotor Tip Speed Ratio. In: Carriveau, R., Ed., *Fundamental and Advanced Topics in Wind Power*. http://dx.doi.org/10.5772/21398

[6] Bird, J. (2007) Engineering Mathematics. 5th Edition, Elsevier Ltd.

Appendix

Symbol	Unit	Description
ρ	Kg/m^3	Density
π	--	Pi(3.14)
A	m^2	Area
λ	--	Tip Speed Ratio
r	m	Radius
C_p	--	Power coefficient

6

Automatic Generation of Water Masks from RapidEye Images

Gideon Okpoti Tetteh, Maurice Schönert

BlackBridge AG, Kurfürstendamm 22, Berlin, Germany
Email: gideon.tetteh@blackbridge.com

Abstract

Water is a very important natural resource and it supports all life forms on earth. It is used by humans in various ways including drinking, agriculture and for scientific research. The aim of this research was to develop a routine to automatically extract water masks from RapidEye images, which could be used for further investigation such as water quality monitoring and change detection. A Python-based algorithm was therefore developed for this particular purpose. The developed routine combines three spectral indices namely Simple Ratios (SRs), Normalized Green Index (NGI) and Normalized Difference Water Index (NDWI). The two SRs are calculated between the NIR and green band, and between the NIR and red band. The NGI is calculated by rationing the green band to the sum of all bands in each image. The NDWI is calculated by differencing the green to the NIR and dividing by the sum of the green and NIR bands. The routine generates five intermediate water masks, which are spatially intersected to create a single intermediate water mask. In order to remove very small waterbodies and any remaining gaps in the intermediate water mask, morphological opening and closing were performed to generate the final water mask. This proposed algorithm was used to extract water masks from some RapidEye images. It yielded an Overall Accuracy of 95% and a mean Kappa Statistic of 0.889 using the confusion matrix approach.

Keywords

Water Mask, Image Threshold, Simple Ratio, Normalized Green Index, Normalized Difference Water Index, Logical and Morphological Operations, RapidEye

1. Introduction

Water, which covers over 70 percent of the earth's surface, sustains all known life forms on earth, making it imperative to monitor the spatial extent and quality of waterbodies over time. Doing this monitoring has been done in the past through direct field measurements, which suffers a draw-back in that it is expensive and lags behind with respect to the frequency at which timely information can be collected. The advent of earth-orbiting satellites over the years has enabled us to remotely monitor waterbodies in a timely manner. This particular research focused essentially on the extraction of the spatial extents of waterbodies from satellite images in an automated fashion.

The automatic extraction of waterbodies from satellite images is well documented. The Normalized Difference Water Index (NDWI), calculated as a ratio of the difference between the green and the NIR band to the sum of those two bands, was first proposed by [1] and was used to extract waterbodies from satellite images. The NDWI was later modified by [2] by simply replacing the NIR band with the MIR band and came up with the name Modified NDWI (MNDWI). Due to the limitations of the NDWI and MNDWI in terms of unstable threshold and the mixture of waterbodies and other dark objects, [3] developed what they call the Automated Water Extraction Index (AWEI) to improve the extraction of waterbodies especially in areas that include shadow and dark surfaces. An automated method for extracting rivers and lakes from Landsat imagery by exploiting the NDWI, MNDWI and AWEI was developed by [4]. The authors then extract waterbodies by applying a threshold value to these indices, extracting both wide and narrow waterbodies and finally combining all the extracted waterbodies.

An automatic object-based waterbody extraction method was developed by [5] based on NDWI using the Locally Excitatory Globally Inhibitory Oscillator Networks (LEGION) as promulgated by [6]. Using a Perceptron Model, [7] developed an automated waterbody extraction routine for Landsat-7 ETM+ by feeding the low reflectance of water in the SWIR band, the relatively lower reflectance in the NIR band to the green band and in the red to the NIR band, and a binary of the summation of NDWI and MNDWI as feature vectors to the model.

The Whole-R Method was developed by [8] to automatically extract waterbodies from Landsat images. This method involves the binarization of the red band based on a threshold value and the spectral angle between the spectral value of the image and that of its spectral library. Two spectral enhancement techniques called Weighted Water Index (WWI) and Weighted Water-Soil Index (WWSI) were introduced by [9] as a means of enhancing waterbodies in satellite images making it easier to extract waterbodies by applying a simple threshold to the enhanced images.

Owing to the limitation of the NDWI and MNDWI in sometimes confusing dark objects like shadow and coal tarred roads with water, [10] came up with an automatic routine to extract a threshold value in the NIR band of some airborne hyperspectral images to mask potential water pixels, and then performs spectral slope analysis to remove possible shadows from the potential water mask to generate the final water mask.

Instead of using a single global threshold value on the NDWI to extract waterbodies, [11] dynamically extracted different threshold values based on the sub-pixel components of each image. Based on empirical analysis of a Landsat TM image of the Huanghe river delta, [12] came up with a decision tree algorithm to automatically extract waterbodies from Landsat image. Through spectral pattern analysis of the green, red, near infrared (NIR) and short wave infrared (SWIR) bands, [13] extracted waterbodies from Spot 5 and Landsat 5 TM images.

In order to come up with an approach which is insensitive to image sensors, [14] introduced an entropy-based method for extracting waterbodies from satellite images based on the justifiable assumption that water normally has a smooth surface in images. Some spectral indices namely Normalized Difference Vegetation Index (NDVI), Normalized Difference Moisture Index (NDMI), Water Ratio Index (WRI), AWEI, NDWI and MNDWI were tested by [15], coming to the conclusion that NDWI gave the best results in extracting waterbodies.

The research presented here adopted the automatic extraction of waterbodies from the NIR band as described by [10], the NDWI as proposed by [1], two proposed simple ratios (SRs), and one newly proposed spectral index called Normalized Green Index (NGI).

2. Materials

2.1. RapidEye Sensor Specifications

The RapidEye satellite system is a constellation of five identical earth observation satellites with the capacity to provide large area, multi-spectral images with frequent revisits in high resolution (6.5 m at nadir). In addition to the blue (440 - 510 nm), green (520 - 590 nm), red (630 - 685 nm) and near-infrared (760 - 850 nm) bands, the sensor has a red-edge (690 - 730 nm) band, which is very suitable for vegetation analysis. The RapidEye Level 3A standard image product covers an area of 25 by 25 km, is radiometrically calibrated to radiance values, and is ortho-rectified and resampled to 5 m spatial resolution.

2.2. Data Used

In order to test the robustness of the developed routine, three geographically varying RapidEye Level 3A images were used. **Table 1** gives a summary of the images used for developing the routines and for the accuracy assessment of the classified waterbodies.

Table 1. Overview of the test images.

Test Site	Location		Water Conditions	Intended test
	Longitude	Latitude		
USA	−89°50'57.17"	29°34'59.71"	Black, shallow, turquoise, yellow, green and brownish water	Different inland and coastal water conditions
Brazil	−36°43'37.19"	−5°5'54.69"	Variety of greenish and turquoise, greyish violet and turquoise waters	Different inland and coastal water conditions
Mali	−4°0'21.4632"	15°18'7.8372"	Greyish-blue, turquoise, black waters and shallow flooded river banks	Flood, different water conditions, flooded AG and vegetation

3. Methodology

The entire methodology adopted in this research is captured in **Figure 1**. The whole process begins by first extracting the solar zenith and acquisition date of each image. These extracted values are used to generate the Top-Of-Atmosphere (TOA) reflectance image from the RapidEye L3A radiance images, which are already corrected for offset. The TOA images serve as base images for all other components of the automation processes.

An initial water mask is created using the low reflectance of water in the NIR band by applying a simple threshold on the NIR band. Water absorbs most of the electromagnetic energy in the NIR region, therefore it has lower reflectance in the NIR region as against other features in an image. The threshold value is extracted automatically from the NIR histogram precisely following the method as proposed by [1]. **Figure 2** shows the typical spectra for six feature types in all five spectral bands of the RapidEye images.

The NIR threshold value is estimated by finding the two local peaks at the left tail of the NIR histogram knowing that the low reflectance surfaces would be at that tail of the NIR histogram. A fifth order polynomial is then fitted on all histogram points between these two peaks, after which the minimum of the polynomial curve is extracted, and finally the corresponding NIR value for this minimum is extracted as the threshold value. This value is then used to threshold the NIR band to obtain the first potential water mask.

Four additional water masks are created using two Simple Ratios (SRs), Normalized Green Index (NGI) and Normalized Difference Water Index (NDWI). Equation (1) to (4) shows the formulas for all the four indices

$$\mathrm{SR1} = NIR / \mathrm{Green} \tag{1}$$

$$\mathrm{SR2} = NIR / \mathrm{Red} \tag{2}$$

$$\mathrm{NGI} = \mathrm{Green} / (\mathrm{Blue} + \mathrm{Red} + \mathrm{RedEdge} + NIR) \tag{3}$$

$$\mathrm{NDWI} = (NIR - \mathrm{Green}) / (NIR + \mathrm{Green}) \tag{4}$$

Sample spectral index values computed for the same aforementioned feature types are shown in **Figure 3**. A close examination of the SRs values shows that the values of both muddy and clear water are lower than all the other feature types and indeed less than one, hence making it easier to extract two additional waterbodies by applying a simple threshold to the two SRs respectively.

The muddy and clear waterbodies had a relatively higher NGI value, 0.25 and above, than all the other feature types, hence that value was used to threshold the NGI image to create the fourth waterbody. An examination of the NDWI values indicated that the waterbodies had the lowest values, 0 and below, making it a straight forward threshold operation to generate the fifth potential waterbodies. There were cases where other dark objects had NGI and NDWI values similar to that of water, hence the next step was to do a *logical intersection* of all the five potential waterbodies to create a single intermediate water mask.

Finally, in order to remove very small waterbodies and any remaining holes within the water mask, morphological opening followed by closing are performed on the intermediate water mask to create the final water mask.

4. Results & Conclusions

The accuracy of the water masking algorithm was assessed using stratified random sampling. 200 points were randomly generated for the water and non-water classes in each classified image. For each point, the ground truth class (non-water and water) was assigned by visual identification with the assistance of Google Earth.

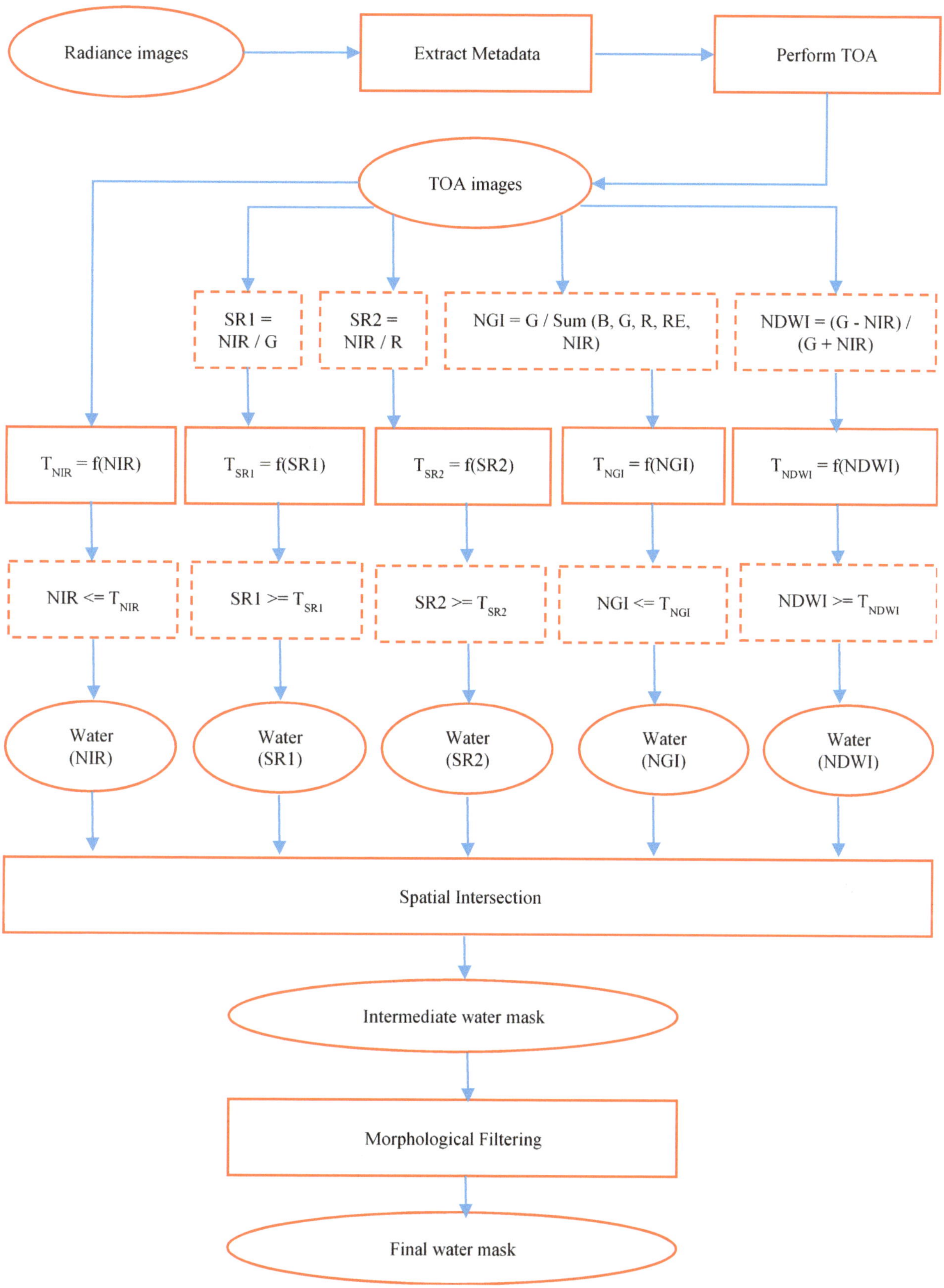

Figure 1. The automated workflow for waterbodies extraction.

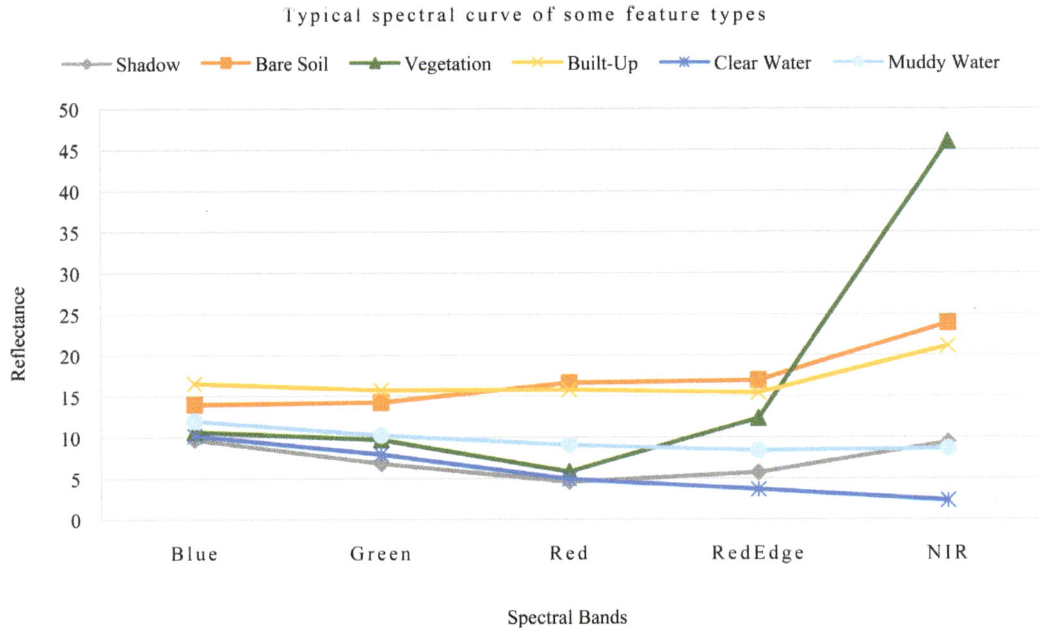

Figure 2. Typical spectral curve of some feature types.

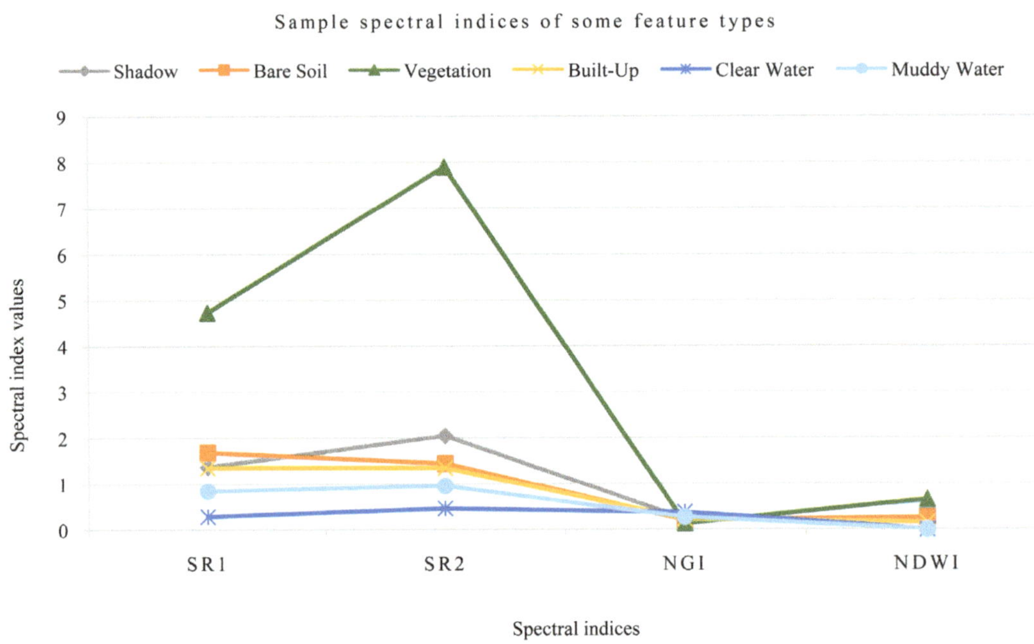

Figure 3. Sample spectral indices of some feature types.

These reference points were compared to those of the generated water masks and the accuracy was assessed using the *Confusion Matrix* approach.

The *Producer's Accuracy*, *User's Accuracy*, *Overall Accuracy* and *Kappa* of the classification result of the implemented algorithm were calculated accordingly. The Producer's Accuracy here refers to the probability that a visually identified and assigned waterbody is classified as such, while the User's Accuracy refers to the probability that a classified waterbody in the image is indeed a waterbody. The Overall Accuracy is the probability of the correctly classified waterbodies. Kappa was used as a measure of agreement between the prediction of the proposed water algorithm and the reality as was done in this case through visual inspection.

The test images, resultant water masks and Confusion Matrices are shown in **Figure 4-6**.

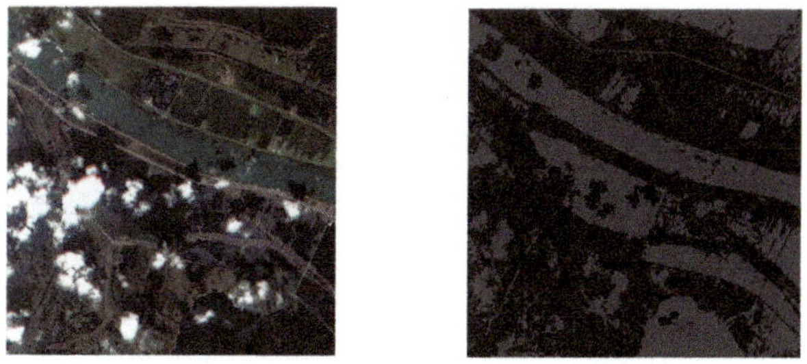

	Non-water	Water	User's Accuracy
Non-water	109	3	97.3
Water	8	80	90.9
Producer's Accuracy	93.2	96.4	
Overall Accuracy = 94.5%			
Overall Kappa = 0.888			

(a) (b) (c)

Figure 4. (a) USA image; (b) Extracted water mask; (c) Confusion matrix.

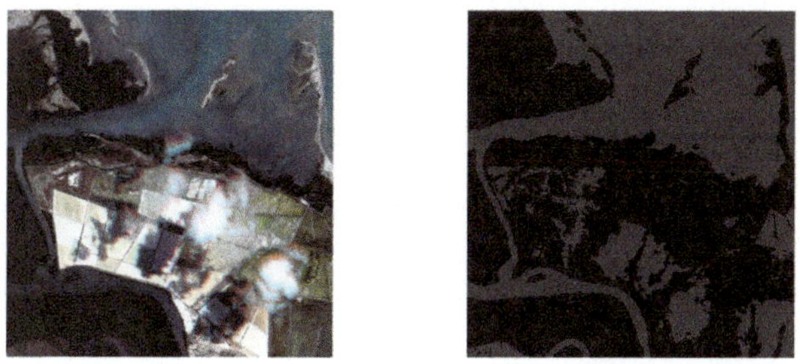

	Non-water	Water	User's Accuracy
Non-water	130	6	95.6
Water	8	56	87.5
Producer's Accuracy	94.2	90.3	
Overall Accuracy = 93.0%			
Overall Kappa = 0.838			

(a) (b) (c)

Figure 5. (a) Brazil image; (b) Extracted water mask; (c) Confusion matrix.

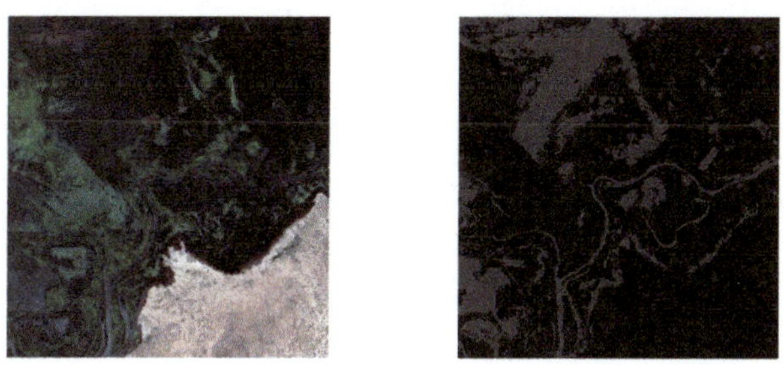

	Non-water	Water	User's Accuracy
Non-water	138	3	97.9
Water	2	57	96.6
Producer's Accuracy	98.6	95.0	
Overall Accuracy = 97.5%			
Overall Kappa = 0.940			

(a) (b) (c)

Figure 6. (a) Mali image; (b) Extracted water mask; (c) Confusion matrix.

In conclusion, the proposed algorithm showed a very good performance on all the water conditions mentioned before. The algorithm detects water with an Overall Accuracy of 95% (mean Kappa of 0.889). For the water class, the mean Producer's Accuracy is 93.9% and the User's Accuracy is 91.7%. The main drawback of this algorithm is that some shadow areas especially the ones over vegetated areas (see **Figure 5**) were also classified as waterbodies. Therefore, this algorithm ought to be further developed to handle those exceptional cases.

References

[1] McFeeters, S.K. (1996) The Use of the Normalized Difference Water Index (NDWI) in the Delineation of Open Water Features. *International Journal of Remote Sensing*, **17**, 1425-1432. http://dx.doi.org/10.1080/01431169608948714

[2] Xu, H.Q. (2006) Modification of Normalised Difference Water Index (NDWI) to Enhance Open Water Features in Remotely Sensed Imagery. *International Journal of Remote Sensing*, **27**, 3025-3033. http://dx.doi.org/10.1080/01431160600589179

[3] Feyisa, G.L., Meilby, H., Fensholt, R. and Proud, S.R. (2014) Automated Water Extraction Index: A New Technique for Surface Water Mapping Using Landsat Imagery. *Remote Sensing of Environment*, **140**, 23-35. http://dx.doi.org/10.1016/j.rse.2013.08.029

[4] Jiang, H., Feng, M., Zhu, Y., Lu, N., Huang, J. and Xiao, T. (2014) An Automated Method for Extracting Rivers and Lakes from Landsat Imagery. *Remote Sensing*, **6**, 5067-5089. http://dx.doi.org/10.3390/rs6065067

[5] Li, M., Xu, L. and Tang, M. (2011) An Extraction Method for Water Body of Remote Sensing Image Based on Oscil-

latory Network. *Journal of Multimedia*, **6**, 252-260. http://dx.doi.org/10.4304/jmm.6.3.252-260

[6] Terman, D. and Wang, D.L. (1995) Global Competition and Local Cooperation in a Network of Neural Oscillators. *Physica D: Nonlinear Phenomena*, **81**, 148-176. http://dx.doi.org/10.1016/0167-2789(94)00205-5

[7] Mishra, K. and Prasad, P.R.C. (2014) Automatic Extraction of Water Bodies from Landsat Imagery Using Perceptron Model. *Journal of Computational Environmental Sciences*, **2015**, 1-9. http://dx.doi.org/10.1155/2015/903465

[8] Nawaz, N., Srinivasulu, S. and Rao, P.K. (2013) Automatic Extraction of Water Bodies Using Whole-R Method. *International Journal of Environmental, Ecological, Geological and Geophysical Engineering*, **7**, 564-567.

[9] Polidorio, A.M., Flores, F.C., Franco, C., Imai, N.N. and Tommaselli, A.M.G. (2010) Enhancement of Terrestrial Surface Features on High Spatial Resolution Multispectral Aerial Images. *Proceedings of the 23rd SIBGRAPI Conference on Graphics, Patterns and Images*, Rio Grande do Sul, 30 August-3 September 2010, 295-300. http://dx.doi.org/10.1109/SIBGRAPI.2010.46

[10] Bochow, M., Heim, B., Kster, T., Roga, C., Bartsch, I., Segl, K. and Kaufmann, H. (2012) On the Use of Airborne Imaging Spectroscopy Data for the Automatic Detection and Delineation of Surface Water Bodies. In: Chemin, Y., Ed., *Remote Sensing of Planet Earth*, InTech, Rijeka, 3-22. http://dx.doi.org/10.5772/34073

[11] Ji, L., Zhang, L. and Wylie, B. (2009) Analysis of Dynamic Thresholds for the Normalized Difference Water Index. *Photogrammetric Engineering and Remote Sensing*, **75**, 1307-1317. http://dx.doi.org/10.14358/PERS.75.11.1307

[12] Fu, J., Wang, J. and Li, J. (2008) Study on the Automatic Extraction of Water Body from TM Image Using Decision Tree Algorithm. *Proceedings of SPIE* 6625, *International Symposium on Photoelectronic Detection and Imaging* 2007: *Related Technologies and Applications*, Beijing, 9 September 2007, 1-9.

[13] Nguyen, D.D. (2012) Water Body Extraction from Multi Spectral Image by Spectral Pattern Analysis. *Int. Arch. Photogramm. Remote Sens. Spatial Inf. Sci.*, **XXXIX-B8**, 181-186. http://dx.doi.org/10.5194/isprsarchives-XXXIX-B8-181-2012

[14] Zhaohui, Z., Prinet, V. and Songde, M. (2003) Water Body Extraction from Multi-Source Satellite Images. *Proceedings of the IEEE International Geoscience and Remote Sensing Symposium*, Toulouse, 21-25 July 2003, 3970-3972. http://dx.doi.org/10.1109/igarss.2003.1295331

[15] Rokni, K., Ahmad, A., Selamat, A. and Hazini, S. (2014) Water Feature Extraction and Change Detection Using Multitemporal Landsat Imagery. *Remote Sensing*, **6**, 4173-4189. http://dx.doi.org/10.3390/rs6054173

7

Assessment of Domestic Wastewater Management Practices in the Communal District I of Maradi City, Niger Republic

Ousmane Laminou Manzo[1*], Hassidou Saidou[2], Salamatou Abdourahamane Illiassou[3], Saoudé Tidjani Idrissa[3]

[1]Department of Agricultural Engineering and Water & Forests, Faculty of Agronomy and Environmental Sciences, Dan Dicko Dankoulodo University of Maradi, Maradi, Niger
[2]Department of Chemistry, Faculty of Science and Technology, Dan Dicko Dankoulodo University of Maradi, Maradi, Niger
[3]Department of Biology, Faculty of Science and Technology, Dan Dicko Dankoulodo University of Maradi, Maradi, Niger
Email: [*]lamine_ous@yahoo.fr, saidou.hassidou@gmail.com, assalamat2@yahoo.com, saoudet@yahoo.fr

Abstract

Wastewater treatment is a problem with much acuity in the city of Maradi in general, and particularly in the Communal District I of Maradi. It is for this reason that the present study was conducted which deals with wastewater sanitation. The main objective of this study conducted through a survey is to investigate in the domestic wastewater sanitation in the Communal District I of Maradi city. Specifically, the study aims to estimate the amount of wastewater produced by households and to analyze the sanitation system at household and at district levels. The wastewater management practices in this district were assessed through a descriptive cross sectional study in which a total of 129 households were selected by sample distribution technique in the district neighborhoods and studied using interviewer and administered questionnaires at communal and household levels. The results of the study showed that the amount of domestic wastewater generated is quite large, about 32.27 m^3 per person per day, and this comes from several sources including showers, laundry, dishwashing and income generating activities. The results also revealed that the unhealthy state of the Communal District I of Maradi is related to the absence of a wastewater treatment system, the methods and practices used in its management, and the population is also largely responsible for this unhealthy state and not willing to participate to a sanitation project.

[*]Corresponding author.

Keywords

Domestic Wastewater, Sanitation, Communal District I of Maradi, Management Practices

1. Introduction

African cities are generally over populated because of the exodus of rural populations in search for employment. In most of these cities, waste management and wastewater in particular is a major concern of the people. The wastewater is a by-product of utilized portable water (domestic wastewater) or industrial process water (industrial wastewater) [1]. Sanitation in these cities is usually dominated by the autonomous sanitation systems (about 60% - 95% of people) [2]. In Africa and Asia, it is estimated that 65% to 100% of homes, which have sewerage systems, are equipped with stand-alone installations that are not connected to a sewer system [3]. This is explained by the fact that the construction of drainage systems is very expensive for these countries.

Practically all the cities in the Niger have the same situation where the autonomous systems are not connected to a sewage network. The city of Maradi is not an exception. Indeed, it is experiencing an accelerated urbanization boom and rapid growth population. These two phenomena are the basis of the increased domestic water demand and consequently a significant increase in the volume of wastewater generated in various forms (excreta, toilet water, dishwashing and laundry, wastewater generated by the activities generating income).

The lack of a complete sewerage system in the different zonal areas makes that the wastewater produced is mostly discharged directly into the environment without any treatment. Indeed, the autonomous sanitation systems used (traditional latrines, septic tanks) generate effluents that are rich in fecal coliforms, helminths, viruses, protozoa, and in other various chemical and physical pollutants. The intrusion of these fecal wastes in aquifers or in water distribution system can be a source of contamination of various diarrheal diseases in communities.

Nevertheless, we note the existence of some culverts for the rainwater drainage in some zonal quarters, but used for the disposal of domestic wastewater. However, in areas where it exists, the malfunction of the system disables the evacuation of these waters, especially in densely populated areas. Urban safety is constantly threatened by stagnant sewage in the empty spaces on the floor, in the streets and drains.

Untreated sewage, lack of a proper environmental sanitation system (excreta management, water drainage) and poor hygiene contribute to a large extent to the deterioration of public health and degradation of the life frame of the populations of Maradi city.

The sanitation system of the city is not, for the people, a guarantee against the risk of disease and degradation of their surroundings because of his failure. Therefore, it is urgent to turn to integrated waste management approach to ensure sustainable sanitation in the city.

Thus, in view of the environmental and health interests of the sanitation sector and also its importance in the urban development process, this study aims to contribute to the improvement of the living in Maradi city, through the case study of Municipal District I of Maradi. The overall objective of this study is to contribute to sustainable domestic wastewater sanitation in the Communal District I of Maradi in order to improve the living conditions of populations.

2. Materials and Methods

2.1. Presentation of the Communal District I of Maradi

The population of the Municipal District I of Maradi is of about 94,122 inhabitants with 50.25% of women [4]. This population is predominantly urban and consists of the following ethnic groups: Hausa (92%), Fulani (4.6%), Tuaregs (1.4%) and others 2%. The density of the population is of about 10 inhabitants per km^2. The average size of a household is of about 7 people. This means that they are about 13,446 households in the Communal District I of Maradi.

The problems of housing and urban planning in the municipality still not taking into account the topography of the area (many high sloppy zones), the presence of bed of old Kori, and the pattern of urbanization, which is lacking for this district, in the habitats construction. In rainy season, the district face dwellings damage (houses collapse), and problems of drainage of rain water and even sewage. Misbehavior, carelessness of the population in compliance with the rules of hygiene and inadequate disposal of household waste materials are all problems that complicate the benefit of management and maintenance of safety in this municipal district I of Maradi.

2.2. Observation Units

A survey was conducted in the Communal district I of Maradi. It is carried out at two levels: communal level (1) and household one (2) and questionnaires were administrated to the population in order to collect socio-demographic variables which are related to the wastewater management and those related to environmental problems.

At the household level, the survey will allow assessing the state of the living environment in these settlements at this level through the different wastewater management modes and practices.

The assessment at the communal level, which constitutes the outside environment or the communal area, will analyze and appreciate the level of the unhealthy state of the neighborhood environment.

2.3. Methods

- **Sample distribution**

The survey was based on the questionnaire mode and carried out on a sample of the studied population. It has affected households. The results obtained are extrapolated to the entire communal district I territory. The sampling frame is a list of geographical units which correspond to geographical areas and the component units are households. The survey type selected is that of the sample survey in which data collection is made for only some of the units usually very small population. The sampling method chosen is simple probabilistic and random type. It is a method of selection in one step that can make inferences about the population, given the observations from the sample and ensures that every possible sample size n has an equal chance of being selected [5] [6].

According to the 2011-2015 Social Development Plan the Urban Community of Maradi, the distribution of the population, and thus the households in neighborhoods is the following : Bouzou Dan Zambadi: 15.9%; Mazadou Jika: 6%; Nouveau carré: 19.1%; Soura Alladey: 6.6%; Soura Bildi: 19.8%; Zaria I: 32.6%.

According to the above percentages, the sample distribution of the study in the Communal District I is of about 129 households and is composed as follow: Bouzou Dan Zambadi: 20 households; Mazadou Jika: 8 households; Nouveau carré: 25 households; Soura Alladey: 8 households; Soura Bildi: 26 households; Zaria I: 42 households.

- **Survey questionnaire**

The questionnaire consisted of twenty six questions which are used in the survey. Questions are addressed and discussed with household head. The main topics of this survey include sociodemographic characteristics of households; wastewater production sources; wastewater management methods and practices; sanitation facilities (structures) used; wastewater disposal practices and; amount of water used for daily needs.

- **Interview**

An interview with officials in charge of hygiene and sanitation and those in charge of urban planning from the city of Maradi was conducted to analyze the sanitation system at municipal level. These interviews were also based on a questionnaire.

2.4. Data Processing and Analysis

The sociodemographic data are related to variables such as type of housing, household size, age, sex, education level and type of employment of household head, Analysis of these is to evaluate the share of households in the wastewater production. Also, the degree of involvement of the population in the environmental management is appreciated.

The analysis of data from variables such as wastewater disposal modes, sanitation structures and wastewater management practices of the population related data help to know the views of people on the state of environment unhealthiness and the possibility of an environmental participatory management.

The analysis of data related to hygiene and sanitation of the population environment, wastewater collection and; impacts of poor sanitation on people allow to assess the actual level of living environment degradation and to analyze the environmental impact caused by wastewater.

Data analysis was carried out in Excel and SPSS11.5 software. The statistical processing with Excel software has allowed drawing tables and charts. SPSS 11.5 software has used to describe data and to calculate descriptive statistical parameters, to analyze correlations of education level of household head and 1) wastewater production, 2) wastewater management modes and 3) practices using Pearson correlation analysis.

One-way analysis of variance (ANOVA) was performed to test the significance of differences in education level of household head and 1) wastewater production, 2) wastewater management modes and 3) practices. The Tukey-Kramer Multiple Comparisons Test was used to compare those parameters and education level.

3. Results

3.1. Sociodemographic Characteristics of Households

Results of the study reveal that three types of house were identified in the communal district I of Maradi: 1) The houses made of clay bricks, which represent 3.1% of the sample, are fully built with clay material and are found in old quarters (the largest percentage is found in Soura Bildi); 2) 34.1% of the houses are semi hard built of clay plastered with cement; and 3) 62.8% of the houses are built entirely with cement and are mainly found in Zaria area (about 97.7%).

The age group that has the largest number of heads of household is 31 - 40 years with 32.5% followed by the class 41 - 50 years (31%). This distribution reflects a certain young population. The responsibility of the head of household, in 93.8% of cases, is masculine against 6.2% held by women.

The current study revealed that a sizeable percentage of number of the respondents, of about 7% had tertiary education, 24.8% secondary school education, 20.1% primary school education, 28.7 koranic education and 19.4 with no formal education.

Among the population of the district, 90% of heads of household have a profession (farmer or raiser, trader, mechanic technician, driver and employee. The others 10% with no activity are composed of pensioners and housewives.

3.2. Wastewater Production

According to the results of the survey, each household head is responsible of an average of 14 people. The results of the survey reveal that domestic wastewater in the study area is generated from four (4) different sources which are the shower, the laundry, the dishwashing and the income generating activities. Then, the estimation of the volume of domestic waste water produced was made considering the number of 1) individuals per household; 2) the number of showers per day for each individual; and the volume of water used per shower 3), the number of laundry performed by each individual per week 4); the volume of water used per laundry 5); le volume of water used for dishwashing per day 6) and the volume of water used for income generating activities daily 7).

The total volume of wastewater daily produced in the communal district I of Maradi is of about 44.95 m^3 by 1393 individuals (**Figure 1**). The largest parts of this quantity are produced in Soura Bildi and I Zaria neighborhoods with 16.86 and 15.39 m^3 respectively.

A descriptive analysis of the wastewater quantities generated, by the households of this district, according to the sources give the results shown in the following **Table 1**.

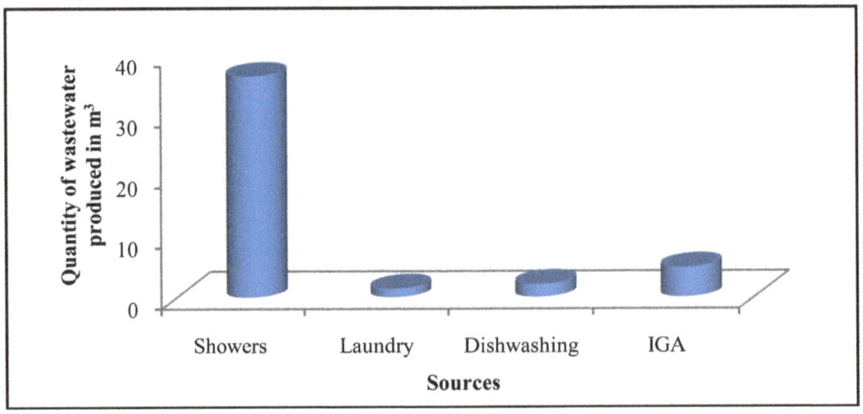

Figure 1. Total volume of wastewater produced in the communal district I of Maradi by production source.

Table 1. Descriptive statistics of wastewater generated in the communal district I of Maradi per production source.

Sources	N	Minimum (in liter)	Maximum (in liter)	Mean	Std. Deviation
Shower	129	25.00	1050.00	282.6550	238.96194
Laundry	129	1.43	60.00	11.1074	9.66967
Dishwashing	129	10.00	40.00	16.6667	6.41450
Income generating activities (IGA)	129	.00	60.00	37.9070	13.44401
Valid N (listwise)	129				

It appears that the wastewater quantity varies depending on the source. The average quantity produced by source and per household are as follow: 282.65; 11.10; 16.66 and 37.90 liters respectively generated by shower, laundry, dishwashing and income generating activities.

3.3. Wastewater Management Practices at Household Level

Wastewater sanitation structures in the study area can be divided into two categories: individual sanitation facilities (latrines and septic tanks) and collective sewerage structures (drainage system of rainwater). In this study, improved latrine means a defecation pit covered with a concrete slab.

In the study area, personal sanitation is the only sanitation mode used because public one is nonexistent. About 14% of surveyed households use modern toilet and have an internal shower and 46% have the system with septic tank.

It also emerged from this study that domestic wastewater is collected by individual facilities (latrines, cesspools, septic tanks). They are the only ones mainly used in the study area (3.9% and 79.9% respectively traditional and improved latrines). Only about 16.2% of surveyed households use modern sanitation facilities (**Figure 2**). This confirms that the majority of people in the communal district I of Maradi do not use improved sanitation facilities.

All households surveyed in the study area are aware of the existence of wastewater drain services. However, only 21.7% of households use their services. As for septic tanks, 75.1% of households empty them manually.

All the households surveyed in the study area are aware of the existence of the wastewater drain services. However, only 21.7% of households use their services. Concerning the septic tanks, 75.1% of households emptied them manually.

Manual pit emptying means that a hole is made next to the septic tank and its content is returned in it and buried. Another method of emptying pits is to pour its contents into the street during the rains to be evacuated by runoff. Among the 78.3% of households that do not use the wastewater drain services, 60.3% do not do it because of its high cost, and 39.7% believe that the drain is usually poorly done or incomplete.

Regarding the wastewater from the laundry, dishwashing and income generating activities, they are removed very often in the street or in the gutters (**Figure 3**). In some households, these waters are discharged even in the courtyard of the house.

Household participation to a sanitation project of wastewater management is a necessity in the context of the sanitation conditions identified in this study. The households interviewed are aware of the magnitude of the consequences on their environment, health and quality of life, due to the mismanagement of the sanitation sector of wastewater. However, 69.8% of households surveyed do adhere to a sanitation project if the services are free. Although the population is aware of its responsibility in the unhealthiness in the study area, only 20.9% of households are willing to invest in a project to clean up their environment and 9.3% do not want it at all.

In the study area, the living environment is continually deteriorated by wastewater. Direct observations reveal that the soil is polluted, the immediate environment of dwellings is deteriorated by septic tanks overflow and wastewater discharge in the streets. The streets are unhealthy and unsightly. The front of 68% of the surveyed households' habitat is invaded by sewage and the air is polluted by the release of noxious odors. The highest percentages of houses whose fronts are unhealthy are found in the districts of Zaria, Soura Bildi and Nouveau carré with respectively 92.8%, 76.9% and 56%, with the presence of sewage.

3.4. Wastewater Management Practices at Communal Level

It is apparent from field observation and interviews with municipal services, responsible for sanitation in the

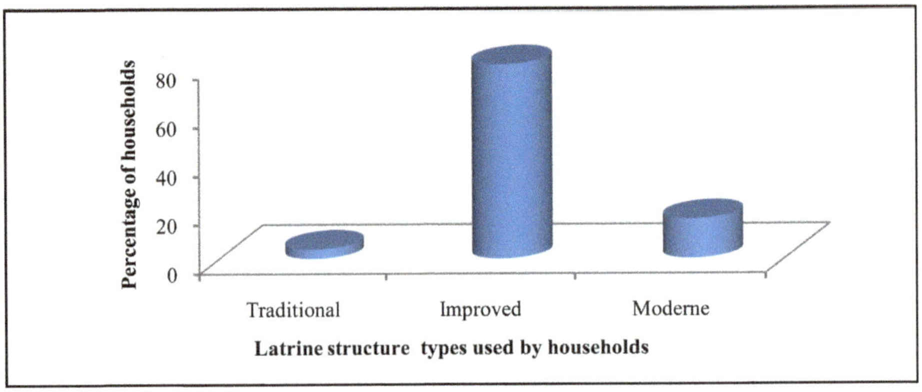

Figure 2. Latrine structures' types used in the communal district I of Maradi.

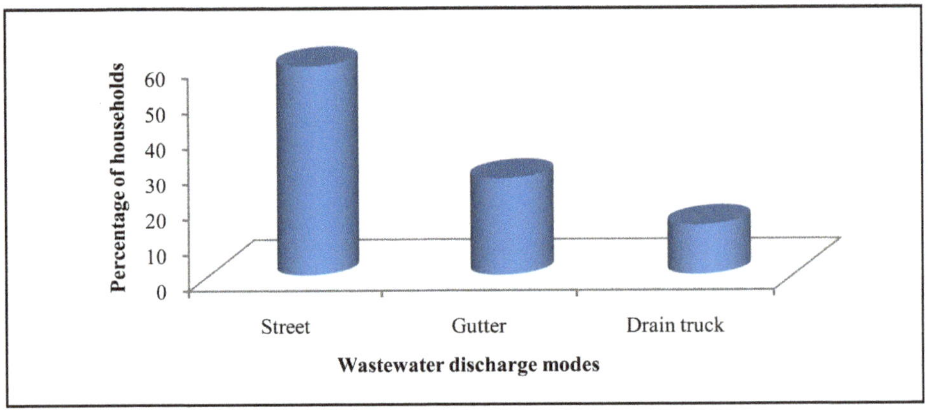

Figure 3. Wastewater disposal modes used in the communal district I of Maradi.

district, that there is no sewage collection network. The sewage system does not exist in the study area. Each inhabitant manages its wastewater at will. Analysis of the results shows that it is individual sanitation that is used in the study area through latrines, cesspools and septic tanks. These structures are often characterized by deficiencies in their design, realization and maintenance.

In addition, the population most often uses the street as a means of wastewater, and even excreta, disposal. Indeed, 75.1% of surveyed households empty their tanks or cesspools manually and discharge the waste directly into the street or in a hole dug next, against 24.9% who use services drain trucks. Wastewater from the laundry is also systematically discharged in the street or in the gutters made for the rainwater evacuation. The result is an almost permanent presence of sewage in the streets as illustrated by the rate of 62.1% of the households having permanent water in front of their houses.

The consequent pollution due to the discharge of wastewater in the streets and their stagnation is important in the study area. About 34.9% of the surveyed households complain about the proliferation of some diseases' vectors like mosquitoes, flies, cockroaches etc., 26.3% of the release of odors and 38.8% fear the disease risks related to the presence of germs and other pathogens in the wastewater.

Also results of the surveys reveal that there is no wastewater management system in the city of Maradi nor in the municipal district Maradi I. Indeed, at the city level, there is no specialized unit responsible for wastewater treatment. Also, there is no sewage system for the collection of these waters, and none other alternative is offered to the population for sanitation. There is also no system for evacuating wastewater. However, this service is partly provided by only two private sewage trucks for all the town of Maradi. Also, these drain services are not subjected to any control by the municipality. There is no wastewater treatment plant, the sewage is discharged in nature without any treatment, any security measures nor monitoring. These operations are not regulated and are made in disregard of the elementary rules of hygiene, health and respect for the environment.

A correlative study was conducted to highlight and analyze any possible links between education level and sewage management type, pits or cesspools emptying practice, the membership in a sanitation project of wastewater.

Table 2. Correlation between education level and 1) wastewater production, 2) wastewater management modes and 3) practices in the communal district I of Maradi.

	Level of education	Volume of wastewater generated per household	Wastewater management mode	Wastewater drain practice	Joining a wastewater sanitation project
Pearson Correlation	1	−0.278**	0.256**	0.237**	−0.069
Sig. (2-tailed)	-	0.001	0.003	0.007	0.437
N	129	129	129	129	129

NB: ** indicate that the statistical difference is highly significant (Tukey-Kramer Multiple Comparisons Test, $p < 5\%$).

The results of the analysis (**Table 2**) show that the level of education has an influence on the quantity of wastewater produced, on the method of wastewater management practice and on the wastewater disposal (pits or cesspools).

The higher the education level of the household head, the lower the amount of wastewater generated is significant ($r = -0.278$; $p = 0.001$), better the household handle the wastewater ($r = 0.256$; $p = 0.003$) and adopt a healthier drain practice ($r = 0.237$; $p = 0.007$).

However, the education level of the household head has no influence on the intention to join a wastewater sanitation project ($r = -0.069$; $p = 0.437$). This supposes that the population is partly responsible for the unhealthiness in the study area.

4. Discussion

Analysis of the results reveals that the amount of waste water produced by the population of the Municipal District I Maradi is important and is generated by several sources. This fact is explained by the population density and the high growth rate of the population. The main sources of domestic wastewater are the shower, laundry, dishwashing and income generating activities.

The results also showed that unhealthiness in the communal District I of Maradi is related to the sanitation system used. Individual and autonomous sanitation system used in the district does not play its role as the sanitation structures are poorly designed, poorly executed and poorly maintained. Failure of many onsite systems is generally not due to inherent flaws in system technology, but rather due to inappropriate siting and design issues or their operation and management are poorly executed [7]. According to a study report on sanitation access, conducted in five African countries (Niger, Sierra Leone, Ghana, Uganda and Rwanda) the Niger's population increased by 7.7 million inhabitants between 1990 and 2010 while only one million of people have access to sanitation during the same period [8]. The report also points out that nearly 91% of Niger's population currently lack access to proper toilets. This could be explained by the low-income population because the construction of a latrine—even traditional ones—requires money to buy cement and rebar for the slab, not counting the labor of work. According to Nassartbaye (2011) [9], in Gamkalé neighborhood in Niamey, more than 23% of the population have no latrine and live in a very unhealthy environment.

The results of our study also established correlations between wastewater management practices, individual behaviors and unhealthiness in the study area. It is clear from these results that the population is partly responsible for the unhealthy environment in the municipal district Maradi I. The study of Younsa (2011) [10] in the "Pays-bas" neighborhood (Niamey), revealed that 52.5% of households discharge their wastewater in the street. Populations have adapted to this system despite all the health and environmental risks to which they are exposed. This unhealthiness environment is common to all the cities of Niger. The impacts associated with onsite wastewater treatment systems has come to the fore in recent years in order to protect public health and the environment from the consequences of poorly performing systems, in particular septic tanks [11] [12].

The study has also highlighted the responsibility of municipalities in the sanitation conditions in the study area. Indeed, there is no wastewater management system, nor a service in charge of wastewater treatment and no control is done in this field of discharge. However, it is well known that the municipal authorities have the responsibility for waste management and wastewater treatment. Winkler (2005) [13] has shown that the sanitation sector is not a concern of municipal authorities. Yet in the context of adoption of innovative practice, sanitation could prove to be an effective tool in the fight against poverty. Under the "Projet d'Assainissement Productif" (Productive Sanitation Project) in Aguié (Maradi), it has been estimated that the annual quantity of fertilizer from sanitation sector produced by a family of nine people correspond to a bag of 50 kg of urea and another of NPK (15:15:15) [14]. The value of this fertilizer can reach 30 to 40,000 FCFA on the local market. It was estimated for the entire population of Niger in 2009, an annual value of about 60 billion FCFA in fertilizers present in the human excreta, the majority is found in the urine.

The correlation study between the different parameters showed that the higher the education level of the household head, the more households adopts healthier practices for wastewater management. This could be explained by the fact that when we are better informed (through mass media), we paid more attention to what may cause health problems to the family. This result is in line with the findings of works done by Adeyemo and Gboyesola (2013) [15] and Adogu *et al.* (2015) [16] on knowledge, attitude, and practices on waste management of people living respectively in the university area of Ogbomoso and Owerri Municipality in Nigeria, which indicated that the respondents were knowledgeable in waste management

However, the study also revealed that the level of education has no influence on adherence to a wastewater sanitation project. Malpractice and recklessness of the population could explain this behavior. Indeed, the population of Maradi does not have the habit of the daily cleaning of the front of houses. However, according to Adogu *et al.* (2015) [16], strict adherence to appropriate waste management practices in any community will insulate the inhabitants from detrimental and hazardous environmental conditions and improve the living standard of the people.

It seems clear that the municipal district I of Maradi does not have a wastewater management system. The sanitation system used through septic tanks, cesspools and latrines does not guarantee the safety of the living conditions of the populations. Nicosia *et al.* (2001) [17] found that failing septic systems are the most frequently reported cause of groundwater contamination. Also, Carroll *et al.* (2006) [18] reported that approximately 50% of waterborne disease outbreaks in the USA were a result of consumption of contaminated groundwater, with septic tanks reported as the most frequent cause of contamination. Populations share the responsibility for the environmental degradation state with municipal authorities. Wastewater management seems not to be a major concern to the municipal authoriyies, probably because of lack of human, financial and material resources. However, Adogu *et al.* (2015) [16] consider that proper waste disposal management is essential to sustain healthy living conditions in any environment. However, it is well known that municipal wastewater can be an important water resource but its use must be carefully planned and regulated to prevent adverse health effects [19].

In view of the results of this study, we could say that unhealthy state of the study area is also similar in the other two municipal districts of the city of Maradi. The findings should be the same concerning the living environment conditions.

5. Conclusions

Unsanitary conditions, in which people live in communal district I Maradi of Maradi, constitute a major problem for the local authorities and the population itself.

The results of this study, conducted in this District, show that this population produces significant quantities of domestic wastewater from showers, laundry, dishwashing and income generating activities. However, there is no sanitation system at the administrative entity level, and no service is set up to handle the wastewater management. This demonstrates a certain lack of involvement of local authorities in this area.

Also, it is demonstrated that the type of sanitation used remains the main cause of unhealthiness in the area, but it is also worth noting that the population is partly responsible for this situation, because of the adoption of malpractice management of wastewater. It is routinely discharged in the street gutter if they exist or even in the courtyard of the concession.

Large numbers of the population are aware of the consequences of the non wastewater management. However, membership household for a possible project of wastewater treatment in the study area did not find a fairly favorable response and participatory. This reflects a kind of wait-and resignation of the population to face its responsibilities in maintaining their lifestyle in a healthy state. As rightly has said Konaté (2012) [20], sanitation should not be regarded as a burden to be borne by the local authorities and population; it is a *sine qua non* condition for sustainable development. Therefore, efforts must be provided at all levels to solve the problem of hygiene and sanitation in the communal District I of Maradi and even in the entire town of Maradi where the urbanization rate and increase of the population continue to rise.

References

[1] Emigilati, M.A., Ishiaku I., Usman, B.Y., Kuta, G.I. and Dangana, K. (2015) Assessment of Effluents Discharged from Textiles Industries in Selected Villages in Kaduna State, Nigeria. *African Journal of Environmental Science and Technology*, **9**, 385-389.

[2] Lacina, C., Diomandé, D., Coulibaly, A. and Gourène G. (2004) Utilisation des ressources en eaux, assainissement et risques sanitaires dans les quartiers précaires de la commune de Port-Bouët (Abidjan; Côte d'Ivoire). *Vertigo*, **5**, Issue 3. http://vertigo.revues.org/3299

[3] Diallo, B.D. (2012) Cours de Gestion écologique des déchets ménagers. Master en Agriculture Durable. Service des Etudes Avancées, Institut Supérieur Agronomique et Vétérinaire de Faranah, Guinée, 104 p.

[4] Institut National de la Statistique (2012) Recensement Général de la Population et de l'Habitat. INS, Ministère des finances, République du Niger, 10 p. http://www.stat-niger.org/statistique/file/rgph2012.pdf

[5] Visticot, C. and Morin, C. (2005) La réalisation d'une enquête, Pédagogie, Économie et gestion, Académie d'Aix-Marseille. http://www.ecogesam.ac-aix-marseille.fr/Resped/Commerce/Outils/

[6] Ministère de l'industrie du Canada (2010) Méthodes et pratiques d'enquête. Statistique Canada. http://www.statcan.gc.ca/pub/12-587-x/12-587-x2003001-fra.pdf

[7] Otis, R.J. and Anderson, D.J. (1994) Meeting Public Health and Environmental Goals: Performance Standards for On-site Wastewater Treatment Systems. *Proceedings of the Seventh National Symposium on Individual and Small Community Sewage Systems*, Atlanta, Georgia, 11-13 December 1994, 1-10.

[8] Ousseini, I. (2013) NIGER: Très peu de gens ont accès à l'assainissement. Inter Press Service News Agency. http://ipsinternational.org/fr/_note.asp?idnews=7474

[9] Nassartbaye, N. (2011) Accès à l'eau potable et à l'assainissement; quels enjeux pour la santé dans les quartiers précaires? Etude appliquée au quartier Gamkallé de la commune IV de Niamey au Niger. MSc Dissertation, Abdou Moumouni University of Niamey, Niamey.

[10] Younsa, H.H. (2011) L'accès à l'eau potable et à l'assainissement dans les quartiers précaires de Niamey: Cas de Pays-bas (commune IV). MSc Dissertation, Abdou Moumouni University of Niamey, Niamey

[11] Geary, P.M. and Whitehead, J.H. (2001) Groundwater Contamination from On-site Domestic Wastewater Management Systems in a Coastal Catchment. In: *Proceedings of the Ninth National Symposium on Individual and Small Community Sewage Systems*, American Society of Agricultural Engineers, Fort Worth, Texas, 11-14 March 2001, 479-487.

[12] Lipp, E.K., Farrah, S.A. and Rose, B. (2001) Assessment and Impact of Microbial Fecal Pollution and Human Enteric Pathogens in a Coastal Community. *Marine Pollution Bulletin*, **42**, 286-293. http://dx.doi.org/10.1016/S0025-326X(00)00152-1

[13] Winkler, S. (2005) Les expériences du CREPA dans la promotion de l'assainissement autonome en Afrique de l'Ouest: Etat des lieux, analyse et perspectives. Mémoire de recherche, Ecole Polytechnique Fédérale de Lausanne, cycle postgrade environnement: Sciences, ingénierie & management.

[14] Linus, D., Kailou, H. and Hadidjatou, I. (2009) L'assainissement productif à Aguié, Niger, Le secteur agricole prend le volant, *Symposium Régional sur l'Assainissement et l'Hygiène en Afrique de l'Ouest*, Accra, Ghana, 3-5 November 2009.

[15] Adeyemo, F.O. and Gboyesola, G.O. (2013) Knowledge, Attitude and Practices on Waste Management of People Living in the University Area of Ogbomoso, Nigeria. *International Journal of Environment Ecology, Family and Urban Studies*, **3**, 51-56.

[16] Adogu, P.O.U., Uwakwe, K.A., Egenti, N.B., Okwuoha, A.P. and Nkwocha, I.B. (2015) Assessment of Waste Management Practices among Residents of Owerri Municipal Imo State Nigeria. *Journal of Environmental Protection*, **6**, 446-456. http://dx.doi.org/10.4236/jep.2015.65043

[17] Nicosia, L.A., Rose, J.B. and Stewart, M.T. (2001) A Field Study of Virus Removal in Septic Tank Drain Fields. *Journal of Environmental Quality*, **30**, 1933-1939. http://dx.doi.org/10.2134/jeq2001.1933

[18] Carroll, S., Goonetilleke, A., Thomas, E., Hargreaves, M., Frost, R. and Dawes, L. (2006) Integrated Risk Framework for Onsite Wastewater Treatment Systems. *Environmental Management*, **38**, 286-303. http://dx.doi.org/10.1007/s00267-005-0280-5

[19] Bouwer, H. (2000) Integrated Water Management: Emerging Issues and Challenges. *Agricultural Water Management*, **45**, 217-228. http://dx.doi.org/10.1016/S0378-3774(00)00092-5

[20] Konaté, S. (2012) Gestion des eaux usées domestiques dans le district de Bamako. Cas de la commune V, MSc Dissertation, University Institute of Territorial Development, Mali.

8

Evaluation of Nutrients and Trace Metals and Their Enrichment Factors in Soil and Sugarcane Crop Irrigated with Wastewater

Mohammed Abdullah Alghobar, Sidduraiah Suresha*

Department of Environmental Science, Yuvaraja's College, University of Mysore, Mysore, India
Email: *sureshakumar12@yahoo.com, alghobar@yahoo.com

Abstract

The study was conducted to investigate enrichment factor of wastewater (WW) irrigation to agricultural soils and the bioconcentration factor of trace metals uptake by sugarcane and to determine the concentrations of nutrients in sugarcane by growing the same on a contaminated soil, during the growth period of crop in Vidyaranyapuram area in the South West of Mysore, Karnataka. The concentrations of macronutrients and trace metals in the soil sampled sugarcane crop were found highly significant in Ca, Mg, K, Na, N, Mn, Cu, Ni and Cr. The enrichment factor was maximum for N and Na, and the values of N and Na were in the significant enrichment category. Whereas the values of following elements (K, P, Ca, SO_4, Cu, Cr, Mg, Mn, Ni, Cd, Fe, Pb, Cl, Co and Zn) in soil irrigated with WW were in moderate enrichment category. The mean concentrations of N, P, K, Ca, Na, Fe, Mn, and Pb in sugarcane leaves in wastewater treatment crops were significantly higher than those with ground water (GW) treated which showed the lowest concentrations. Calculation of the bioconcentration factor (BCF) indicated that the accumulation of heavy metals in sugarcane crop was in the order Cd > Zn > Co > Pb > Mn > Ni > Cu > Cr > Fe. Pearson's correlation coefficient indicated the relationship between different elements content in soil and plant. The results showed that the elements, Ca, Mg, Na, K, N, P, Mn, Zn and Cr, were significantly positively correlated, between the soil and the plant.

Keywords

Heavy Metals, Wastewater, Sugarcane, Enrichment Factor

*Corresponding author.

1. Introduction

There is increasing use of treated and untreated sewage water for irrigation, especially in the agricultural regions of India, and there is need to adequately assess the possible environmental impacts of such applications and agricultural practices, to ensure sustainable agricultural development. There is the possibility toxic of effects of trace elements in soil and their bioaccumulation in plants and vertical leaching/transport into shallow groundwater systems [1]. Irrigation with contaminated water contributes as a source of contamination of trace elements in agricultural soils and is harmful especially for vegetable production [2]. In agricultural soils, the presence of metals is of increasing concern because they have the potential to get accumulated in less soluble forms, get transferred into soil solution, and subsequently deteriorate the groundwater and crop quality [3]. Increasing soil heavy metal concentrations may result in increased uptake by plants and potentially cause serious health problems such as stomach ache, renal dysfunction and pulmonary emphysema [4]. Trace metal mobility depends on soil characteristics including soil pH and texture, the type and quantity of the oxyhydroxides present, the content of organic matter, and soil main constituents such as carbonates, phosphates, and clays which are responsible for toxic metal and metalloid sorption [5]. Increase of pH, EC and availability of N, P, K, Fe, Mn, Zn, Cu, Cd, Co, Pb, Ni and Cr in surface layer of paper industry effluent irrigated soil, when compared to well water irrigated soils [6]. Mean concentrations of Cu, Fe, and Cd in irrigation water and Cd content in soil were much above the recommended levels. Accumulation of heavy metals in vegetables studied was lower than the recommended maximum tolerable levels proposed by the Joint FAO, with the exception of Cd which exhibited elevated level. Uptake and translocation pattern of metals from soil to edible parts of vegetables were quite distinguishable for almost all the elements examined [7]. Effect of irrigation of forage crops with treated sewage water provided factors other than N and P for improving crop production such as improving soil fertility and soil organic matter [8]. Soils irrigated with wastewater accumulated heavy metals like Zn, Pb, Cd, Ni and Cr in surface soil [9]. Sewage water irrigated soils showed higher Cd, Ni and Cu levels than the desirable levels, while the levels of Hg, Pb, and Zn were lower than the maximum threshold values recommended for crop production. Comparing between the crops, it was found that Hg and Cd levels were highest in soils under maize and declined linearly from maize to spinach to olive to tomato and control site [10]. Long term irrigation with sewage water led to the deposition of heavy metals (Ni, Pb, Cd, Cr and Cu) in soil and also bio-accumulation in the plants and tubers of potato crop [11]. Continuous use of sewage and industrial effluents for irrigation recorded improvement in water retention, hydraulic conductivity, organic C and build-up of available N, P, K and micronutrient status. The EC although increased due to sewage irrigation, it was within the tolerance limit to cause any soil salinity hazard. The toxic metals like Cd, Cr and Ni were found to get accumulated in the soil and plant due to long-term use of sewage irrigation. Hence, Cd, Cr and Ni are more likely to be the elements that may become health hazard for consumers of the crops grown in sewage irrigated soils. The concentrations of these metals were greater in leafy vegetables than in grain crops [12]. Understanding and prediction of the accumulation of heavy metals and metalloids in crops grown on contaminated croplands were motivated to secure crop quality and food safety [13]. The present study was conducted to investigate enrichment factor of sewage water irrigation to agricultural soils and the bioconcentration factor of trace metals uptake by sugarcane and to determine the concentrations of nutrients in sugarcane by growing the same on a contaminated soil.

2. Materials and Methods

2.1. Description of Study Area

The study area is located in the suburban area in the south western part of Mysore city, Karnataka, India, near Vidyaranyapuram sewage treatment plant. Locations were selected to gain understanding of the effect that typical wastewater (WW) creates on soil and sugarcane crop in Mysore city. The present study also covers the physico-chemical characteristics of water samples collected from Vidyaranyapuram sewage treatment plant station. More than fifty percent of the wastewater handled by in Mysore city is received by Vidyaranyapuram sewage treatment plant. The total sewage generation of sewage treatment plant is 67.75 million liters per day. It is a biological treatment plant situated next to the solid waste disposal area at the foothills of Chamundi Hills; the treated wastewater of Vidyaranyapuram sewage treatment plant crosses the Dalvai Lake and reaches drinking water source that is the Kabini River. The treated sewage water is pumped out after sewage treatment to field channels for direct use as irrigation water; also the farmers use this untreated wastewater for irrigating various crops.

2.2. Sampling and Analysis of Water

Wastewater (WW), treated wastewater (TWW) and ground water (GW) samples. Manual sampled water was carried collected in plastic bottles of 2000 ml capacity which were thoroughly cleaned and were rinsed with the water being sampled. The randomly collected water samples were studied and analyzed for the physicochemical parameters in accordance with procedures outlined in Standard Methods (APHA) [14]. To estimate heavy metals concentration an aliquot consisting of 500 mL of water samples, ach digest was made up to 50 ml with the addition of double distilled water. Pb, Zn, Cu, Co, Cr, Cd, Fe, Mn and Ni were determined in the digest by ICP-OES (Perkin Elemer model 8000 DV).

2.3. Soil Sampling and Analysis

The survey was conducted around Vidyaranyapuram sewage treatment plant area of Mysore city during 2012-2013. In these areas wastewater both (treated and untreated) are being used by farmers for irrigation of sugarcane crop. Soil samples were routinely collected from fields to obtain a representative soil sample. In the fields, soil samples were collected from three depths (0 - 20, 20 - 40 and 40 - 60 cm), by using auger of 3 cm diameter. Soil samples from the farmer's fields were randomly collected and bulked together to form a composite sample and transported to the laboratory. Samples were air-dried, crushed, passed through 2 mm sieve and saved in plastic bottles. Soil samples were analyzed as per the standard procedures of APHA [14] and ISI [15].

2.4. Crop Sampling and Analysis

Sugarcane crop samples were collected randomly in triplicate from fields irrigated with WW, as well as GW. In the laboratory, sugarcane leaves were washed with distilled water to remove soil particles, were dried in a hot air oven at 70˚C - 80˚C for 24 h, crushed and powdered for metals determination. For the digestion of sugarcane leaves samples, wet digestion method was used. For each analysis 0.5 g of the sample was accurately weighed and digested with 1 ml of perchloric acid ($HClO_4$) and 4 ml of Nitric acid (HNO_3). The samples were allowed to cool and the contents were filtered off using Whatman No. 42 filter paper. The filtrate was made up to 25 ml with distilled water. Blank solution was made using the same procedure except the addition of leaf sample. Heavy metals (Fe, Mn, Cu, Zn, Pb, Cd, Ni, Co and Cr) concentrations in crop samples were estimated on ICP-OES (Perkin Elmer, Model 8000 DV).

3. Evaluation Indices and Data Analyses

3.1. Enrichment Factor (EF)

EF is the relative abundance of a chemical element in a soil compared with the background [16]. In this study, the EF method was used to estimate the level of possible contamination of the soil. Elements accumulation in soil amended with wastewater on a contaminated site with respect to soil on uncontaminated soil (control soil) gives the EF. The EF is calculated according to the following equation [17]:

EF = concentration of metals in the amended soil/concentration of metals in the control soil

The enrichment factor is used to assess soil contamination (enrichment), and its interpretation is as follows: EF < 2 – depletion to minimal enrichment, EF 2 < 5 – Moderate enrichment, EF 5 < 20 – Significant enrichment, EF 20 < 40 – very high enrichment EF > 40 – extremely high enrichment [18] [19]. EF can also be used to evaluate element depletion from soil [20].

3.2. Bioconcentration Factor (BCF)

BCF is a parameter used to describe the transfer of trace elements from soil to plants. It was calculated as the ratio between the concentration of particular elements in the plant and the same in the corresponding soil as well; all based on (dry weight) and for each crop calculated separately [21]. The BCF of Ca, Mg, Na, K, N, P, Fe, Mn, Cu, Zn, Cd, Ni, Pb, Co, and Cr and Pb in the plant samples were calculated as follows:

BCF = Concentration of heavy metal in plant/Concentration of heavy metal in soil

3.3. Statistical Analyses

The results of soil and crop analyses were subjected to analyses of variance (ANOVA) as applicable to a completely randomized block design .The statistical analyses were performed for each parameter, and were compared as for Duncan's least significant difference test at $p < 0.05$. All statistical analyses were carried out using the SAS program, Version 9.1 [22]. Pearson's correlation coefficients between soil elements and elements in crops were calculated to evaluate the relation between these parameters in the soil and the crop. Correlation analysis was computed using the software package SPSS (version 19.0).

4. Results and Discussion

4.1. Physical and Chemical Parameters of Wastewater Effluent

The results of this study showed that the levels of pH were 7.50, 8.13, 9.19 and 8.30 in WW, TWW and GW, respectively. Generally WW treatment shows the lowest value. Domestic wastewater before treatment typically has a pH of 6.5 to 8.5, according to FAO [23]. Electrical conductivity of water is a useful and easy indicator of its salinity or total salt content. The conductivity values were 1032, 1225, 906 and 1099 μS/cm, for WW, TWW and GW, respectively. The conductivity of water is useful as an indicator of the salinity or total salt content which is high in the wastewater. This result is in consistent with [24], who observed wastewater from domestic sewage often contained high amounts of dissolved salts. The concentration of TDS is 560, 624 and 696 for WW, TWW, and GW respectively. These values obtained for TDS for all the water samples were saline for irrigation waters purpose but lower than FAO [23] standard according to which it is 2000 mg/l for water samples for irrigation. So when used this type of water, do not use the more saline waters in this class on soils with restricted drainage. Even with adequate drainage, best management practice for salinity may be required, and the salt tolerance of the plants to be irrigated also must be considered [25]. The concentrations of COD for WW, TWW, and GW, were 964, 145 and 16 mg/l, respectively. COD value is greater than that prescribed by FAO as shown in **Table 1**. in all different wastewaters. TDS cause toxicity through increased salinity, changes in the ionic composition of the water and toxicity of individual ions thereby resulting in low COD removal efficiencies [26]. While concentration of BOD in WW as higher than that of FAO values of 100 mg/l. High COD and BOD concentration observed in the wastewater might be due to the use of chemicals, which are organic or inorganic [24].

The results of concentration of major ions, macronutrients and micronutrient in water samples are presented in **Table 2** and **Figure 1**. Cation concentrations are within the permissible limit for irrigation water and anion concentrations of wastewater in all different water samples could be rated as marginal. Based on the results and on the standards given by FAO [23]-[27] for using the wastewater and for discharging them on land for irrigation it may be stated that the wastewater, is of less hazardous. Macro and Micronutrient status and the concentrations of heavy metals in are given in **Figure 1**. N and P concentrations of WW and TWW treatment are higher than permissible limits by FAO. The heavy metals (Fe, Mn, Cu, Zn, Cd, Ni, Pb, Co, and Cr) analysed in irrigation water are not at elevated levels except Cd, which is higher in wastewater and groundwater (**Figure 1**). On average, the levels of Mn were higher in the irrigation water sources compared to the recommended 0.1 mg/l; the concentration of Cu was below detection limits. Related studies on wastewater irrigation in two cities in India have shown similar trends. In Varanasi, [28] reported equally low levels in wastewater. Continuous application of irrigation from wastewater could contribute to heavy metal accumulation in soils and crops, even if their levels are low in irrigation water source.

4.2. Effect of Wastewater Irrigation on Soil Properties

Soil properties greatly influence the availability of nutrients and elements to plants [29]. The solubility of metallic

Table 1. Physiochemical parameters of water samples.

Water quality	T°C	pH	EC μs/cm	DO mg·L^{-1}	COD mg·L^{-1}	BOD mg·L^{-1}	TDS mg·L^{-1}
WW	25	7.50	1032	Nil	964	650	560
TWW	25	8.13	1225	2.3	145	30	624
GW	25	8.30	1099	6.9	16	2	696
Standard Limit for irrigation [23]-[27]		6.5 - 8.5	<3000	<9	80 - 500	100	2000

Table 2. Concentration of major ions ($mg \cdot L^{-1}$) in different water samples.

Water quality	Cation mgL^{-1}				Anion mgL^{-1}			
	Na^+	K^+	Ca^{2+}	Mg^{2+}	HCO_3^-	CO_3^{2-}	Cl^-	SO_4^{2-}
WW	48	24	43.37	27.01	296	Nil	93	24
TWW	60	20	62.64	28.89	392	Nil	115	20
GW	56	20	56.22	68.50	544	40	17	52
Standard Limit for irrigation [23]-[27]	900	-	400	60	600	6	1100	1000

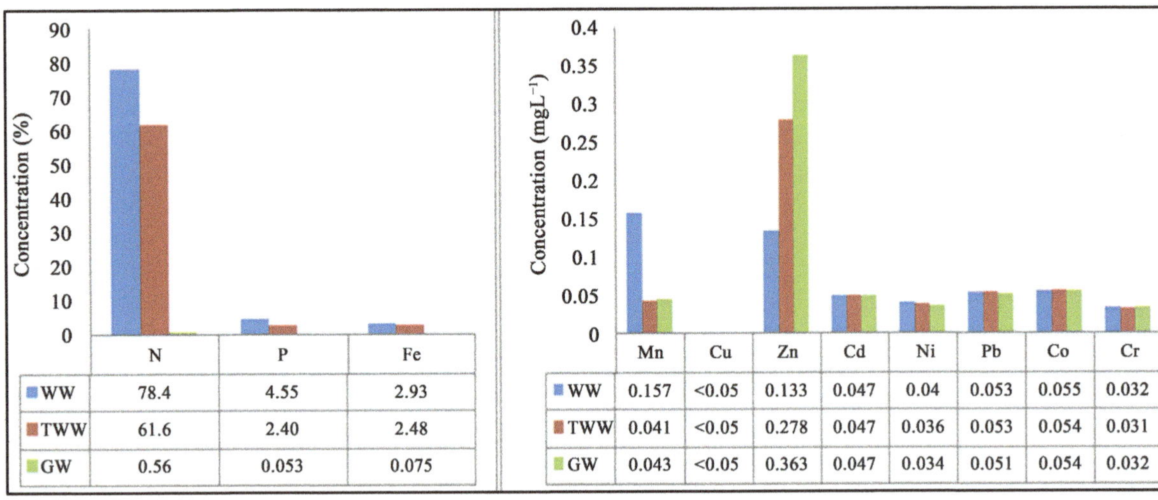

Figure 1. Concentration of macronutrients and micronutrient in water samples.

ions depends on various soil physiochemical factors, such as the pH [30]. In the present study, analysis of soils irrigated with different wastewater was performed to study the effect of soil properties on enrichment factor and availability of nutrients in soil. The data of soil analysis are shown in **Table 3**. Soils sampled with WW showed acidic pH against alkaline pH of TWW and GW soils, the values were 6.67 for WW, 7.56 and 7.92 respectively. Data presented in **Table 3** also show that the soil EC of experimental field was significantly influenced by WW irrigation, which are 259 and 214 μs/cm for WW and TWW irrigated soil respectively. The exchangeable cations like Ca, Mg, K and Na showed highly significant difference between WW, TWW and GW, the mean value content recorded of WW, TWW and GW in the present study were 0.90%, 0.81% and 0.50% for Ca, 0.44%, 0.39% and 0.34% for Mg, 0.31%, 0.22% and 0.059% for Na, 0.44%, 0.33% and 0.13% for K. N content showed highly significant value of WW for 0.53% followed 0.46% for TWW and lowest of 0.038% was for GW. The low organic nitrogen content of the untreated olive mill wastewaters increased the soil C/N ratio to as much as 50 [31]. In the case of P, the highly significant value was 0.061% for WW compared to lower value of 0.032% for control GW. The WW contained appreciable amounts of useful major plant nutrients, such as N, P, K, and Ca, which are reflected in the appreciable build-up of these nutrients in the sewage-irrigated soils [32]. Soil concentrations of heavy metals were found highly significant values in the Mn, Cu, Ni and Cr in the field experiment soils due to the application of wastewater are given in **Table 3**. Heavy metal concentration in WW, TWW and GW irrigated sites ranged between 238, 224 and 204 mg/kg for Mn, 266, 243 and 177 mg/kg for Cu, 58, 54 and 51 mg/kg for Ni and 107, 99 and 80 mg/kg for Cr. According to [33] although the concentration of heavy metals in sewage effluents are low, long-term use of WW on agricultural lands often results in the buildup of elevated levels of heavy metals in soil. In WW irrigated soil Mn, Cu, Ni and Cr contents were significant high Fe, Cd, Pb and Co were equal and Zn was lower on compared to GW irrigated soil.

4.3. Assessment of Trace Metals Enrichment Factor (EF) in Soils

Enrichment factor is an indication of the anthropogenic impact on soil or enrichment factor can give an insight into differentiating an anthropogenic source from a natural origin [34]. The enrichment factor of the elements in

Table 3. Nutrients concentrations and estimated enrichment factors for soil.

Parameters	Soil of WW		Soil of TWW		Soil of GW (Control)
	Treatment	EF	Treatment	EF	
pH	6.67 c	-	7.56 b	-	7.92 a
EC µs/cm	259 a	-	214 b	-	187 c
Ca^{2+} %	0.90 a	1.80	0.81 ab	1.60	0.50 c
Mg^{2+} %	0.44 a	1.30	0.39 ab	1.15	0.34 b
Na^{+} %	0.31 a	5.25	0.22 b	3.72	0.059 c
K^{+} %	0.44 a	3.38	0.33 b	2.54	0.13 c
Cl^{-} %	0.002 a	1.00	0.002a	1.00	0.002a
N %	0.53 a	13.95	0.46 a	12.11	0.038 b
P %	0.061 a	1.97	0.041 b	1.28	0.032 b
SO_4^{2-} %	0.0017 a	1.70	0.0017 a	1.07	0.0010 a
Fe^{2+} %	3.63 a	1.02	3.53 a	0.99	3.55 a
Mn^{2+} mg/kg	238 a	1.16	224 b	1.09	204 c
Cu^{2+} mg/kg	266 a	1.50	243 b	1.37	177 d
Zn^{2+} mg/kg	341 d	0.67	404 c	0.80	503 a
Cd^{2+} mg/kg	12 a	1.09	12 a	1.09	11 a
Ni^{2+} mg/kg	58 a	1.13	54 b	1.05	51 b
Pb^{2+} mg/kg	81 a	1.02	80 a	1.01	79 a
Co^{2+} mg/kg	16 a	1.00	16 a	1.00	16 a
Cr^{2+} mg/kg	107 a	1.34	99 b	1.24	80 d

Different letters in the same row indicate significant difference among means as determined by Duncan's multiple-range test ($p < 0.05$).

soil were in the sequence of N (13.95) > Na (5.25) > K (3.83) > P (1.97) > Ca (1.80) > SO_4 (1.70) > Cu (1.50) > Cr (1.34) > Mg (1.30) > Mn (1.16) > Ni (1.13) > Cd (1.09) > Fe, Pb (1.02)> Cl, Co (1.00) > Zn (0.67) in the WW irrigated soil (**Table 3**). However, values of N and Na were in the enrichment categories of significant enrichment, while values of K, P, Ca, SO_4, Cu, Cr, Mg, Mn, Ni, Cd, Fe, Pb, Cl, Co and Zn in the enrichment categories of moderate enrichment. The EF values of elements in soil irrigated with TWW was in the order N (12.11) > Na (3.72) > K (2.54) > Ca (1.60) > Cu (1.37) > P (1.28) > Cr (1.24) > Mg (1.15) > Mn, Cd (1.09) > SO_4 (1.07) > Ni (1.05) > Pb (1.01) > Cl, Co (1.00) > Fe (0.99) > Zn (0.80). There are significant differences in enrichment factor values among the elements; value of N is in the enrichment category of significant enrichment. While values of Na, K, Ca, Cu, P, Cr, Mg, Mn, Cd, SO_4, Ni, Pb, Cl, Co, Fe and Zn are in the moderate enrichment category. When enrichment value is high (>1) it indicates higher availability and distribution of metals in soil irrigated with contaminated water and thereby increasing the average heavy metal concentration in major crops, vegetables and weeds in wastewater irrigated area with respect to their reference values [35]. [36] reported the enrichment factor values were lower than the values of heavy metals in soil irrigated with wastewater generated from the sponge iron industries which were in the moderate enrichment category.

4.4. Nutrient Concentrations in Plants

Mean concentrations of macronutrients, and heavy metals in sugarcane leaves are presented in **Table 4**. Higher concentrations of macronutrients (N, P, K, Ca, and Na) in plants were observed in treatment irrigated with wastewater; the constant trend in their concentrations in decreasing order was as follows: WW, TWW, and GW, The mean values were 1.94%, 1.84% and 0.46%, for N, 0.16%, 0.13% and 0.10% for P, 1.25%, 1.18% and 0.93% for K, 0.64%, 0.63% and 0.51% for Ca, 0.27%, 0.25% and 0.24% for Na. Irrigation with raw or diluted wastewater resulted in higher concentrations of most nutrients in the millet plant tissues, as expected, this higher accumulation in plant paralleled higher nutrient concentrations in the irrigation water [37].

Table 4. Mean concentrations of macronutrients and heavy metals in sugarcane leaves.

Parameters	WW	TWW	GW
Ca^{2+} %	0.64 a	0.63 ab	0.51 c
Mg^{2+} %	0.34 a	0.37 a	0.39 a
Na^{+} %	0.27 a	0.25 b	0.24 b
K^{+} %	1.25 a	1.18 a	0.93 b
N %	1.94 a	1.84 a	0.46 c
P %	0.16 a	0.13 b	0.10 c
Fe^{2+} mg/kg	410 a	370 ab	340 b
Mn^{2+} mg/kg	72 a	66 b	58 c
Cu^{2+} mg/kg	38 a	36 a	35 a
Zn^{2+} mg/kg	220 c	283 b	332 a
Cd^{2+} mg/kg	13 a	13 a	13 a
Ni^{2+} mg/kg	12 a	12 a	11 a
Pb^{2+} mg/kg	30 a	27 ab	25 b
Co^{2+} mg/kg	8 a	8 a	8 a
Cr^{2+} mg/kg	15 a	14 a	13 a

Different letters in the same row indicate significant difference among means as determined by Duncan's multiple-range test ($p < 0.05$).

Table 5. Bioconcentration factors for heavy metals in sugarcane crop.

Parameters	WW	TWW	GW
Fe^{2+} mg/kg	0.01	0.01	0.01
Mn^{2+} mg/kg	0.30	0.29	0.28
Cu^{2+} mg/kg	0.14	0.14	0.19
Zn^{2+} mg/kg	0.64	0.70	0.66
Cd^{2+} mg/kg	1.08	1.08	1.18
Ni^{2+} mg/kg	0.20	0.22	0.21
Pb^{2+} mg/kg	0.37	0.33	0.35
Co^{2+} mg/kg	0.50	0.50	0.50
Cr^{2+} mg/kg	0.14	0.14	0.16

Concentration of heavy metals (Fe, Mn, Pb) in sugarcane leaves in WW treatments were significantly higher compared to GW treatments which had the lowest concentration. While the mean values for WW, TWW, and GW were 410, 370 and 340 mg/kg, 72, 66 and 58 mg/kg, 30, 27 and 25 mg/kg for Fe, Mn and Pb, respectively. This is in consistent with reports of higher concentration of heavy metals in vegetables from wastewater irrigated areas as compared to tube well water irrigated areas [38] [39] analyzed plants for Cd, Cr, Ni, Pb, Co, Zn, Cu, Fe and Mn in sewage contaminated area. Plants grown on polluted soil irrigated with wastewater recorded higher level of heavy metals, although the levels of heavy metals in the wastewater are low, the irrigated soil shows much higher concentrations of heavy metals, this may be due to the accumulation of heavy metals in the soil due to long-term irrigation, this high concentration of heavy metals in soil irrigated with wastewater and thereby increasing the average heavy metal concentration in cultivated vegetables in wastewater-irrigated area with respect to their reference values.

4.5. Bioconcentration Factor (BCF) for Heavy Metals in Sugarcane Crop

Bioconcentration factors are shown in **Table 5**. It was calculated for elements transfer from soil to sugarcane. The Bioconcentration factors for elements in soil irrigated with WW were in the descending order of Cd (1.08) > Zn (0.64) > Co (0.50) > Pb (0.37) > Mn (0.30) > Ni (0.20) > Cu, Cr (0.14) Fe (0.01), for TWW the elements were Cd (1.08) > Ca(0.77) > Zn(0.70) > Co (0.50) > Pb (0.33) > Mn (0.29) > Ni (0.22) > Cu, Cr (0.14) Fe (0.01),

Table 6. Pearson correlation coefficient between soil nutrient and nutrient in sugarcane plant.

	Ca	Mg	Na	K	N	P	Fe	Mn	Cu	Zn	Cd	Ni	Pb	Co	Cr
Ca	0.830**														
Mg	−0.566	−0.284													
Na	−0.234	−0.221	−0.148												
K	0.761**	0.639*	0.850**	0.807**											
N	0.893**	0.702*	0.897**	0.903**	0.972**										
P	0.703*	0.664*	0.887**	0.864**	0.779**	0.825**									
Fe	0.250	0.526	0.211	0.412	0.522	0.502	0.263								
Mn	0.830**	0.836**	0.843**	0.926**	0.819**	0.815**	0.423	0.835**							
Cu	−0.042	0.300	0.167	0.164	0.160	0.182	0.392	0.391	0.069						
Zn	−0.821**	−0.752**	−0.852**	−0.869**	−0.908**	−0.879**	−0.599*	−0.736**	−0.814**	0.961**					
Cd	−0.033	0.264	−0.003	0.298	0.180	0.159	0.368	0.368	0.109	−0.188	0.383				
Ni	0.312	0.487	0.343	0.363	0.574	0.345	0.369	0.176	0.326	−0.523	0.037	0.458			
Pb	0.111	0.531	0.108	0.259	0.411	0.421	0.530	0.202	0.152	−0.447	0.193	0.456	0.189		
Co	0.100	0.016	0.219	0.218	0.400	0.017	−0.017	−0.058	0.170	−0.334	0.376	0.162	−0.204	0.522	
Cr	0.696*	0.515	0.722**	0.588*	0.550	0.567	0.438	0.666*	0.564	−0.620*	−0.383	0.509	0.218	−0.183	0.608*

**: Correlation is higher significant at the 0.01 level (2-tailed); *: Correlation is significant at the 0.05 level (2-tailed).

whereas in the control site GW, it was in the order of Cd (1.18) > Zn (0.66) > Co (0.50) > Pb (0.35) > Mn (0.28) > Ni (0.21) > Cu (0.19) > Cr (0.16) Fe (0.01). There was no a significant difference in BCF values among the WW, TWW and GW sampling sites. Relatively high BCF values were found for Mn and Pb in WW site grown sugarcane crop, whereas the lowest BCF values were found for control site (GW). When BCF is ≤1, this denotes that the plant only absorbs but does not accumulate heavy metals; when BCF > 1, this indicates that plant accumulates metals [40]. Bioconcentration factor values of heavy metals were less than 1 in the sugarcane except for Cd which was >1. The results indicate that metal bioavailability was low at WW, TWW and control site (GW).

4.6. Pearson's Correlation between Elements in Soil and Plant

The relationship between different elements content in soil and plant were analyzed by Pearson's correlation coefficient. The correlation analysis is a bivariant method which is applied to describe the relation between two different parameters. The high correlation co-efficient (near +1 or −1) means a good relation between two variables, and its concentration around zero means no relationship between them at a significant level of 0.05% level, it is strongly correlated, if r > 0.7, as where r values between 0.5 and 0.7 show moderate correlation between two different parameters [41]. Pearson's correlation at 5% and 1% levels of significance was calculated for finding relationship between elements content in soil and plant. Significant positive correlation was observed between different elements in soil and plant suggesting the correlation of elements irrigated with wastewater (**Table 6**). The variation of elements content in soils was well correlated to plant. The results showed that the relationship among the elements Ca, Mg, Na, K, N, P, Mn, Zn and Cr, were significant correlations, between the soil and the plant, showed strong positive correlation; this may be due to their common source in the wastewater. While there were no correlations observed with Fe, Cu, Cd, Ni, Pb and Co. The translocation of elements from soil to plant parts (transfer factor) was calculated to determine the relative uptake of heavy metals by the plants with respect to soil. The ratio of metals between soil and plant parts (TF) is an important criterion for the contamination assessment of soils with high level of heavy metals. The ratio ">1" means higher accumulation of metals in plant parts than soil [42].

5. Conclusion

This study was carried out to assess the enrichment factor as a result of wastewater irrigation to agricultural soil and also the bioconcentration factor by trace metals uptake by sugarcane and to determine the concentrations of nutrients in sugarcane crop by growing on contaminated sites in Vidyaranyapuram area. The high manurial values in the form of macro nutrients like N, P, K, Ca and Mg were provided by wastewater, which is the effect on soils and also as an alternate source of water for agriculture, for the crops. However, the results of the study showed high significance for Mn, Cu, Ni and Cr contents in soil and Fe, Mn, Zn, Cu, Pb, Ni, Cd, Co and Cr concentrations were highly significant in sugarcane tissue, which was the soil irrigated with wastewater continuously soil for 12 years. Generally, the average application rate of wastewater should not be in excess of the normally allowable application rates. Hence the wastewater should be used judiciously for irrigation purpose in agriculture.

References

[1] Cui, Y., Zhu, Y.G., Zhai, R., Huang, Y., Qiu, Y. and Liang, J. (2005) Exposure to Metal Mixtures and Human Health Impacts in a Contaminated Area in Nanning, China. *Environment International*, **31**, 784-790. http://dx.doi.org/10.1016/j.envint.2005.05.025

[2] Huang, B., Shi, X., Yu, D., Oborn, I., Blomback, K., Pagella, T.F., Wang, H., Sun, W. and Sinclair, F.L. (2006) Environmental Assessment of Small-Scale Vegetable Farming Systems in Peri-Urban Areas of the Yangtze River Delta Region, China. *Agriculture, Ecosystems and Environment*, **112**, 391-402. http://dx.doi.org/10.1016/j.agee.2005.08.037

[3] Kelepertzis, E. (2014) Accumulation of Heavy Metals in Agricultural Soils of Mediterranean: Insights from Argolida Basin, Peloponnese, Greece. *Geoderma*, **221-222**, 82-90. http://dx.doi.org/10.1016/j.geoderma.2014.01.007

[4] Yeung, A. and Hsu, C. (2005) Electrokinetic Remediation of Cadmium Contaminated Clay. *Journal of Environmental Engineering*, **131**, 298-304. http://dx.doi.org/10.1061/(ASCE)0733-9372(2005)131:2(298)

[5] Marcussen, H., Holm, P.E., Strobel, B.W. and Hansen, H.C.B. (2009) Nickel Sorption to Goethite and Montmorillonite in Presence of Citrate. *Environmental Science and Technology*, **43**, 1122-1127. http://dx.doi.org/10.1021/es801970z

[6] Roy, R., Prasad, J. and Joshi, A.P. (2008) Changes in Soil Properties Due to Irrigation with Paper Industry Wastewater. *Journal of Environmental Science and Engineering*, **50**, 277-282.

[7] Ahmad, J.U. and Abdul Goni, M. (2010) Heavy Metal Contamination in Water, Soil, and Vegetables of the Industrial Areas in Dhaka, Bangladesh. *Environmental Monitoring and Assessment*, **166**, 347-357. http://dx.doi.org/10.1007/s10661-009-1006-6

[8] Munir, J.M. and Ayadi, M. (2005) Forage Yield and Nutrient Uptake as Influenced by Secondary Treated Wastewater. *Journal of Plant Nutrition*, **27**, 351-365. http://dx.doi.org/10.1081/PLN-120027659

[9] Mishra, V.K., Upadhyay, A.R. and Tripathi, B.D. (2009) Bioaccumulation of Heavy Metals and Two Organochlorine Pesticides (DDT and BHC) in Crops Irrigated with Secondary Treated Wastewater. *Environment Monitoring and Assessment*, **156**, 99-107. http://dx.doi.org/10.1007/s10661-008-0466-4

[10] Dikinya, O. and Areola, O. (2010) Comparative Analysis of Heavy Metal Concentration in Secondary Treated Wastewater Irrigated Soils Cultivated by Different Crops. *International Journal of Environmental Science and Technology*, **7**, 337-346. http://dx.doi.org/10.1007/BF03326143

[11] Bharose, R., Lal, B.S., Singh, K.S. and Srivastava, K.P. (2013) Heavy Metals Pollution in Soil-Water-Vegetation Continuum Irrigated with Ground Water and Untreated Sewage. *Bulletin of Environmental and Scientific Research*, **2**, 1-8.

[12] Antil, R.S. (2012) Impact of Sewage and Industrial Effluents on Soil-Plant Health, Industrial Waste. InTech. http://www.intechopen.com/books/industrial-waste/impact-of-sewer-water-and-industrial-wastewaters-on-soilplant-health

[13] Adamsa, M.L., Zhao, F.J., Mc Gratha, S.P., Nicholsonb, F.A. and Chambersb, B.J. (2004) Predicting Cadmium Concentrations in Wheat and Barley Grain Using Soil Properties. *Journal of Environmental Quality*, **33**, 532-541. http://dx.doi.org/10.2134/jeq2004.5320

[14] APHA (1998) Standard Methods for Examination of Water and Wastewater. 20th Edition, American Public Health Association, Washington DC.

[15] Indian Standards (1982) Tolerance Limits for Inland Surface Water Subject to Pollution. IS: 2296-1082.

[16] Hernandez, L., Probst, A., Probst, J.L. and Ulrich, E. (2003) Heavy Metal Distribution in Some French Forest Soils: Evidence for Atmosphere Contamination. *Science of the Total Environment*, **312**, 195-210. http://dx.doi.org/10.1016/S0048-9697(03)00223-7

[17] Al-Hwaiti, M. and Al-Khashman, O. (2015) Health Risk Assessment of Heavy Metals Contamination in Tomato and Green Pepper Plants Grown in Soils Amended with Phosphogypsum Waste Materials. *Environmental Geochemistry and Health*, **37**, 287-304.

[18] Kartal, S., Aydın, Z. and Tokalıoğlu, S. (2006) Fractionation of Metals in Street Sediment Samples by Using the BCR Sequential Extraction Procedure and Multivariate Statistical Elucidation of the Data. *Journal of Hazardous Materials*, **132**, 80-89. http://dx.doi.org/10.1016/j.jhazmat.2005.11.091

[19] Sutherland, R.A. (2000) Bed Sediment-Associated Trace Metals in an Urban Stream. *Oahu, Huwaii, Environmental Geology*, **39**, 611-627. http://dx.doi.org/10.1007/s002540050473

[20] Blaser, P., Zimmermann, S., Luster, J. and Shoty, K.W. (2000) Critical Examination of Trace Element Enrichment and Depletions in Soils: As, Cr, Cu, Ni, Pb and Zn in Swiss Forest Soil. *Science of the Total Environment*, **249**, 257-280. http://dx.doi.org/10.1016/S0048-9697(99)00522-7

[21] Liu, W., Li, X., Li, H.H. and Wang, Y.W. (2006) Heavy Metal Accumulation of Edible Vegetables Cultivated in Agricultural Soil in the Suburb of Zhengzhou City, People's Republic of China. *Bulletin of Environmental Contamination and Toxicology*, **76**, 163-170. http://dx.doi.org/10.1007/s00128-005-0903-9

[22] SAS (2004) Statistical Analysis System. SAS Institute, Inc., Cary.

[23] FAO (1992) Wastewater Treatment and Use in Agriculture. Irrigation and Drainage Paper 47, FAO, Rome.

[24] Singh, S.N., Srivastav, G. and Bhatt, A. (2012) Physicochemical Determination of Pollutants in Wastewater in Dheradun. *Current World Environment*, **7**, 133-138.

[25] Environment Protection Authority (1991) Guidelines for Wastewater Irrigation. Publication 168, Environment Protection Authority, Melbourne.

[26] Kargi, F. and Dincer, A.R. (1996) Effect of Salt Concentration on Biological Treatment of Saline Wastewater by Fed-Batch Operation. *Enzyme and Microbial Technology*, **19**, 529-537. http://dx.doi.org/10.1016/S0141-0229(96)00070-1

[27] FAO (1985) Guidelines: Land Evaluation for Irrigated Agriculture. Soils Bulletin 55, Food and Agriculture Organization of the United Nations, Rome.

[28] Singh, A., Sharma, R.K., Agrawal, M. and Marshall, F.M. (2010) Risk Assessment of Heavy Metal Toxicity through Contaminated Vegetables from Wastewater Irrigated Areas in Varanasi, India. *Tropical Ecology*, **51**, 375-387.

[29] Singh, O.V., Labana, S., Pandey, G., Budhirajaand, R. and Jain, R.K. (2003) Phytoremediation: An Overview of Metallic Ion Decontamination from Soil. *Applied Microbiology and Biotechnology*, **61**, 405-412. http://dx.doi.org/10.1007/s00253-003-1244-4

[30] Magnuson, M.L., Ketty, C.A. and Kelty, K.C. (2001) Trace Metal Loading on Water-Borne Soil and Dust Particles Characterized through the Use of Spilt-Flow Thin-Cell Fractionation. *Analytical Chemistry*, **73**, 3492-3496. http://dx.doi.org/10.1021/ac0015321

[31] Di Serio, M.G., Lanza, B., Mucciarella, M.R., Russi, F., Iannucci, E., Marfisi, P. and Madeo, A. (2008) Effects of Olive Mill Wastewater Spreading on the Physico-Chemical and Microbiological Characteristics of Soil. *International Biodeterioration & Biodegradation*, **62**, 403-407. http://dx.doi.org/10.1016/j.ibiod.2008.03.006

[32] Rattan, R.K., Datta, S.P., Chhonkar, P.K., Suribabu, K. and Singh, A.K. (2005) Long-Term Impact of Irrigation with Waste Water Effluents on Heavy Metal Content in Soils, Crops and Groundwater—A Case Study. *Agriculture, Ecosystems and Environment*, **109**, 310-322. http://dx.doi.org/10.1016/j.agee.2005.02.025

[33] Ghosh, A.K., Bhatt, M.A. and Agrawal, H.P. (2012) Effect of Long-Term Application of Treated Sewage Water on Heavy Metal Accumulation in Vegetables Grown in Northern India. *Environmental Monitoring and Assessment*, **184**, 1025-1036. http://dx.doi.org/10.1007/s10661-011-2018-6

[34] Chopra, A.K. and Pathak, C. (2013) Enrichment and Translocation of Heavy Metals in Soil and *Spinacea oleracea* Grown in Sugar Mill Effluent Irrigated Soil. *Sugar Tech*, **15**, 77-83. http://dx.doi.org/10.1007/s12355-012-0194-5

[35] Kisku, G.C., Barman, S.C. and Bhargava, S.K. (2000) Contamination of Soil and Plants with Potentially Toxic Elements Irrigated with Mixed Industrial Effluent and Its Impact on the Environment. *Water, Air, and Soil Pollution*, **120**, 121-137. http://dx.doi.org/10.1023/A:1005202304584

[36] Gupta, S., Nayek, S., Saha, R.N. and Satpati, S. (2008) Assessment of Heavy Metal Accumulation in Macrophyte, Agricultural Soil, and Crop Plants Adjacent to Discharge Zone of Sponge Iron Factory. *Environmental Geology*, **55**, 731-739. http://dx.doi.org/10.1007/s00254-007-1025-y

[37] Marschner, H. (1986) Mineral Nutrition of Higher Plants. Academic Press, London.

[38] Kawatra, B.L. and Bakhetia, P. (2008) Consumption of Heavy Metal and Minerals by Adult Women through Food in Sewage and Tube Well Irrigated Area around Ludhiana City (Punjab, India). *Journal of Human Ecology*, **23**, 351-354.

[39] Raj, B.G., Patnaik, M.C., Babu, S.P., Kalakumar, B., Singh, M.V. and Shylaja, J. (2006) Heavy Metal Contaminants in Water-Soil-Plant Animal Continuum Due to Pollution of Musi River around Hyderabad in India. *Indian Journal of Animal Sciences*, **76**, 131-133.

[40] Singh, J., Upadhyay, S.K., Pathak, R.K. and Gupta, V. (2011) Accumulation of Heavy Metals in Soil and Paddy Crop (*Oryza sativa*), Irrigated with Water of Ramgarh Lake, Gorakhpur, UP, India. *Toxicological and Environmental Chemistry*, **93**, 462-473. http://dx.doi.org/10.1080/02772248.2010.546559

[41] Rakesh Sharma, M.S. and Raju, N.S. (2013) Correlation of Heavy Metal Contamination with Soil Properties of Industrial Areas of Mysore, Karnataka, India by Cluster Analysis. *International Research Journal of Environment Sciences*, **2**, 22-27.

[42] Barman, S.C., Sahu, R.K., Bhargava, S.K. and Chaterjee, C. (2000) Distribution of Heavy Metals in Wheat, Mustard and Weed Grown in Field Irrigated with Industrial Effluents. *Bulletin of Environmental Contamination and Toxicology*, **64**, 489-496. http://dx.doi.org/10.1007/s001280000030

9

Hyporheic Zone Hydrochemistry of the Mine-Polluted River

D. Ciszewski, P. Bijata

AGH-University of Sciences and Technology, Kraków, Poland

Email: ciszewski@geol.agh.edu.pl

Abstract

Intensity of stream waters mixing with groundwaters and lateral extent of these processes in the hyporheic zone were investigated in a near-bank sandbar and an adjacent floodplain through the comparison of groundwaters and stream water chemistry of the Biała Przemsza River in southern Poland. The stream waters were polluted by the discharge of mine waters from "Bolesław" lead and zinc mine. The investigated waters were several times more mineralized than the natural spring waters of the river valley. The concentration of: potassium, sodium, and the pH, as well as cadmium, lead, and zinc decreased in the hyporheic zone towards the stream bank, whereas conductance, calcium, magnesium, sulphates, as well as silica contents were the highest on the floodplain, diminishing towards the stream. The changes observed in the chemical composition of groundwaters were apparent in mixing stream waters below the depth of 2 m with shallow groundwaters draining the valley slope. Hyporheic mixing also takes place in the 10-meter-wide, marginal zone of the sandbar, whereas in the 5-meter-wide stream-side zone of the sandbar groundwaters represent weakly transformed stream water.

Keywords

Hyporheic Zone, Groundwaters, Pollution, Water Chemistry, Metal Mining

1. Introduction

The hyporheic zone is an area beneath and adjacent to the stream where stream waters mix with groundwaters. This zone contains from several per cent up to 100% of river waters which penetrate alluvia over distances from centimetres to kilometres from the channel and several meters below the channel bottom (Woessner, 2000) [1]. Intensive investigations of the hyporheic zone during two most recent decades revealed their important ecological role. It was found that a channel bottom constitutes a shelter for hyporheic organisms which are unique for this specific ecotone. This ecotone can act as a physical, chemical and biological filter capable of transforming nutrients and pollutants (Bourg & Bertin, 1993) [2]. Microbial biofilms, in particular, mediate the transformation of water chemistry and affect the nitrogen cycling within the river system. Typically, along the hyporheic flow path well-oxygenated stream waters infiltrating through the bottom lose oxygen during nitrification, followed by denitrification in anoxic conditions, before unaerated waters reenter the stream (Hancock, Boulton & Humphreys, 2005) [3]. High oxygen, temperature and flow velocity gradients in the hyporheic zone, together with the microbial activity, favour the transformation of contaminants transported by the streams affected by mining. Studies of gravel alluvia also recognized the precipitation of manganese and iron oxides on coarse grains as an important mechanism controlling the removal of heavy metals from river waters (Fuller & Bargar, 2014) [4].

Whereas the hyporheic zone is extensively studied on gravel-bed rivers and gravel bars (considered as part of this zone), the processes of contaminant transformation are much less recognized in the sandy alluvia. Sandy river beds are more mobile and this mobility was found to affect the storage and transformation of solutes and fine particles (Harvey *et al.*, 2012) [5]. The aim of the study was to recognize the relative importance of the hyporheic flow in the sand bar of the Biała Przemsza River channel, cut in the fine grained alluvia, which are typical for the Pleistocene sediments in Central Europe. The river channel is polluted by the discharge of mine waters from "Bolesław" lead and zinc mine in southern Poland. The chemistry of the river water is markedly altered due to the discharged waters, thus their comparison to groundwater chemistry was used to reveal the lateral extent of the hyporheic zone.

2. Method

The Biała Przemsza River is the left-bank tributary of the Vistula River, which is the longest and largest river of Poland. The Biała Przemsza, with the length of over 60 kilometres, drains the upland area of the most industrialized region of Poland called Upper Silesia. The river valley is several hundred meters wide when it crosses cuesta sections, built of carboniferous rocks, and is several kilometres wide in inter-cuesta basins filled with fluvioglacial sandy deposits, with a thickness of as much as 25 m. In the studied middle river reach the valley gradient decreases and the channel reaches the width of up to 50 m because the sandy sediments accumulate, forming mid- and side channel bars (**Figure 1**). The tendency for the accretion of bed material has been observed here since the middle of the 19th century when weirs protecting the railway bridge and later the water supply plant had been constructed about 2 km downstream. These constructions, together with the valley narrowing by the high Pleistocene terrace in this river section, induced the lateral channel stabilization over the course of the 20th century. This stabilization favoured the development of the turf grass cover over most of the sand bars in this river reach as their surface became closer to the average level of the water table. The average discharge of the Biala Przemsza River in this reach equalled 4 m^3/s with a half of it originating from one of the largest lead and zinc mines in Europe—the "Bolesław" Mine and Metallurgical Plant (ZGH). The plant discharges mine water with higher-than-natural mineralization, the content of zinc, lead, cadmium, and turbidity, which has decreased since the beginning of the mine operation (Ciszewski, 1998) [6].

Seven sediment cores were collected from the 20-metre-long transect across the side channel bar and the floodplain of the Biała Przemsza River at Maczki (**Figure 1**). Four sediment cores were drilled to the depth of 4 meters and three profiles—up to 1 meter deep. In the core sites, 4- and 1-meter long piezometers were installed. Sediment cores were split into sections about 30 cm long and transported to a laboratory for analyses. Samples of groundwaters and stream waters were collected 10 times over the period of one year. The values of the pH and conductance were established using a Thermo Scientific device. The samples were filtered before the analy-

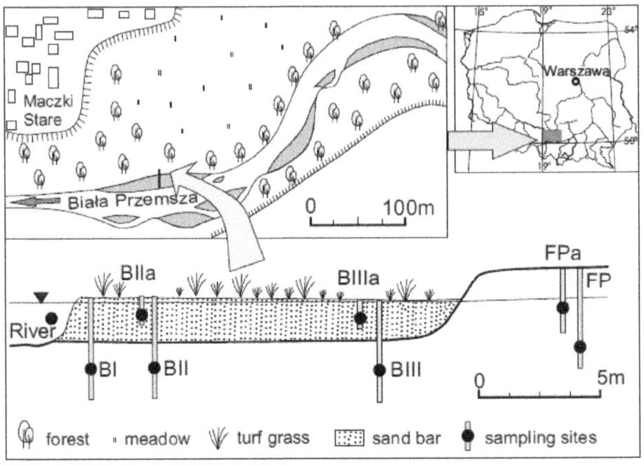

Figure 1. Location of sampling sites of river water and groundwater in the channel bar (B) and in the floodplain (FP) (black dots).

sis through 0.45 µm filters and concentrations of Cl^-, NO_3^-, SO_4^{2-}, PO_4^{3-}, CO_3^{2-} and macroions (Ca, Mg, Na, K) were determined within 48 h using Ion Chromatography (DIONEX 1000). Standard reference materials (Canadian waters Hamilton-20) were employed to determine the accuracy of anion analyses. For analyses of Zn, Cd, Pb, Fe, Ba and Si, the waters were filtered on site through a syringe filter 0.45 µm, acidified to a pH below 2 with ultrapure HNO_3 and measured with an inductively coupled plasma-mass spectrometer (Perkin Elmer ELAN 6100) in the certified Hydrogeochemical Laboratory according to the standard certified analytical quality control procedure (PN-EN ISO 17294-1:2007). In the collected sediments, losses on ignition at 550°C and content of 0.2 mm fraction were analyzed after drying at 105°C.

3. Results

Groundwater levels in the sandbar varied within less than 10 cm over the study period and were never recorded below 20 cm of the bar surface. The groundwater depths on the floodplain varied between 68 and 78 cm within the year. No head gradient was observed within the channel bar. Generally, the average conductance of stream waters was high and equalled 885 µS (**Table 1**). Very similar values were in groundwaters within 5 meters from the bar edge and they increased markedly at the river bank. These values were higher at the surface than at greater depths and on the floodplain.

The opposite distribution pattern was observed for the pH. Whereas stream waters were alkaline, with the pH of 8, groundwaters became more acidified further away from the stream, reaching 6.5 on the floodplain (**Figure 2**). In the same sampling site (II, III or FP) more acidified waters occurred closer to the surface than at the depth

Table 1. River and ground waters characteristics from the sampling sites in the channel bar (B) and in the floodplain (FP) of the Biała Przemsza river at Maczki (Median values) [mg/dm³]; n.d.: not detected.

Water chemistry characteristics	River	B I depth 3 m	B II depth 3 m	B IIA depth 0.6 m	B III depth 3 m	B IIIA depth 0.6 m	FP I depth 3 m	FP IA depth 1 m	Natural spring
Conductivity (µS)	885	880	875	904	831	979	961	915	228
pH	8.1	7.5	7.5	6.8	7.3	6.9	7.0	6.5	6.5
Ca	118	120	115	129	118	141	135	134	30
Mg	33	32	32	30	33	35	37	36	7
Na	22	26.3	23	21	17	11	21	9	3
K	5.6	6.8	5.1	4.0	4.8	4.5	3.0	0.8	2
NH_4	0.72	0.02	0.47	0.16	0.72	0.51	0.19	0.44	0.01
HCO_3	278	268	273	341	273	390	311	297	76
SO_4	198	199	210	181	179	191	248	211	35
Cl	35	31	33	28	28	29	33	31	8
NO_3	8.4	1.53	nd.	nd.	nd.	nd.	nd.	5.2	2.4
NO_2	0.22	0.21	nd.	nd.	nd.	nd.	nd.	0.67	nd.
PO_4	0.39	0.01	nd.	nd.	nd.	nd.	nd.	nd.	nd.
F	0.74	0.269	0.37	0.35	0.46	0.270	0.16	0.308	0.1
Ba (µg/dm³)	91	111	211	129	290	127	259	148	80
Cd (µg/dm³)	3.5	<0.1	<0.1	<0.1	<0.1	<0.1	<0.1	3.3	nd.
Pb (µg/dm³)	36	5.4	2.1	5.0	2.8	6.2	11.5	32	3.1
Zn (µg/dm³)	878	284	14	681	11	21	11	950	30
Fe (µg/dm³)	148	59	798	6795	3042	12,006	4804	940	300
Si (µg/dm³)	8792	7982	7890	9470	8437	10,471	8357	16,231	12,650

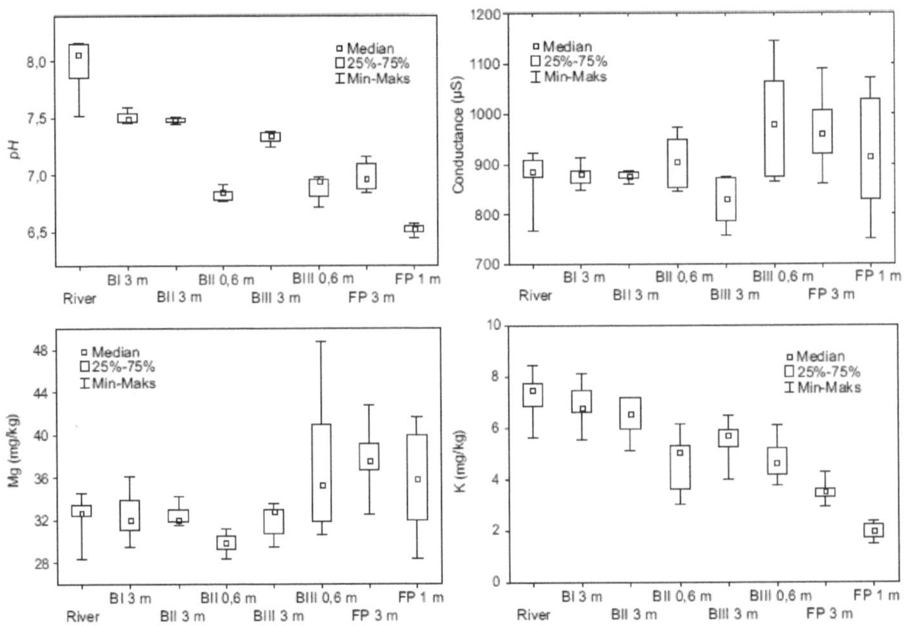

Figure 2. Variability of the pH, conductance and Mg and K concentrations in waters sampled from the river and from different depths of the channel bar (B) and of the floodplain (FP).

of 3 m. Very similar distribution pattern was observed for potassium as the maximum concentrations of this element occurred in the stream waters and there were markedly lower values in the floodplain; the decrease of this element content from the stream was progressive (**Figure 2**). Higher concentrations in stream waters than in floodplain groundwaters were observed also for sodium and its changes were closely associated with those of potassium (**Table 1**). Particularly low sodium values were recorded in shallow floodplain waters, whereas at the depth of 3 m the content of both elements was much higher or even similar to that in the river.

In contrast with sodium and potassium, magnesium exhibited the distribution pattern similar to that of the conductance. Differences between waters of the floodplain and those of the channel bar were not very high, but statistically significant. Intermediate values characterized shallow waters from the marginal zone of the channel but they varied markedly over the investigated period (BIIIA, **Figure 2**). Variability of calcium was also similar to that of magnesium; calcium content in the surface waters was generally lower than in the sandbar close to the stream (BI, BII) and increased at the channel margin and on the floodplain.

River waters were markedly polluted with sulphates to the level of about 200 mg/l, but their content did not differ significantly from the values for the channel bar. Only values at the deep horizon of the floodplain were markedly higher than the rest of the investigated points. Hydrocarbonates with the content of about 280 - 300 mg/l were more uniformly distributed in the investigated transect with significantly (statistically) higher values only in shallow waters of the sandbar. Their concentration in floodplain waters was only slightly higher than that in the river. Concentration of chlorides was also weakly differentiated with the highest values, on average, in the stream. Detectable values of nitrates, nitrites and phosphates occurred solely in the stream, in close proximity to the stream (BI) and at the floodplain surface (FPI). Concentrations of all these compounds were at least several times higher than in the natural spring water draining the Pleistocene terrace about 3 km upstream the sampling site (**Table 1**).

The sediments of the sandbar were composed almost exclusively of uniform, medium grained sands. The content of the fine fraction of sand and silt did not exceed several per cent in almost all strata of sediment cores. The highest content of the fine fraction was found at the surface of the sandbar. The surface 0.3 m strata contained also up to 10% of organic matter. However, some lenses of more organic and finer grained deposits were also observed at lower depths of the profiles.

4. Discussion

There are several methods of investigation of groundwater and surface waters interactions in the hyporheic zone.

In addition to the direct flow measurements, pumping tests, temperature measurements or solute tracing, contaminants may be used as markers for stream water or groundwater which mix beneath the stream (Kalbus, Reinstorf & Schirmer, 2006) [7]. Hydrochemical methods are particularly effective where chemical properties of stream water and groundwater differ markedly (Woessner, 2000; Hoehn & Scholtis, 2011) [1] [8]. In the presented studies this difference was used in order to identify the relative intensity of mixing of the two types of water. Two different distribution patterns of chemical compounds were identified. In one of them the concentrations decreased from the highest in the stream towards the lowest in the floodplain. It is represented by potassium, sodium and the pH, as well as by cadmium, lead, and zinc. The opposite distribution, with overall higher values in the floodplain and lower in the stream and in the sandbar, is represented by conductance, calcium, magnesium, and sulphates, as well as silica. These differences are less visible for chlorides, hydrocarbonates, barium, and iron and seem to represent an intermediate distribution pattern. The observed distribution patterns indicate that shallow groundwaters in the floodplain differ markedly from the rest of the waters because they drain a large area of the valley slope and have a prolonged contact with the subsurface. This is particularly visible in the highest content of silica. The groundwaters at lower levels were for some compounds more closely related to groundwaters which drain the floodplain and some concentrations were more similar to that in the sandbar at the channel margin under the influence of water from the stream. High variability also characterizes concentrations of many compounds in the subsurface waters at the channel margin (BIIIA). The influence of water draining the floodplain is higher during rainfall or snow-melting periods and only the channel margins are the zone of most intensive mixing of ground and stream waters. The proportion of stream waters at the depth of 3 m on the floodplain seems to be quite high, considering the chlorides content is as high as in the stream. Nevertheless, it is clearly visible that the proportion of stream waters is the highest closer to the stream. In particular, the concentrations of most of the compounds in BI and BII sites *i.e.* in the marginal zone of the sandbar about 5 metres wide are almost the same as in the stream. It seems that in this zone 100% of stream water flow is parallel or the intensive exchange with the stream and lateral groundwater inflow from the floodplain occurs at the channel margin, where stream waters mix with groundwaters in variable proportions. This zone is characterized by the highest average conductance. The data enabled the identification of the hyporheic zone extending over the channel bar and outside the river channel below the depth of 2 m.

Investigations of hyporheic exchange processes in partially submerged bed structures use observations of stream stage, head data in piezometers, artificial tracer additions or variations in natural electrical conductivity and were attempted mainly in gravel bed rivers (Trauth *et al.*, 2015) [9]. The electrical conductivity and the other characteristics of water in the sandbar were used to show high complexity of the morphological factors controlling the exchange in the sandy alluvia. These investigations indicated that sand bars, particularly at channel margins, not actively reworked in the channel, may not have a uniform structure. The sediments accreted during numerous flood episodes contain some stratification which modify the flow velocity in the hyporheic zone. The bars formed at the channel margin shift out of the bank both the flow and the zone of the flow parallel to the channel, sometimes called the *parafluvial zone* (Boulton et al. 1998) [10]. As a result, hyporheic mixing processes occur as well within the channel in a zone proportional to the width of the channel bar and their grain size variability.

The mixing processes seem to be connected to the conservative behaviour of chemical compounds. However, some substances are biologically transformed in the hyporheic zone. Nitrates and nitrites, abundant in the stream water, had to have already disappeared in the BI point, localized a few decimetres from the bar edge (**Table 1**). The same substances are also present only in the surface waters of the floodplain. This suggests that microbial reduction of nitrogen load is intensive and takes place within the decimetres of the flow in the hyporheic zone. The widespread absence of oxygen in the sandbar or its very low content (<0.5 mg/l) only in the BI site at the bar edge, demonstrates its rapid drop in the fully oxygenated (10 mg/l) river waters infiltrating the sand bar. Some of the other chemical compounds can also be transformed within the channel bar. The loss of phosphates from the stream waters is suggested by its rapid decrease already at the bar margin. Moreover, processes of dissolution of iron hydroxides are responsible for very high iron content in the subsurface waters on the sandbar. These processes are characteristic of the reducible dissolution of iron compounds in the presence of organic matter and also occur in the swampy areas of the Biała Przemsza River valley.

5. Conclusion

Stream waters polluted by the discharge from "Bolesław" lead and zinc mine are more abundant in most of the chemical compounds investigated than in the natural spring waters of the Biała Przemsza River valley. The concentrations of potassium, sodium, and the pH, as well as cadmium, lead, and zinc decreased in the hyporheic zone towards the stream banks, whereas conductance, calcium, magnesium, and sulphates, as well as silica contents were higher on the floodplain and decreased towards the stream. The observed changes in the chemical composition of the groundwater exhibit the inflow of shallow groundwater from the catchment mixing with stream water below the depth of 2 m. Hyporheic mixing extends along the bank and within the approximately 10-metre-wide marginal zone of the sandbar, whereas in the 5-metre-wide marginal zone of the sandbar, groundwaters represent weakly transformed stream waters flowing along the stream.

Acknowledgements

The authors gratefully acknowledge the National Science Center of Poland for funding the project No. 2012/05/B/ST10/00235.

References

[1] Woessner, W.W. (2000) Streams and Fluvial Plain Ground Water Interactions: Rescaling Hydrogeologic Thought. *Ground Water*, **38**, 423-429. http://dx.doi.org/10.1111/j.1745-6584.2000.tb00228.x

[2] Bourg, A.C.M. and Bertine, C. (1993) Biogeochemical Processes during the Infiltration of River Water into an Alluvial Aquifer. *Environmental Science and Technology*, **27**, 661-666. http://dx.doi.org/10.1021/es00041a009

[3] Hancock, P.J., Boulton, A.J. and Humphreys, W.F. (2005) Aquifers and Hyporheic Zones: Towards an Ecological Understanding of Groundwater. *Hydrogeological Journal*, **13**, 98-111. http://dx.doi.org/10.1007/s10040-004-0421-6

[4] Fuller, Ch.C. and Bargar, J.C. (2014) Processes of Zinc Attenuation by Biogenic Manganese Oxides Forming in Hyporheic Zone of Pinal Creek, Arizona. *Environmental Sciences and Technology*, **48**, 2165-2172. http://dx.doi.org/10.1021/es402576f

[5] Harvey, J.W., Drummond, J.D., Martin, R.L., McPhillips, L.E., Packman, A.I., Jerolmack, D.J., Stonedahl, S.H., Aubenau, A.F., Sawyer, A.H., Larsen, L.G. and Tobias, C.R. (2012) Hydrogeomorphology of the Hyporheic Zone: Stream Solute and Fine Particle Interactions with Dynamic Streambed. *Journal of Geophysical Research*, **117**, 1-20. http://dx.doi.org/10.1029/2012JG002043

[6] Ciszewski, D. (1998) Channel Processes as a Factor Controlling Accumulation of Heavy Metals in River Bottom Sediments: Consequences for Pollution Monitoring (Upper Silesia, Poland). *Environmental Geology*, **36**, 45-54. http://dx.doi.org/10.1007/s002540050319

[7] Kalbus, E., Reinstorf, F. and Schirmer, M. (2006) Measuring Methods for Groundwater-Surface Water Interactions: A Review. *Hydrology and Earth System Science*, **10**, 873-887. http://dx.doi.org/10.5194/hess-10-873-2006

[8] Hoehn, E. and Scholtis, A. (2011) Exchange between a River and Groundwater, Assessed with Hydrochemical Data. *Hydrology and Earth System Science*, **7**, 9023-9042. http://dx.doi.org/10.5194/hessd-7-9023-2010

[9] Trauth, N., Schmidt, Ch., Vieweg, M., Oswald, S.E. and Fleckenstein, J.H. (2015) Hydraulic Controls of In-Stream Gravel Bar Hyporheic Exchange and Reactions. *Water Resources Research*, **51**, 2243-2263. http://dx.doi.org/10.1002/2014WR015857

[10] Boulton, A.J., Findlay, S., Marmonier, P., Stanley, E.H. and Valet, H.M. (1998) The Functional Significance of the Hyporheic Zone in Streams and Rivers. *Annual Review of Ecological Systems*, **29**, 59-81. http://dx.doi.org/10.1146/annurev.ecolsys.29.1.59

10

Hydrological Mann-Kendal Multivariate Trends Analysis in the Upper Yangtze River Basin

Lei Ye[1,2], Jianzhong Zhou[1,2*], Xiaofan Zeng[1,2], Muhammad Tayyab[1,2]

[1]College of Hydropower and Information Engineering, Huazhong University of Science and Technology, Wuhan, China
[2]Hubei Key Laboratory of Digital Valley Science and Technology, Huazhong University of Science and Technology, Wuhan, China
Email: *jz.zhou@hust.edu.cn

Abstract

Hydrological events should be described through several correlated variables, so multivariate HFA has gained popularity and become an active research field during recent years. However, at present multivariate HFA mainly focuses directly on fitting the frequency distribution without confirming whether the assumptions are satisfied. Neglecting testing these assumptions could get severely wrong frequency distribution. This paper uses multivariate Mann-Kendal testing to detect the multivariate trends of annual flood peak and annual maximum 15 day volume for four control hydrological stations in the Upper Yangtze River Basin. Results indicate that multivariate test could detect the trends of joint variables, whereas univariate tests can only detect the univariate trends. Therefore, it is recommended to jointly apply univariate and multivariate trend tests to capture all the existing trends.

Keywords

Trend Analysis, Multivariate Mann-Kendal Test, Hydrological Variable, Upper Yangtze River Basin

1. Introduction

Hydrological frequency analysis (HFA), foundation of the hydraulic engineering design and water resources management, is mainly used for purposes of improving hydrological extreme events prediction. Researchers can estimate possible hydrological variable value of certain frequency through analyzing frequency distribution of the hydrological variable. [1] summarized the researches of HFA and concluded that HFA is composed of four main steps: 1) carrying out the descriptive and exploratory analysis and outlier detection, 2) checking the basic assumptions including stationarity, homogeneity and independence, 3) selecting a probability distribution model and estimating the parameters, and 4) risk evaluation and analysis. One of the most important steps is confirm-

*Corresponding author.

ing whether the stationarity assumption of the hydrological variable is satisfied, which is the statistical property of hydrological variable remains the same in the past, present and future.

For univariate HFA, extensively studies have been devoted to treat the above steps [2]-[4]. Hydrological events generally need be described through several correlated characteristics, such as volume, peak for floods. The importance and justification of jointly considering all variables characterizing an event has been realized by the researchers and it has become an active and attractive research area in hydrology [5]. However, very limited attention has been paid to step b for multivariate HFA.

In fact, existing research has shown that, due to the impact of climate change and human activities, the physical mechanism of rainfall runoff process has changed a lot, the hydrological stationarity assumption has been challenged. Therefore, before step c it is necessary to apply trend analysis to the hydrological variables of interest. If the hydrological variables have significant tendency, researchers should use nonstationary HFA method, otherwise can go to step c directly.

For hydrological multivariate frequency analysis, if one only uses trend testing for each variable separately, he/she can only test whether there exists trend for the single variable and can't check out whether there exists changing trend of the correlation between joint variables. Ignoring the testing step in multivariate HFA may lead to very inaccurate or wrong results and hence to inappropriate decisions. Therefore, this paper introduces the widely used multivariate trend testing method in water quality data to the multivariate hydrological events in order to test the stationarity of joint variables prior to selecting a probability distribution model and estimating the parameters.

2. Methodologies

For the univariate trend analysis, Mann-Kendal test [6] [7] is the most widely used nonparametric hydro-meteorological time series method and is highly recommended by the World Meteorological Organization as standard nonparametric procedures when testing for trend [8]. MK statistical test can detect any kind of monotonic trend, whether of linear or nonlinear behavior, and it is also not influenced by the interference of a few outliers, thus making it suitable for hydrological data which is usually not normal distributed. For multivariate trend analysis, this paper applies multivariate MK trend analysis to test whether there exist significant trends in joint variables.

2.1. Univariate Mann-Kendal Test

First the MK statistic S is calculated as:

$$S = \sum_{i=1}^{N-1} \sum_{j=i+1}^{N} \operatorname{sgn}(X_j - X_i) \tag{1}$$

where X_j and X_i are the hydrological data values in years j and i, respectively, with $j > i$, N is the total number of years and sgn() is the sign function:

$$\operatorname{sgn}(X) = 1 \text{ if } X > 0; =0 \text{ if } X = 0; =-1 \text{ if } X < 0 \tag{2}$$

The statistics S is approximately normally distributed, with mean zero and variance given by:

$$E[S] = 0 \tag{3}$$

$$Var(S) = N(N - 1)(2N + 5)/18 \tag{4}$$

The standard normal variable Z is then formulated as:

$$Z = \frac{S_i}{\sqrt{Var(S_i)}} \tag{5}$$

If Z is positive, the hydrological time series has an increase trend, otherwise it has a decrease trend. The null hypothesis is rejected at significance level α if $|Z| > Z_{1-\alpha/2}$, indicating the hydrological time series has significant changing trend.

2.2. Multivariate Mann-Kendal Test

[9] developed an improved the multivariate nonparametric MK test for trend detection in water quality time series, as the extension of univariable MK test. [1] firstly introduced the multivariate MK test to hydrology field

and summarized the procedures of multivariate MK test calculation.

First univariate MK test statistic is calculated according to Equation (1) for each variable, then constructing a vector using these univariate MK test statistic as:

$$S_{multi} = [S_1, S_2, \ldots, S_d] \quad (6)$$

where d is the total number of joint variables. S_{multi} is asymptotically d-dimensional normal with zero mean and covariance matrix $C_M = (c_{u,v})_{u,v=1,\ldots,d}$, with $c_{u,v} = \mathrm{cov}(S_u, S_v)$. The covariance matrix is calculated as:

$$\hat{c}_{u,v} = \frac{t_{u,v} + r_{u,v}}{3} \quad \text{when } u \neq v \quad (7)$$

in which:

$$t_{u,v} = \sum_{1 \le i < j \le N} \mathrm{sgn}((X_j^u - X_i^u)(X_j^v - X_i^v))$$
$$r_{u,v} = \sum_{i,j,k=1}^{N} \mathrm{sgn}((X_k^u - X_j^u)(X_k^v - X_i^v)) \quad (8)$$

when $u = v$, the variance is calculated according to Equation (4).

2.2.1. The Covariance Inversion Test

The first multivariate extension of the MK test was proposed by [10]. Here we call it the Covariance Inversion Test (CIT) as proposed by [9]. C_M^{-1} is the inverse matrix of C_M when it has full rank, otherwise a generalized inverse of C_M. The test statistic D is given as:

$$D = SC_M^{-1}S^T \quad (9)$$

D is asymptotically $\chi^2(q)$-distributed in which q is the rank of C_M. if the value of D exceeds the significance level according to the $\chi^2(q)$ distribution quantile, the joint variables are considered have a significant trend.

2.2.2. The Covariance Sum Test

[11] developed a generalization of the multivariate MK test similar to (9) in a model which assumed independent seasons, by considering each season as a variable and thus obtaining a multivariate setting [1] [9] call this modified seasonal MK test the Covariance Sum Test (CST), as it is referred to in this paper. The test statistic H is then:

$$H = S_1 + S_2 + \cdots + S_d \quad (10)$$

The statistic H is approximately normally distributed, with mean zero and variance given by:

$$Var(H) = \sum_{u=1}^{d} S_u + 2\sum_{v=1,u=1}^{d,v-1} c_{u,v} \quad (11)$$

in which $c_{u,v}$ is calculated according to Equation (7). For a given significant level, the covariance sum test is the same as the univariate MK test, constructing standard normal variable and then comparing the value with the threshold value.

3. Case Study

3.1. Study Area and Data

This paper analyzes the annual flood peak (Q) and annual maximum 15 d flood volume (V) multivariate trend in the Upper Yangtze River Basin by taking four control hydrological stations (Pingshan, Lijiawan, Beibei and Yichang stations) into consideration. Data series of flood peak and flood volume (V) are available for the period 1955-2005 corresponding to the four stations. **Figure 1** presents the locations of the upper Yangtze River basin and the control hydrological stations

3.2. Univariate MK Test Analysis

Time series of the two hydrological variables Q and V corresponding to the selected four hydrological stations

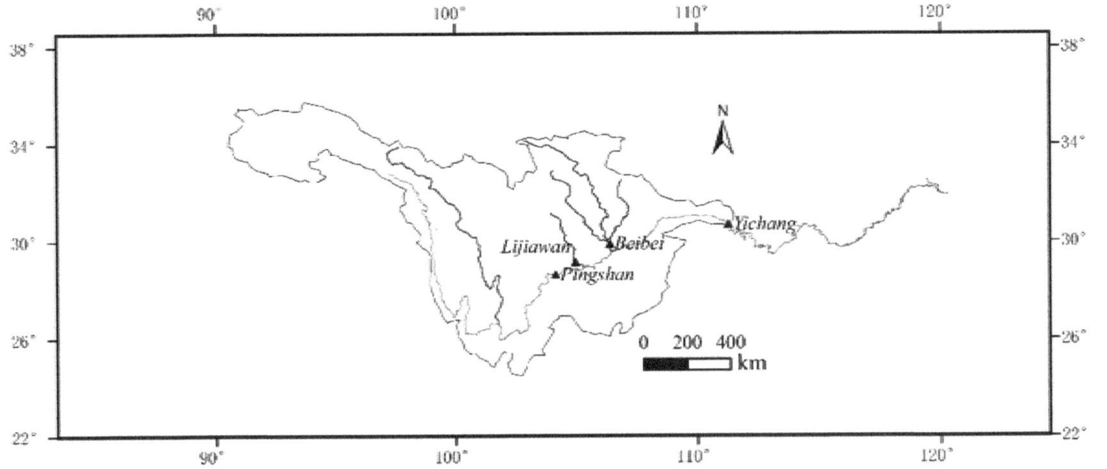

Figure 1. Locations of the upper Yangtze River basin and the controlling hydrological stations.

for the period 1955-2005 altogether 8 time series data are plotted in **Figure 2**. **Figure 2** also presents the linear regression line of the time series data.

According to the univariate MK test described in Section 2.1, we applied the trend analysis for the Q and V corresponding to the four stations. Given the significant level $\alpha = 0.05$, the threshold value is 1.96. When the absolute Z value for the variable is higher than 1.96, the variable has a significant changing trend. The results are shown in **Table 1**.

It can be seen from **Table 1** that the absolute Z values of both the Q and V for Pingshan and Yichang stations did not reach 1.96, indicating no significant changing trend for the both variables. The absolute Z value of only V for Beibei station passed 1.96 and meanwhile the Z value was negative, which showed significant decreasing trend for V. For Lijiawan station, Z values for both Q and V passed the threshold value, indicating significant decreasing trend for both variables.

3.3. Multivariate MK Test Analysis

According to the multivariate MK test described in Section 2.3, we applied the CIT and CST trend analysis for the joint variables (Q and V) corresponding to the four stations. Given the significant level $\alpha = 0.05$, when the absolute Z value for the variable is higher than threshold, the joint variables have a significant changing trend. The results are presented in **Table 2**.

It can be seen from **Table 2** that both the CIT and CST values did not pass the threshold values for Pingshan and Yichang stations, indicating no significant changing trend for the joint variables. For Beibei station, the CIT value passed threshold and the CST value did not pass the threshold. The differences generated by two different multivariate MK test methods encourage us to combine univariate test results to judge the change trend of Beibei station. Both multivariate tests detected significant decreasing bivariate trend for Lijiawan station.

3.4. Discussion

From the results of univariate and multivariate MK test we can find that the no significant change trend existed in annual flood peak, annual maximum 15 d flood volume and the joint variables for Pingshan and Yichang stations. Therefore, traditional HFA method could be used to fit the frequency distribution function of annual flood peak, annual maximum 15 d flood volume and the joint variables.

The annual maximum 15 d flood volume for Beibei station has significant decreasing trend while no significant trend was detected for annual flood peak. Meanwhile, only the CIT test detected significant change trend for joint variables, which indicated that 15 d flood volume dominated the changing trend of multivariate, but the correlation between annual flood peak and annual maximum 15 d flood volume has not changed significantly. Therefore, traditional HFA method could be used to fit the frequency distribution function of annual flood peak and the joint variables. As to annual maximum 15 d flood volume, nonstational HFA method should be applied.

The annual flood peak, annual maximum 15 d flood volume and the joint variables for Lijiawan station have

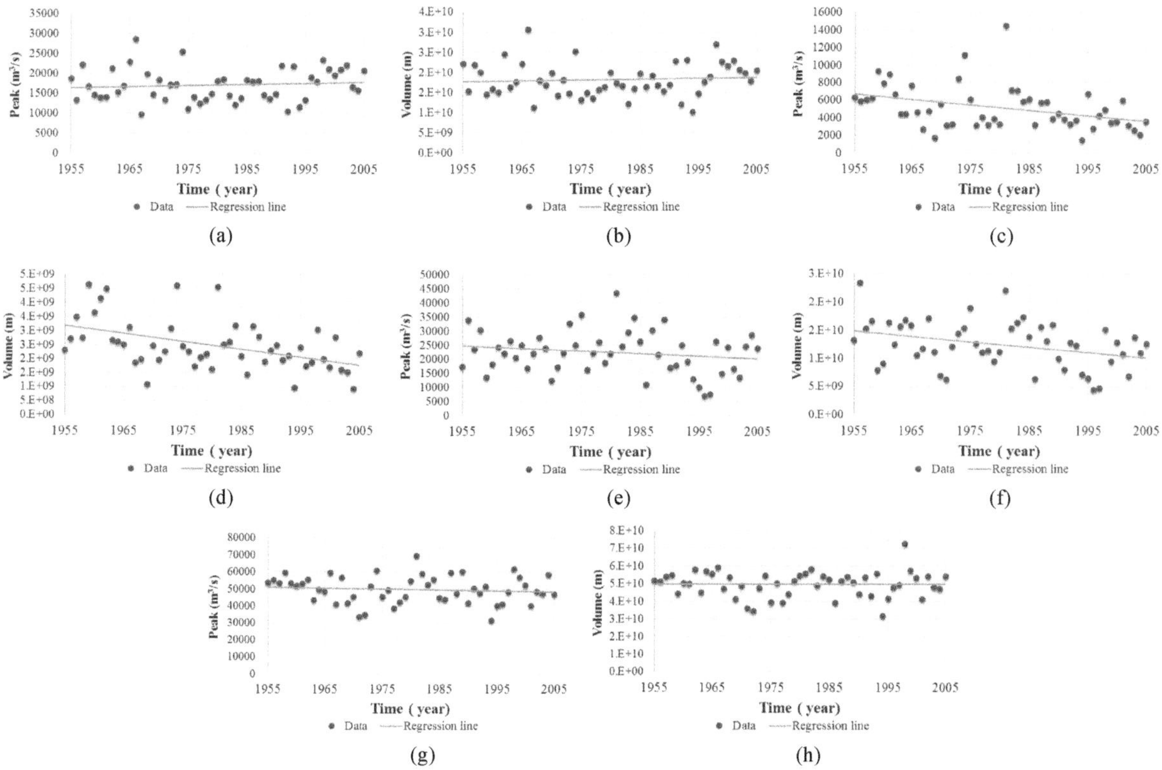

Figure 2. Scatterplot of annual peak and maximum 15 day flood volume as well as linear regression line. (a) Peak from Pingshan station; (b) Volume from Pingshan station; (c) Peak from Lijiawan station; (d) Volume from Lijiawan station; (e) Peak from Beibei station; (f) Volume from Beibei station; (g) Peak from Yichang station; (h) Volume from Yichang station.

Table 1. Results of univariate MK testing.

Station	Variable	*Z* value
Pingshan	*Q*	0.93
	V	1.13
Lijiawan	*Q*	−3.14
	V	−3.60
Beibei	*Q*	−1.09
	V	−2.28
Yichang	*Q*	−1.01
	V	−0.35

Table 2. Results of multivariate MK testing.

Stations	CIT	Threshold	CST	Threshold
Pingshan	1.35	5.99	1.05	1.96
Lijiawan	13.03	5.99	−3.52	1.96
Beibei	7.73	5.99	−1.75	1.96
Yichang	1.35	5.99	−0.73	1.96

been detected with significant decreasing trend. Therefore, nonstationarity should be taken into consideration when applying HFA to Lijiawan Station including the nonsationarity for univariate and the correlation (described by copula function).

4. Conclusion

This paper highlights the importance of testing for trend in multivariate hydrological event and briefly introduces two multivariate Mann-Kendal test methods. Four control hydrological stations in the Upper Yangtze River basin are selected for univariate and multivariate MK test for flood event. Results indicate that significant changing trends exist in some stations including univariate or multivariate trend. For the stations with only univariate trend, we could apply nonstational hydrological frequency analysis method, while still using the traditional copula function. But for the stations with both univariate and multivariate trends, we should consider the nonsationarity for the univariate and the correlation between the variables. These tests insure the validity of the HFA results and can help to guide the selection of the appropriate multivariate distribution (margins and copula).

Acknowledgements

This work was supported by the State Key Program of National Natural Science of China (No. 51239004) and the National Natural Science Foundation of China (No. 51309105).

References

[1] Chebana, F., Ouarda, T.B. and Duong, T.C. (2013) Testing for Multivariate Trends in Hydrologic Frequency Analysis. *Journal of Hydrology*, **486**, 519-530. http://dx.doi.org/10.1016/j.jhydrol.2013.01.007

[2] Cunnane, C. (1988) Methods and Merits of Regional Flood Frequency Analysis. *Journal of Hydrology*, **100**, 269-290. http://dx.doi.org/10.1016/0022-1694(88)90188-6

[3] Sarhadi, A., Soltani, S. and Modarres, R. (2012) Probabilistic Flood Inundation Mapping of Ungauged Rivers: Linking GIS Techniques and Frequency Analysis. *Journal of Hydrology*, **458**, 68-86. http://dx.doi.org/10.1016/j.jhydrol.2012.06.039

[4] Viglione, A., Merz, R., Salinas, J.L. and Blöschl, G. (2013) Flood Frequency Hydrology: 3. A Bayesian Analysis. *Water Resources Research*, **49**, 675-692. http://dx.doi.org/10.1029/2011WR010782

[5] Zhang, L. and Singh, V. (2006) Bivariate Flood Frequency Analysis Using the Copula Method. *Journal of Hydrologic Engineering*. http://dx.doi.org/10.1061/(ASCE)1084-0699(2006)11:2(150)

[6] Kendall, M. (1975) Rank Correlation Methods. Charles Griffin, London.

[7] Mann, H.B. (1945) Nonparametric Tests against Trend. *Econometrica*: *Journal of the Econometric Society*, 245-259. http://dx.doi.org/10.2307/1907187

[8] Mitchell, J., Dzerdzeevskii, B., Flohn, H., Hormeyr, W., Lamb, H., Rao, K. and Wallen, C. (1966) Climate Change. WMO Tech. Note, 79. WMO.

[9] Lettenmaier, D.P. (1988) Multivariate Nonparametric Tests for Trend in Water Quality1. Wiley Online Library.

[10] Dietz, E.J. and Killeen, T.J. (1981) A Nonparametric Multivariate Test for Monotone Trend with Pharmaceutical Applications. *Journal of the American Statistical Association*, **76**, 169-174.

[11] Hirsch, R.M. and Slack, J.R. (1984) A Nonparametric Trend Test for Seasonal Data with Serial Dependence. *Water Resources Research*, **20**, 727-732. http://dx.doi.org/10.1029/WR020i006p00727

11

Major and Trace Element Characterization of Shallow Groundwater in Coastal Alluvium of Chidambaram Town, Cuddalore District, South India

Thangasamy Jeyavel Raja Kumar, Chinadurai Dushiyanthan, Boopalan Thiruneelakandan, Radhakrishnan Suresh, Shanmugam Vasanth Raja, Mailare Senthilkumar

Department of Earth Sciences, Annamalai University, Chidambaram, India
Email: tjeyavel@rediffmail.com

Abstract

Hydrogeochemical investigation of groundwater of Chidambaram town has been carried out to assess the suitability for drinking uses and the source for the ions acquisition process. Geologically, the area comes under the alluvial zone and is followed by tertiary formation, includes sands, sand stone, laterite, clayey black and stiff clay. The water level varies from 6 m to 10 m bgl. Twenty groundwater samples are collected and analysed for pH, Electrical Conductivity (EC), Total Dissolved Solids (TDS), major cations Ca^{2+}, Mg^{2+}, Na^{+}, K^{+} and anions HCO_3^-, Cl^-, SO_4^{-2}. The trace elements of Zn, Pb, Cr, Fe, Cd, Cu and Mn are determined. Study results reveal that groundwater in study area is a fresh water type. The pH, EC and TDS are noticed high and sequence of abundance of the major ions is $Ca^{2+} > Na^{+} > Mg^{2+} > K^{+}$ and $Cl^- > SO_4^{-2} > HCO_3^-$. The Stufzand chloride classification indicates that the groundwater is derived from precipitations and Scholler classification shown type I. Hydrochemical facies of groundwater are $Ca^{2+} > Na^{+} > Cl^-$ and saturation index shows that nearly all the water samples are under saturated with respect to carbonate (calcite, dolomite and aragonite) and sulfate (gypsum and anhydrite) minerals. Except Zn, other elements of Fe, Pb, Cr, Mn, Cu, and Cd are found high and above the permissible limit of drinking water standard. The ionic concentration present in the groundwater of the study area can be from wastes of commercial activities, domestic sewage and automobiles rather than by ionic exchange and geological formation.

Keywords

Groundwater Quality, Hydrogeochemical Processes, Saturation Index, Trace Elements

1. Introduction

Groundwater is a precious natural resource for several vital functions such as public, industries and agricultural water supply. In recent years, the quality of the groundwater is getting higher priority, due to the increasing trend of contamination level. At the same time, safe drinking water is a vital requirement to human being, and its availability is important for the overall socioeconomic development of a nation [1]. For the purpose of use, particularly drinking, determination of groundwater composition is of utmost importance from the point of view of its suitability. In common, rock water interaction, anthropogenic and industrial effluents are the primary source for groundwater composition and contamination; however, over exploitation of groundwater due to urbanization and addition of pollutants have rendered these resources to be useless in many instances. In addition to these processes water borne pathogens, toxic and nontoxic pollutants are the major water quality degradation parameters which are transported from recharge area to discharge area through aquifers by groundwater motion. In this scenario, it is essential that the study on quality of the groundwater is important not only for determining the level of concentration but also to learn about the contaminating sources of elements to the aquifers. In above all, undesirable and soluble constituents in the water cannot be controlled after entering the ground [2] [3]. Several comprehensive studies of groundwater contamination and sources have carried out in south India [4]-[7].

1.1. Need of the Study

In this study, an attempt has been made to understand the groundwater chemistry and quality of shallow aquifers of alluvium in Chidambaram town. The holy town Chidambaram has expanded very fast in the recent decades, and the running population also increased to worship and visit the temple. The requirement of fresh water has been increased, which drove to drill very deeper tube wells for the domestic usage due to most of the existing shallow tube wells of freshwater lens being dried during the summer season. The limited available quantity of potable water deteriorates due to salinity and pollution generated by sea water intrusion, agricultural activities in Chidambaram and its environs is at the verge of reaching an alarming situation [8]. Moreover, during the last decades, many cases of groundwater pollution in the coastal track due to seawater intrusion and anthropogenic inputs were reported [9] [10].

1.2. Study Area

The present study area Chidambaram town lies between longitude 79˚40' to 79˚45'E and latitude 11˚20' to 11˚25'N and located in the survey of India (SOI) topo sheet No. 58 M/15 as shown in **Figure 1**. It is geographically located in the Cuddalore district of Tamil Nadu state, which is about 40 km east of Neyveli Lignite Mines, Neyveli. Geologically, the area comes under the alluvial zone which is the youngest in age and followed by tertiary formation, includes sands, sand stone, laterite, clayey black and stiff clay are pronominal (**Figure 2**). The water level varied from 6 m to 10 m bgl. The elevation is approximately about + 5.75 m above the mean sea level. The maximum and minimum temperature of the study area is 37.2˚C and 21.6˚C. The precipitation recorded is mainly from the north east monsoon. The maximum and minimum rainfall received is 1400 mm and 1200 mm. The higher rates of relative humidity (about 85%) is observed during the north east monsoon period and in summer the relative humidity is low, about 60%.

The occurrence of groundwater in the study area is on two formations, the unconsolidated Tertiary and Quaternary sediments. Fresh groundwater has been observed in the tertiary semi-consolidated sand stone and sands on the Cuddalore series. Most of the aquifers in the Cuddalore series are extensive and interconnected with aquifer above and below, and due to this water can move to greater distances through them. The transmissivity of the alluvium is 98 m^2/d and permeability is 19.7 m/d [11].

2. Materials and Methods

In order to understand the hydrogeochemistry and assess quality, twenty (20) tube well samples were collected (**Figure 3**) during the month of August, 2011 using standard procedure for the sampling techniques discussed by various authors [12] [13]. The analysis has been carried out in the Department of Earth Sciences, Annamalai University for pH, Electrical Conductivity (EC), Total Dissolved Solids (TDS), major cations Ca^{2+}, Mg^{2+}, Na^{+}, K^{+} and anions HCO_3^-, Cl^-, SO_4^{-2}.

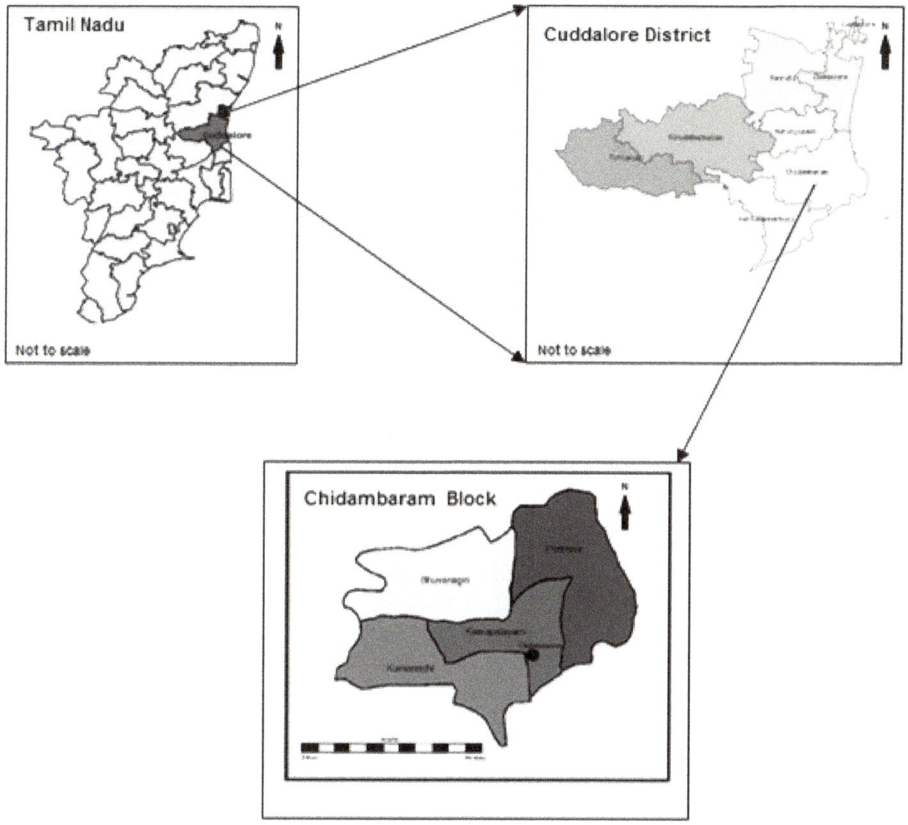

Figure 1. Location map of the study area.

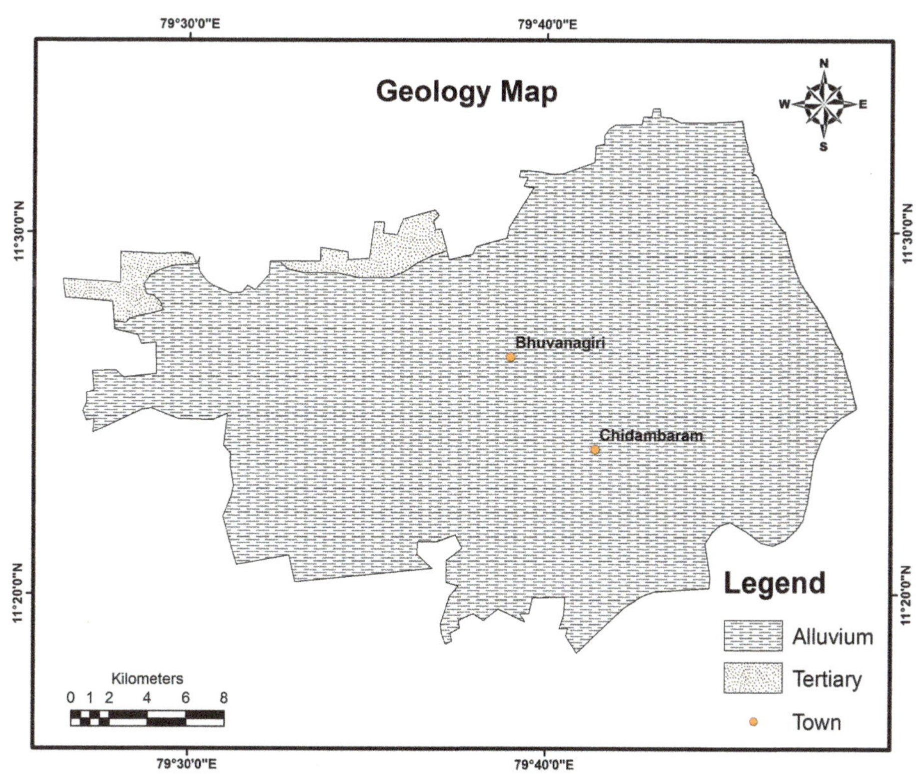

Figure 2. Geology map of the study area.

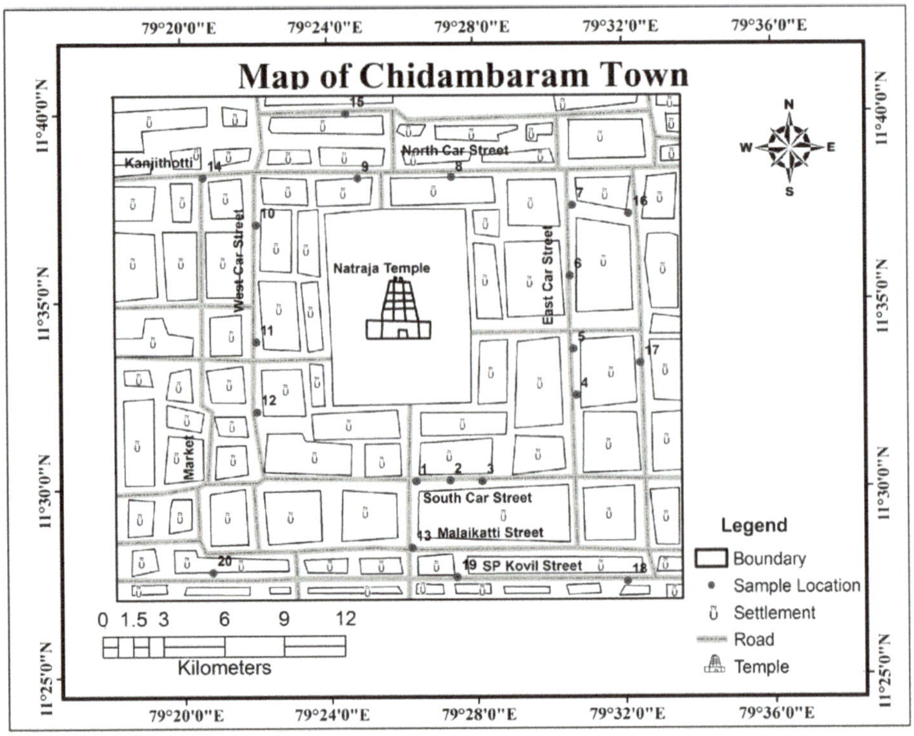

Figure 3. Sample location map.

Analytical Techniques

In the present study, the pH was measured using the instrument model Systronic-335. The TDS was measured using Weight Balance Schimedge T4231 muffle furnace technique, and the EC was measured using the model Systronic-304. The Ca^{2+}, Mg^{2+} and HCO_3^-, Cl^-, were measured by the volumetric method whereas the Na^+, K^+ was measured using flame photo meter model CL 378. Further, the SO_4^{-2} was measured using spectro photometer model SL-171 minispec. Except pH, other parameters are expressed in Mg/L whereas the EC value is expressed in mohs/cm at 25˚C. The trace elements of Zn, Pb, Cr, Fe, Cd, Cu and Mn were determined using AAS model Elico-SL-176. The data of major ions has analysed and interpreted using software WATCLAST [14]. The groundwater sample analysis data of the study area is presented in **Table 1**. Similar studies have carried out extensively by many authors on groundwater quality [15]-[18].

3. Results and Discussion

3.1. General Hydrogeochemistry

The collected groundwater samples shown colourless in most of the locations and few samples were shown pale yellowish colour. The yellowish colour of the samples is due to the presence of suspended solids and silts. Analytical results indicate that pH of groundwater in the study area ranges from 7.1 to 8.6. It is observed that all the samples are above the permissible limit of 6.5, and within the desirable limit of 8.5 suggested by World Health Organization [19] standard for drinking. The pH of the water is influenced by atmospheric contribution, decomposition of organic matter and anthropogenic causes.

Electrical Conductivity (EC) and Total Dissolved Solids (TDS) of the groundwater varies from 171 to 6358 microsimens/cm at 25˚C and 386.3 mg/L to 1491 mg/L respectively. The EC observed to exceed the permissible limit of 750 microsimens/cm (WHO, 2006) in 16 locations. TDS found above the permissible limit of 500 mg/L in 11 locations and above the excessive limit of 1000 mg/L in 5 locations. Generally, water containing more than 500 mg/L of total dissolved solids is not considered potable. However, it is observed that highly mineralized water is also used in the absence of quality water. According to Freeze and Cherrry [20] classification, 15 samples are fresh water category (TDS <1000 mg/L) and 5 samples are brackish water category (TDS >1000 mg/L).

Table 1. Analyzed hydrogeochemical parameters of groundwater of the study area.

S.No	Ca^{2+}	Mg^{2+}	Na^{+}	K^{+}	Cl^{-}	HCO_3^{-}	SO_4^{-2}	pH	EC	TDS	SAR	Zn	Pb	Cr	Fe	Cd	Cu	Mn
1	67.6	51.7	38.5	8.4	62	47.2	37.5	7.4	1875	965	1.092	0.09	1.09	0.77	0.61	0.02	2.12	0.26
2	67.8	91.5	28.9	6.5	52	42.6	13.4	8.0	1294	735	1.079	0.05	1.09	0.77	0.61	0.02	2.24	0.26
3	74.4	42.5	38.5	6.4	91	71.5	29.1	8.1	1414	805	0.992	0.03	0.83	1.1	0.91	BD	2.24	0.26
4	65.1	51.9	43.6	5.2	91	39.6	26.3	8.1	1231	750	1.078	0.01	0.06	0.45	0.76	0.02	1.94	0.33
5	94.7	32.5	41.5	7.6	122	56.2	45.6	7.8	263	1491	0.761	0.06	0.32	0.77	0.76	0.02	2.35	0.12
6	71.8	51.6	42.9	3.3	58	41.2	30.8	8.1	786	463	4.992	0.05	1.09	0.45	0.16	0.07	0.59	0.04
7	68.8	85.5	41.9	3.3	64	39.5	26.8	8.1	1072	620	3.414	0.06	0.83	0.13	0.76	0.02	1.12	0.12
8	64.5	21.5	42.5	2.9	58	21.5	45.2	8.1	6358	386	6.575	0.05	0.83	0.19	0.16	0.05	1.24	0.12
9	74.2	82.6	93.4	4.8	139	57.5	43.6	7.9	1998	924	3.082	0.09	0.57	0.52	0.46	0.02	1.36	0.12
10	74.2	92.5	38.5	6.3	91	24.7	34.5	8.5	766	412	3.566	0.06	0.83	0.52	0.46	0.05	1.18	0.19
11	71.2	47.2	61.5	7.1	127	54.2	53.1	8.0	1793	753	3.307	0.13	1.09	0.84	0.61	0.02	1.71	0.26
12	73.2	56.5	41.5	6.4	81	41.5	24.1	8.0	1301	987	3.926	0.11	0.46	0.84	1.21	BD	1.77	0.19
13	66.4	56.9	40.7	8.5	52	54.5	48.1	8.0	171	978	4.147	0.06	0.2	1.16	0.91	BD	1.88	0.12
14	74.9	74.2	87.5	6.3	113	51.5	48.4	8.1	1618	878	3.814	0.06	0.06	1.48	0.91	0.02	2.06	0.26
15	68.4	71.9	41.7	5.8	101	23.9	22.5	8.6	728	451	4.855	0.09	0.32	1.16	0.91	BD	2.06	0.19
16	72.5	65.5	86.5	10.2	136	85.5	12.4	7.9	2318	1385	3.276	0.09	0.32	1.81	1. 36	0.02	1.94	0.26
17	70.2	48.8	101.5	10.4	139	69.7	25.7	7.1	243	1263	3.793	0.09	0.57	1.48	1.21	0.02	2.06	0.26
18	68.4	39.1	84.9	8.9	141	66.5	24.9	7.1	2302	1271	3.285	0.09	0.32	1.16	1.21	BD	2.06	0.26
19	69.4	36.5	95.3	2.6	123	41.7	38.5	8.0	1012	579.4	3.904	0.08	0.32	1.48	1.51	0.02	1.06	0.4
20	65.1	38.5	71.5	7.9	113	51.5	32.6	8.0	1732	1097	3.948	0.1	0.32	1.81	0.91	0.05	2.06	0.4

*Except pH, EC, and SAR others are expressed in mg/L. EC is in microsiemens/cm at 25°C.

3.2. Major Cations Chemistry

The Ca^{2+} ionic concentration is found to be low as 64.5 mg/L in location 8 whereas it is observed as high concentration of 94.7 mg/L in location 5. The permissible limit of Ca^{2+} for drinking water is specified as 75 mg/L (WHO, 2006) and except sample number 5, all other samples are within the permissible limit. The Mg^{2+} concentration varies from 21.5 mg/L to 92.5 mg/L and the limit of Mg^{2+} for drinking water is 30 mg/L (WHO, 2006), and except sample number 8, all other are above the permissible limit. The high concentration of Mg^{2+} is observed in location number 10 whereas, low concentration is observed in sample location 8. The Na^{+} concentration varies from 28.9 mg/L to 101.5 mg/L. The permissible limit of Na^{+} for drinking water is specified as 200 mg/L (WHO, 2006), and all are within the permissible limit. The higher value is observed in locations 17 and low value is noticed in locations 2. The K^{+} concentration varies from 2.6 mg/L to 10.4 mg/L. The permissible limit of K^{+} for drinking water is specified as 50 mg/L (WHO, 2006) and all samples are below the permissible limit. The higher value of 10.4 mg/L is observed in location19 and low value of 2.6 mg/L noticed in location 17.

Generally, concentration of Ca^{2+}, Mg^{2+}, Na^{+} in the groundwater from clay minerals such as montmorillionite, illeite and chlorite [21]. In this area, Na^{+} and K^{+} ions present relatively higher concentrations in groundwater is particularly derived from leaching of clay minerals, gypsum and anhydrites, whereas the calcium ion is also derived from cation exchange process. Sodium has its source from silicate minerals, atmospheric precipitation and halite deposits. In the study, the major concentrations except Mg^{2+} are low and below the permissible limit of the standards for drinking water. Since the level of concentration is very low, the source could be from cation exchange process and from pollution sources, which have the indication of Na/Cl ratio >1. Sewage, industrial effluents, the use of sodium compounds for corrosion control and water softening process have contributed to Na^+ concentration in groundwater of the study area.

3.3. Major Anions Chemistry

The Bicarbonate concentration varies from 21.5 mg/L to 85.5 mg/L. The HCO_3^- concentration is rather high in location number 16, and it is observed low in location number 8. The Cl^- concentration varies from 52 mg/L to 141 mg/L. The permissible limit of Cl^- concentration for drinking water is specified as 250 mg/L and excessive limit of 1000 mg/L (WHO, 2006). All samples are within the permissible limit and high concentration of Cl^- is noticed in location number 18 whereas low concentration is found in sample location 2 & 13. The Sulphate concentration is found varying from 12.4 mg/L to 53.1 mg/L. The permissible limit for drinking water is specified as 200 mg/L (WHO, 2006) and all are within the permissible limit.

The major source, however, is due to chemical weathering of rocks, whereas most of the pollutants could be of anthropogenic origin. All rocks do not weather at the same rate. Sulfate ion concentrations are probably derived from weathering of sulfate and gypsum bearing sedimentary rocks [5] [6]. The most significant source of sulphate is rock weathering and pollution. Dissolved sulphate has its origin from dissolution of gypsum, oxidation of pyrites and contaminated sulphate from industrial discharges. Sulphate is also a major constituent in the aerosols which is from smoke coming from industries and automobiles. In the study area, higher sulfate might have also been derived from the leaching of the evaporate sediments rich in gypsum abhydrite and sodium sulfate [8]. Health concerns regarding sulphates in drinking water have been raised because of reports of diarrhea associated with ingestion of water with high levels of sulphates [22].

3.4. Quality of Groundwater Based on Chloride

High concentration of chloride gives an undesirable taste to water. Infants and children may suffer if they consume water high in chloride as their tissues may be damaged by the high osmotic pressure brought about by the presence of high concentration of salts. In drinking water, chloride concentration above 250 mg/L causes on objectionable salty taste and makes a deleterious effect on metallic pipes, structures and agriculture crops.

To assess the water based on Chloride, Stuyfzand [23] classification is widely followed and has been classified into 8 main types as given below:

Chloride type	Cl concentration in epm	Number of samples
Extremely Fresh	<0.141	Nil
Very Fresh	0.141 - 0.846	Nil
Fresh	0.846 - 4.231	05
Fresh Brackish	4.321 - 8.462	09
Brackish	8.462 - 28.206	06
Brackish-Salt	28.206 - 282.064	Nil
Salt	282.064 - 564.127	Nil
Hyperhaline	>564.127	Nil

It is observed that the most of the samples were fell in fresh to fresh brackish water category. The brackish water type shown in six locations and it could be mixing of in situ saline water from bottom aquifers. Hence, the water could have infiltrated from recent precipitation and level of concentration may increase from marine sediments water interaction by means of gradual ionic exchange.

3.5. Schoeller Classification

Schoeller [12] proposed a water type classification based on the concentration of groundwater within the sub surface which is as follows:

$$rHCO_3 > rSO_4\text{---------Type I } (r = epm)$$

as the duration of stay increases, the above relation changes to

$$rSO_4 > rHCO_3\text{---------Type II}$$

with further increase of duration, the ionic concentration changes to

$$rCl > rSO_4 > rHCO_3\text{---------Type III}$$

in the final stage the ionic concentration reaches

$$rCl > rSO_4 > r\,HCO_3 \text{ along with}$$

$$rNa > rMg > rCa\text{----------Type IV}$$

The changes from type I to type IV depend on the residence time of water in the subsurface and the extent of rock water interaction.

The samples collected from present study area demonstrate that all the 20 groundwater samples have come under the water type I. As the present study was carried out in shallow level aquifers, it is observed that the strata sediment characterization and interaction are responsible for the present level of concentration in the groundwater. In general, the saturated thickness of groundwater could be maintained for certain level in the subsurface lithounits due to hydrostatic pressure, which could lead to release of soluble elements and responsible for the present ionic concentration in groundwaters.

3.6. Chemical Relationship

Piper's Trilinear diagrams are very useful in bringing out the chemical relationships of ground waters in more definite terms [24]. It identifies hydrochemical facies by plotting the chemical composition of water in a central diamond-shaped diagram in which all the four sides are defined by the proportions of major ions in water. In the present study, the major cations and anions of water samples have been plotted with Piper Trilinear diagram [25] and is shown in **Figure 4**. The plotted samples were distributed in sector 6 of the diamond-shaped field which indicates that Non-carbonate exceeds 50%. The hydrogeochemical facies of the water samples are as $Ca^{2+} > Na^{+}$ and $Cl^{-} > SO_4^{2-}$ facies.

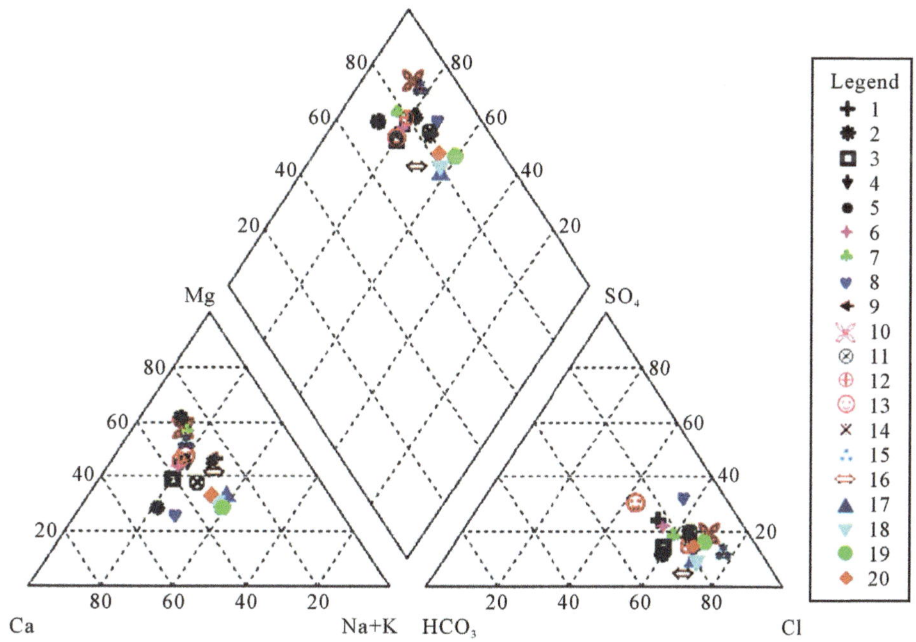

Figure 4. Piper plot.

3.7. Corrosivity Ratio

Corrosivity Ratio is defined as the corrosive tendency of water due to the proportional occurrence of alkaline earths and saline salts along with sulphides in the composition of the water. [17] [26] [27] have used the ratio to evaluate corrosive tendency of groundwater on metallic pipes of various regions. According to Balasubramanian [28] the corrosivity value less than one is considered to be safe for transport of water in any pipe. Similarly, corrosivity higher than one is to as not safe for transport of water. The effects of corrosion are losses in hydraulic capacity of pipes. It is observed that 15 samples were greater than 1.0 and hence the study area groundwater cannot be transported through the pipe for drinking purpose for long duration activities. The release on ions from the flow path lithology and anthropogenic could be the main feeder into the groundwater.

3.8. Saturation State

Saturation indexes are used to evaluate the degree of equilibrium between water and minerals. It is a convenient parameter to evaluate the proximity to equilibrium for various solubility reactions, describing the tendency of the solution to dissolve or precipitate a set of minerals [29]. Changes in the saturation state are useful to distinguish different stages of hydrochemical evolution and help identify which geochemical reactions are important in controlling water chemistry [29]-[31]. The saturation index of a mineral is obtained from equation SI = log (IAP/Kt) where the IAP is the ion-activity product of the mineral-water reaction, and Kt is the thermodynamic equilibrium constant. The saturation for a mineral is indicated when the SI is between −1.0 and +1.0. The index describes the deviation of water from equilibrium with respect to dissolved minerals quantitatively [32] [33]. If the SI is greater than zero, a state of saturation or super saturation is achieved, and minerals would precipitate. If SI is less than zero it means the conditions for mineral dissolution are achieved. The SI around zero indicates equilibrium conditions; therefore, dissolution or precipitation of a certain mineral phase controls the level of the dissolved components contained in that phase.

In this study saturation index has been studied for the Calcite and Sulphate minerals as shown in **Figure 5**. The Calcite group consisting of Magnesite, Dolomite, Calcite and Aragonite and Sulphate group consisting of Aragonite and Gypsum minerals. The index of both the calcite and sulphate minerals are undersaturated state. It indicates the insufficient amount of the mineral for solution or short residence time.

4. Trace Elements

Early concerns over the quality of groundwaters were focused mainly on dissolved mineral salts. However, recently, organics, heavy and trace metals have become a focus of concern for several reasons, including enhanced ability to detect chemicals in the microgram and nano gram per liter concentration levels and an increasing awareness of potentially hazardous risks to public health and the environment. Many trace elements are essential nutrients however certain trace elements such as As, Cd, and Hg are known to be persistent environment contamination and toxic to most form of life. Trace elements are generally present in small concentration in the natural water system. Trace elements are contributed to groundwater from a variety of natural and anthropogenic sources. Once liberated to groundwater, element distributions are continually modified by complex geochemical and biological processes [34].

The trace element data may help to assess the level of heavy metal concentration in groundwater of the study area and possible source. Selected trace metals of Fe, Pb, Cr, Mn, Cu, Zn and Cd were studied in groundwater samples of the study area. The minimum and maximum of trace elements of the analysed groundwater samples of the study area is given in **Table 2** and also the variation is presented as a graph in **Figure 6**.

4.1. Iron (Fe)

Fe is presented in fully aerated water of normal pH in small amounts (as 0.5 mg/L) and acid water may contain ferric iron in large amounts. Microorganisms are causes for dissolving and precipitating iron from water. Rarly, the concentration over 50 mg/L may occur in waters with a pH of 5 to 8 and concentration upto 10 mg/L is uncommon. Morethan about 0.5 mg/L stains clothes and utensils. Iron is objectionable in food processing industry [35]. In the study area, Fe concentration varied from 0.16 to 1.51 mg/L where the higher concentration found in location 19 and low in location 6. The standard for the drinking is 0.3 mg/L where 18 locations are found above the permissible limit and in locations 12, 16, 17, 18 & 19 are found the excessive limit of 1 mg/L. The high

Table 2. Comparison with drinking water standards.

Paramters	Min	Max	Mean	WHO, 2006		Above permissible limit	BIS, 2003		Above permissible limit
				Permissible	Excessive		Desirable	Permissible	
pH	7.1	8.6	7.9	7.0 - 8.5	>8.5	2	6.5	8.5	2
EC	171	6358	1672	750	-	17	500	-	17
TDS (mg/L)	386	1491	867	500	1000	16	500	2000	16
Calcium (Ca^{2+}) (mg/L)	64.5	94.7	71.9	75	100	01	75	200	Nil
Magnesium (Mg^{2+}) (mg/L)	21.5	92.5	57.0	30	100	19	30	100	Nil
Sodium (Na^{+}) (mg/L)	28.9	101.5	58.8	200	-	Nil	-	200	Nil
Potassium (K^{+}) (mg/L)	2.6	10.4	6.4	50	150	Nil	-	-	
Sulphate (SO_4^{-2}) (mg/L)	12.4	53.1	33.1	200	400	Nil	200	400	Nil
Chloride (Cl^{-}) (mg/L)	52	141	97.6	250	1000	Nil	250	1000	Nil
Zn	0.01	0.13	0.07	5	15				
Pb	0.06	1.09	0.57	NA	0.05				
Cr	0.13	1.81	0.94	NA	0.05				
Fe	0.16	1.51	0.79	0.3	1.0				
Cd	BD	0.07	0.02	0.01	0.01				
Cu	0.59	2.35	1.7	0.05	1.5				
Mn	0.04	0.4	0.22	0.05	0.5				

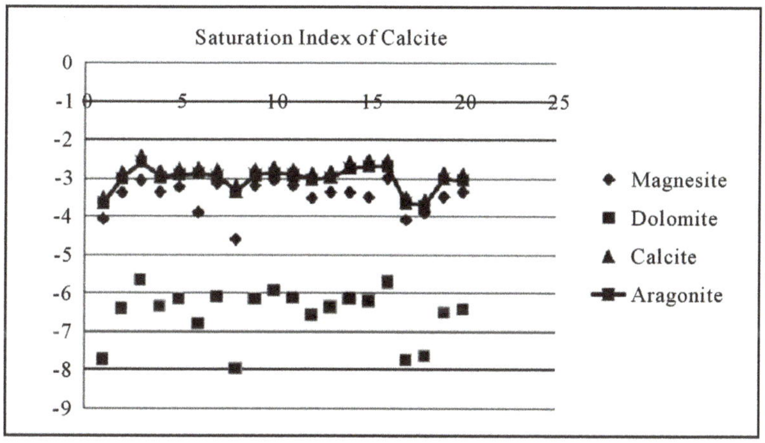

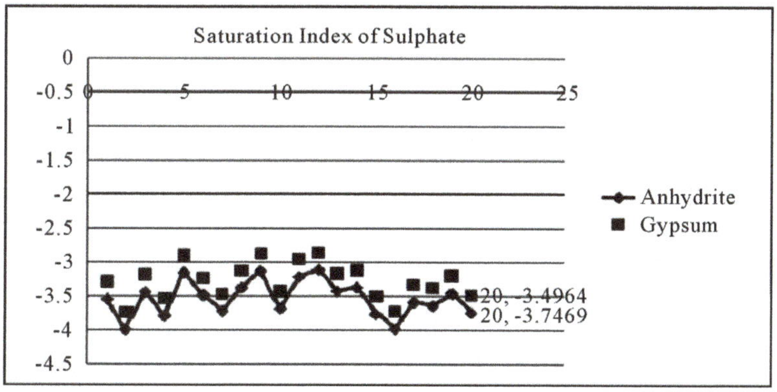

Figure 5. Saturation index of calcite and sulphate minerals.

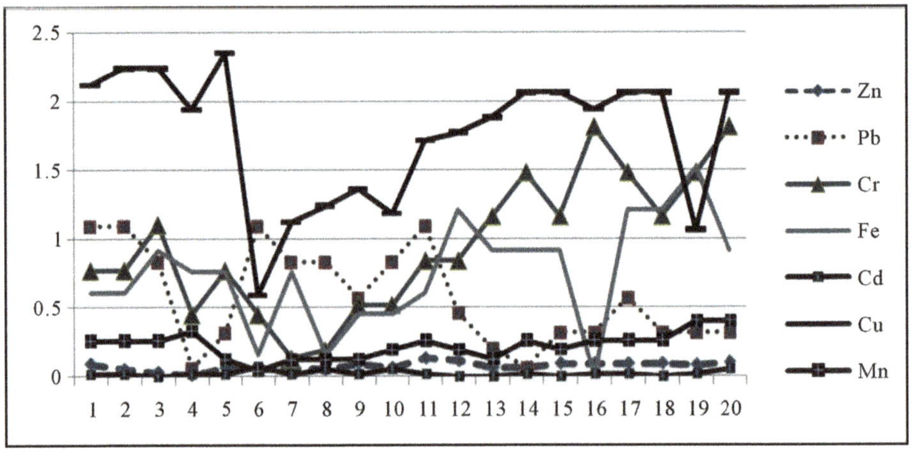

Figure 6. Variation of trace elements.

values are could be attributed to the corrosion effect of borehole materials in water, the redox environment in groundwater, and drainage from domestic activities.

4.2. Lead (Pb)

Lead concentration in natural water is mainly attributed to anthropogenic activities as it is extensively used in some pesticides such as lead arsenate. Besides, oil, grece, smoke from vehicle movements, drainage could be the important sources. Lead enters into the environment from industry, mining, plumbing, gasoline, coal, and as a water additive [36]. In the study area, Pb varied from 0.06 to 1.09 mg/L and all the samples are found with above the permissible limit of 0.05 mg/L WHO (2006) standard.

4.3. Chromium (Cr)

Presence of high concentration of Chromium (III) and heavy doses of Chromium salts even though are rapidly eliminated from the human body, could corrode the intestinal tract [37]. Chromium enters the environment from old mining operations runoff and leaching into groundwater, fossil-fuel combustion and waste incineration. The concentration of the Chromium in the study area ranged from 0.13 - 1.81 mg/L with average concentration of 0.94 mg/L. The groundwater quality standard of Chromium desirable and maximum permissible limit (WHO) is 0.05. Higher concentrations were observed in all locations when compare to WHO standards. Highest concentrations of 1.81 mg/L were observed at locations 16, 20.

4.4. Manganese (Mn)

Manganese occurs naturally as a mineral from sediment and rocks or from mining and industrial waste. The excess Mn is due to the organic circulation by plants and leaves [38]. Excess concentrations of manganese make water distasteful to drinking with no specific toxic effects [39]. The concentration of the Manganese in our study area ranged from 0.04 - 0.40 mg/L. The higher concentration found in location 19, 20 and low in location 6. The groundwater quality standard of Manganese desirable and maximum permissible limit (WHO) is 0.05 mg/L. The manganese concentration is found less than the permissible limit of 0.05 mg/L in sample location 6 and found above the permissible limit in 19 locations but below the excessive limit of 0.5 mg/L.

4.5. Copper (Cu)

The concentration of the Copper in our study area ranged from 0.59 - 2.35 mg/L with an average of 1.7 mg/L. The groundwater quality standard of Copper desirable limit (WHO) is 0.05 mg/L and maximum permissible limit (excessive) is 1.5 mg/L. In the study area 14 locations were found above desirable limit and 6 locations found above the excessive limit. Copper enters into water bodies mostly anthropogenic activities like metal plating, industrial and domestic waste, mining, and mineral leaching were main sources for the presences in environment.

4.6. Zinc (Zn)

Zinc is found naturally in water, most frequently in areas where it is mined and from industrial waste, metal plating, and plumbing, and is a major component of sludge. The concentration of the Zinc in our study area ranged from 0.01 - 0.13 mg/L with an average of 0.7 mg/L. The higher concentration found in location 11 and low in location 4. The ground water quality standard of Zinc desirable limit (WHO) is 5.0 mg/L and maximum permissible limit is 15.0 mg/L. In the study area; all sampling locations are within the desirable limit (WHO).

4.7. Cadmium (Cd)

Cadmium found in low concentrations in rocks, coal, and petroleum and enters the ground and surface water may enter the environment from industrial discharge, mining waste, metal plating, water pipes, batteries, paints and pigments, plastic stabilizers, and landfill leachate.

The concentration of the Cadmium in our study area ranged from below the detected level to 0.07 mg/L with an average of 0.02 mg/L. The higher concentration of 0.07 mg/L found in location 6, and below the detected level noted in locations 3, 12, 13, 15 & 18. The groundwater quality standard of Cadmium desirable and maximum permissible limit (WHO) is 0.01 mg/L. Above the desirable and maximum permissible limit of WHO was noted remaining 15 sampling locations.

5. Comparison with Standards

To estimate whether the groundwater of the study area is suitable for drinking purposes, it has been compared with the standards of Beauro of Indian standards [19] [40]. The samples were analysed and compared with the prescribed standards, and the result are presented in **Table 2**. Among the samples, considerable numbers of samples were identified with the desirable limits of the standards of drinking purposes. Nevertheless, more than 80% of the samples have shown above the permissible limit of BIS and WHO standard for drinking water.

6. Conclusions

Interpretation of hydrogeochemical analysis reveals that the groundwater in study area is fresh water type and needs simple filtering techniques. The pH, EC and TDS are noticed high in most of the locations, and major ionic concentrations are quite below the drinking water standard. The sequence of abundance of the major ions is $Ca^{2+} > Na^{+} > Mg^{2+} > K^{+}$ and $Cl^{-} > SO_4^{-2} > HCO_3^{-}$... The chloride classification elucidates fresh and fresh brackish water type, which indicates the groundwater is derived from precipitations and residence time in the sub surface strata is less. The short residing time is well depicted from the Scholler classification as it has shown type I. In the study area, the dominant hydrochemical facies of groundwater are $Ca^{2+} > Na^{+} > Cl^{-}$. The results of calculation saturation index show that the nearly all of the water samples are undersaturated with respect to carbonate (calcite, dolomite and aragonite) and sulfate (gypsum and anhydrite) minerals. The trace elements' concentration represents that except Zn, other elements of Fe, Pb, Cr, Mn, Cu, and Cd are found high and above the permissible limit of drinking water standard.

Since the groundwater is received from recent precipitations, ionic concentration by ionic base exchange and geological formation can be less. However, anthropogenic particularly wastes from commercial activities and domestic sewage can be rendered remarkable extent. The result also reveals that there is no sea water mixing in the shallow aquifers, and it can have a chance at deeper aquifers in the study area. As for as trace elements are concerned, industrial and agricultural activities are unlikely to be the major contamination sources because of the absence of these activities. The contamination of groundwater is also due to domestic waste water and sewers through subsurface drainage systems from high density residential and commercial buildings. Pollutants accumulated on road surfaces can be washed by storm or street cleansing as road runoff and finally are collected into the nearby sewage drainage.

References

[1] Ramon, L. (1993) All of Us, Environmental Education Dossiers. Vol. 4, Centre UNESCO de Catalunya, Mallorca, 285.

[2] Johnson, C.C. (1979) Land Application of Waste—An Accident Waiting to Happen. *Groundwater*, **17**, 69-72. http://dx.doi.org/10.1111/j.1745-6584.1979.tb03277.x

[3] Sastri, J.V.C. (1994) Groundwater Chemical Quality in River Basins. Hydrogeochemical Modeling, Lecture Notes-Refresher Course, School of Earth Sciences, Bharathidasan University, Tiruchirapalli.

[4] Subba Rao, N., Gurunadha Rao, V.V.S. and Gupta, C.P. (1998) Groundwater Pollution Due to Discharge of Industrial Effluents in Venkatapuram Area, Visakhapatanam, Andhra Pradesh, India. *Environmental Geology*, **33**, 289-294. http://dx.doi.org/10.1007/s002540050248

[5] Elango, L., Kannan, R. and Kumar, S. (2003) Major Ion Chemistry and Identification of Hydrochemical Processes of Groundwater in a Part of Kancheepuram District, Tamil Nadu, India. *Journal Environmental Geosciences*, **10-14**, 157-166.

[6] Krishnakumar, S. (2004) Quality Characterization and Geochemical Characteristics of Lower Gadilum River Basin, Tamil Nadu, India. Dissertation, University of Madras.

[7] Jeevanandam, M., Kannan, R., Srinivasalu, S. and Rammohan, V. (2006) Hydrogeochemistry and Groundwater Quality Assessment of Lower Part of the Ponnaiyar River Basin, Cuddalore District, South India. *Environmental Monitoring and Assessment*, **132**, 263-274.

[8] Mala, V. (1997) Study of the Environmental Hydrogeochemical and Sedimentological Aspects between Puduchatram to Chidambaram Area, Cuddalore District, East Coast of India. Phil Dissertation, Annamalai University, 146.

[9] Chandrasekaram, D. (1989) Hydrogeochemical Study of Coastal Groundwaters of Kerala. *Indian Academy of Sciences* (*Earth and Planetary Sciences*), **98**, 287-295.

[10] Ramanathan, A.L., Mani, R., Manoharan, K., Kesavan, S. and Kathiresan, R.M. (1998) Hydrogeochemical Studies of Coastal Groundwaters from Cuddalore to Puduchatiram. *Indian Journal of Marine Sciences*, **27**, 167-172.

[11] PWD (2000) Groundwater Perspectives: A Profile of Cuddalore District, Tamil Nadu. State Ground and Surface Water Resources Data Centre, Water Resources Organisation in Public Work Department, Government of Tamil Nadu, Chennai, 89.

[12] Schoeller, H. (1967) Qualitative Evaluation of Groundwater Resources (In Methods and Techniques of Groundwater Investigation and Development. *Water Resource Series*, **33**, 44-52.

[13] Palmquist, W.M. (1973) Sampling for Groundwater Quality Investigations. *Groundwater Studies*, **125**, 14.

[14] Chidambaram, S., Ramanathan, A.L., Srinivasamoorthy, K. and Ananthan, P. (2004) WATCLAST—A Computer Program for Hydrogeochemical Studies, Recent Trends in Hydrogeochemistry (Case Studies from Surface and Subsurface Waters of Selected Countries). Capital Publishing Company, New Delhi, 203-207.

[15] Prasad, B.V. and Rameshchandra, P. (1997) Groundwater Quality in an Industrial Zone. *Pollution Research*, **16**, 105-107.

[16] Tripathi, J.K. (2003) Groudwater Hydrochemistry in and around Bhanjabihar, Ganjam District (Orissa). *Pollution Research*, **22**, 185-188.

[17] Aravindan, S., Manivel, M. and Chandrasekar, S.V.N. (2004) Ground Water Quality in the Hard Rock Area of Gadilam River Basin, Tamil Nadu. *Journal of Geological Society of India*, **63**, 625-635.

[18] Hedge, G.V. (2006) Evaluation of Chemical Quality of Groundwater Resource in Dharwad District, Karnataka. *Journal of Geological Society of India*, **67**, 47-58.

[19] WHO (2006) Guidelines for Drinking Water Quality. Volume 1, Recommendations, World Health Organisation, Geneva.

[20] Freeze, A. and Cherry, J. (1979) Groundwater. Prentice-Hall, Englewood Cliffs, 604 p.

[21] Garrels, R.M. (1976) A Survey of Low Temperature Water Mineral Relations. In: *Interpretation of Environmental Isotope and Hydrogeochemical Data in Groundwater Hydrology*, International Atomic Energy Agency, Vienna, 65-84.

[22] Prasad, R.N., Chandra, R. and Tiwari, K.K. (2008) Status of Groundwater Quality of Lalsot Urban Area in Dausa District, Rajasthan. *Nature Environment and Pollution Technology*, **7**, 377.

[23] Stuyfzand, P.T. (1989) A New Hydrochemical Classification of Water Types with Examples of Application. *AHS*, **184**, 89-98.

[24] Walton, W.C. (1970) Groundwater Resource Evaluation. McGraw Hill Book Co., New York, 664 p.

[25] Piper, A.M. (1944) A Graphical Procedure in the Chemical Interpretation of Ground Water Analyses. *Transactions, American Geophysical Union*, **25**, 914-923. http://dx.doi.org/10.1029/TR025i006p00914

[26] Ryzner, J.W. (1944) A New Index for Determining Amount of Calcium Carbonate Scale Formed by a Water. *Journal of American Water Works Association*, **36**, 472-486.

[27] Sankar, K. (1995) Hydrological Studies in the Kanyakumari District, Tamil Nadu. Unpublished PhD Thesis, University of Madras, Chennai, 90-91.

[28] Balasubramanian, A. (1986) Hydrogeological Investigations in the Tambaraparani River Basin, Tamilnadu. Unpub-

lished PhD Thesis, University of Mysore, Mysore, 348 p.

[29] Langmuir, D. (1997) Aqueous Environmental Geochemistry. Prentice-Hall, Inc., Englewood Cliffs, 601.

[30] Drever, J.I. (1997) The Geochemistry of Natural Waters. Prentice-Hall, Englewood Cliffs, 436.

[31] Coetsiers, M. and Walraevens, K. (2006) Chemical Characterization of the Neogene Aquifer, Belgium. *Hydrogeology Journal*, **14**, 1556-1568. http://dx.doi.org/10.1007/s10040-006-0053-0

[32] Garrels, R.M. and Mackenzie, F.T. (1971) Evolution of Sedimentary Rocks. W.W. Norton and Company, Inc., New York, 397 p.

[33] Stumm, W. and Morgan, J.J. (1981) Aquatic Chemistry: An Introduction Emphasizing Chemical Equilibria in Natural Waters. Wiley-Interscience Publication, John Wiley and Sons, New York, 780 p.

[34] Alabdula'aly, A.I., Al Zarah, A.I. and Khan, M.A. (2011) Assessment of Trace Metals in Groundwater Sources Used for Drinking Purposes in Riyadh Region. *International Journal of Water Resources and Arid Environments*, **1**, 5-9.

[35] Ramakrishnan, S. (1998) Groundwater. Ramakrishnan Publishing, Chennai, 761.

[36] Reddy, T.B., Ramana, C.V., Bhaskar, C. and Chandrababu, P.J. (2012) Assessment of Heavy Metal Study on Ground Water in and Around Kapuluppada MSW Site, Visakhapatnam, AP. *International Journal of Science and Nature*, **3**, 468-471.

[37] Akoteyon, S.I. and Soladoye, O. (2011) Statistical Study of Physicochemical Characteristics of Wastewater in Odogunyan, Lagos, Nigeria. *E-Journal of Chemistry*, **8**, 2010-2015. http://dx.doi.org/10.1155/2011/787839

[38] Longe, E.O. and Enekwechi, L.O. (2007) Investigation on Potential Groundwater Impacts and Influence of Local Hydrogeology on Natural Attenuation of Leachate at a Municipal Landfill. *International Journal of Environmental Science and Technology*, **4**, 133-140. http://dx.doi.org/10.1007/BF03325971

[39] BIS (2003) Indian Standard Drinking Water Specifications IS 10500:1991, Edition 2.2 (2003-2009). Bureau of Indian Standards, New Delhi.

[40] Offiong, O.E. and Edet, A.E. (1998) Water Quality Assessment in Akpabuyo, Cross River Basin, South-Eastern Nigeria. *Environmental Geology*, **34**, 167-174. http://dx.doi.org/10.1007/s002540050268

12

Irrigation Demand VS Supply-Remote Sensing and GIS Approach

Ch. Ramesh Naidu[1], M. V. S. S. Giridhar[2]

[1]GVP College of Engineering (Autonomous), Visakhapatnam, India
[2]Jawaharlal Nehru Technological University, Hyderabad, India
Email: chrnaidu@rediffmail.com, mvssgiridhar@gmail.com

Abstract

To determine the irrigation requirements of rice crop on different soils, an integrated approach is used using Remote Sensing and GIS techniques. Depending on the type of soil, climate and the crop acreage, the water requirement for paddy fields is derived. This study is focused on estimating the water demand for rice crop in Rabi season. Crop evapo-transpiration and soil percolation losses account more in rice fields especially in hot climate like Rabi season. In addition to evapo-transpiration and percolation losses, the conveyance losses are also accountable in the case of unlined canals. Satellite data is used to estimate the rice and fallow lands. In conjunction to satellite interpreted data, climate and soil data are also integrated in GIS platform. CROPWAT model is used to determine the crop evapo-transpiration (ET_c). There are 11 Water User's Associations (WUA) in the command area and under which 13 canal blocks are delineated. These blocks are again delineated in to 212 sub blocks. This study indicates that there exists a 5% to 20% of water deficiency in some WUAs and also water surplus in some WUAs varying from 15% to 40%.

Keywords

GIS, Water Users Association, Evapo-Transpiration, Remote Sensing, Irrigation

1. Introduction

The estimation of irrigation demand [1] [2] is an important component for managing the water effectively in the canal command area. To meet the present water demand, irrigation management plays an important role. One of the reasons for low efficiency is the un-accounted volume of water loss due to percolation in the rice fields and conveyance and seepage losses occurred during the water distribution through canals to fields [3] [4]. These losses are accounts in calculating total irrigation demand in addition to crop evapo-transpiration. Traditional way of calculating irrigation demand and the absence of scientific approach and adoption of new technologies is another reason for low irrigation efficiency. The selected study area lacks complete information about the command area including its water resources, distribution system details, land use, cropping patterns, soil, geology, climate and socio economic factors and also lacking regular evaluation and working knowledge regarding proper water management. This can be attributed to the dearth of spatially related information available regarding the canal network system and command area. The hard copy printed maps which are currently available are not in scale followed by lack of integration between the maps for the purpose of evaluation. The water allocation in command area is marked by blocks and sub-blocks. These blocks and sub-blocks come under the purview of Water Users' Association (WUA). Proper delineation of block boundaries is essential in order to evaluate water allocation and its usage in terms of actual or theoretical water demand. GIS technology [5] in conjunction with Remote Sensing [6] has proved to be effective for land use and water management. For effective decision making on water use, stresses the need of generation of spatial and non-spatial database in GIS platform [7]. Multispectral and multi-temporal remote sensing data are very useful as input for finding out rice crop acreage and with this, cartographic and data overlaying capability of GIS coupled with its dynamic linking ability plays a vital role in decision making process. Resulting disparities are existed in the availability of water between head-reach and tail end farms and between large and small farms that reflect some portions of the command area posing water shortage and water logging problems due to under and over utilization of canal water. This demands the calculation and analysis of irrigation water requirements of the command area in detail up to water users association, canal block level by using the latest available technologies for optimum utilization of irrigation water.

2. Materials and Methods

The selected study area is under block number five left main canal in Nagarjuna Sagar Project [8]. The area is located between 16°39'2.84" and 16°56'40.81" N latitude and 79°25'16.01" and 79°40'52.90" E longitude. The study area is bounded by Musi River in the East, Krishna River in the South, Lalbahadur canal in the North and Tungapadu vagu in the West. In the command area reserved forest occupies 18 to 20 percent. The command area extends to a total of 26,725 ha, distributed across three mandals namely Mirialguda, Dameracherla and Tripuraram. The methodology adopted for the study involves the use of satellite imagery in conjunction with soil, crop and climate data. Water losses from the field due to seepage and percolation, evapo-transpiration and conveyance losses were estimated. The crop information and statistical reports extracted from satellite imagery were integrated to crop, soil and climate data and the results were analyzed on GIS platform. The flowchart in **Figure 1** depicts overall methodology.

Since rice is dominant crop in the study area, multi-spectral and multi-temporal satellite imagery on the basis of crop calendar for Rabi season were procured from National Remote Sensing Centre (NRSC), Hyderabad, India. Imageries in the initial crop development stage and the late mid season stage were selected for the present study. This includes IRS P6 LISS III [9] [10] in the month of February and March. Survey of India (SOI) topo-maps in 1:25,000 scale were collected to use as a reference. WUA maps were collected from the Walamtari, Hyderabad and integrated to canal blocks by setting the common projection and converted to GIS format. **Figure 2** and **Figure 3** shows the maps in GIS platform depicting WUAs in command area and canal distribution on different soils. Collateral data such as soil data, daily meteorological data, and climate and rainfall data were collected from different sources to study the parameters of evapo-transpiration, percolation/seepage loss and conveyance loss. Daily meteorological data was collected from the nearest and most representative meteorological station. Mandal wise daily rainfall data was collected from the District Information Center. Some essential information were collected from the field that includes crop and crop variety, first and last planting date and first and last harvesting date. The satellite imageries after being geo-referenced with respect to SOI topo maps were digitally enhanced to extract appropriate information for rice crop acreage estimates. The crop acreage reports were then generated WUA wise in GIS platform to find out the water demands for each WUA. For this study CROPWAT version 8.0 [11] [12] was used. ETc and Net Irrigation Water Requirements (NIWR) were estimated by using CropPWAT Software. The inputs used at various stages of the software are rainfall, temperature, wind speed, sunshine, sowing and harvesting dates of Rabi, crop factors, rooting depth and soil characteristics of the study area [13] [14]. The effective rainfall and the crop water requirement decide the amount of irrigation water that has to be applied. The effective rainfall is subtracted from the crop water requirement to calculate the net irrigation water requirements (NIWR). This formula is taken into consideration while running the CROPWAT model using the daily rainfall data in the program. The net irrigation water requirements have been computed on daily and 10-day basis for each WUA and block/canal [15]. Calculation of the NIWR is carried out by pooling up sequentially the appropriate climate and rainfall data sets together with the crop files and the corresponding sowing/planting dates. In case of NIWR calculation of rice, soil data is also required. NIWR is determined by using the formula: NIWR = ETc − Re. Re is the effective rainfall. The Gross Irrigation Water Requirements (GIWR) [16] required for irrigation gift for the entire command area is equal to the sum of the net irrigation water requirements, water losses due to percolation and seepage and canal conveyance losses.

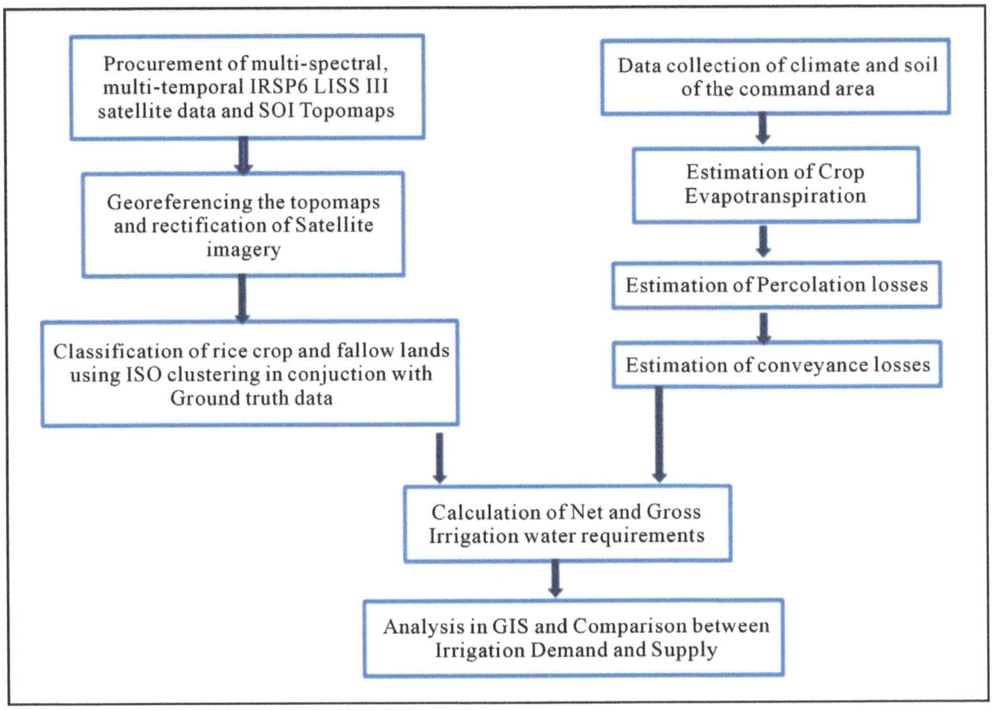

Figure 1. Flow Chart showing the study methodology.

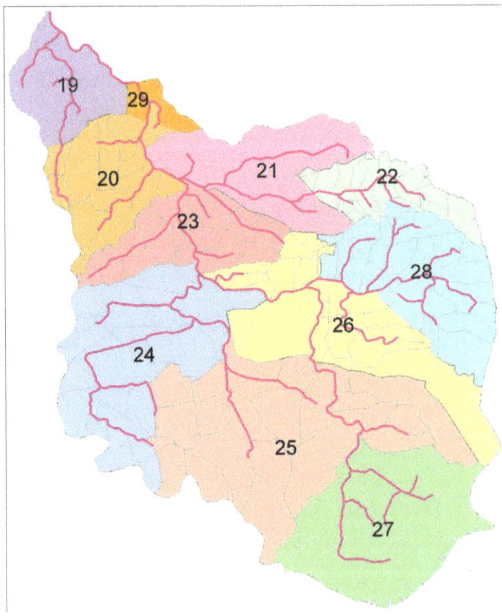

Figure 2. WUA boundaries of command area.

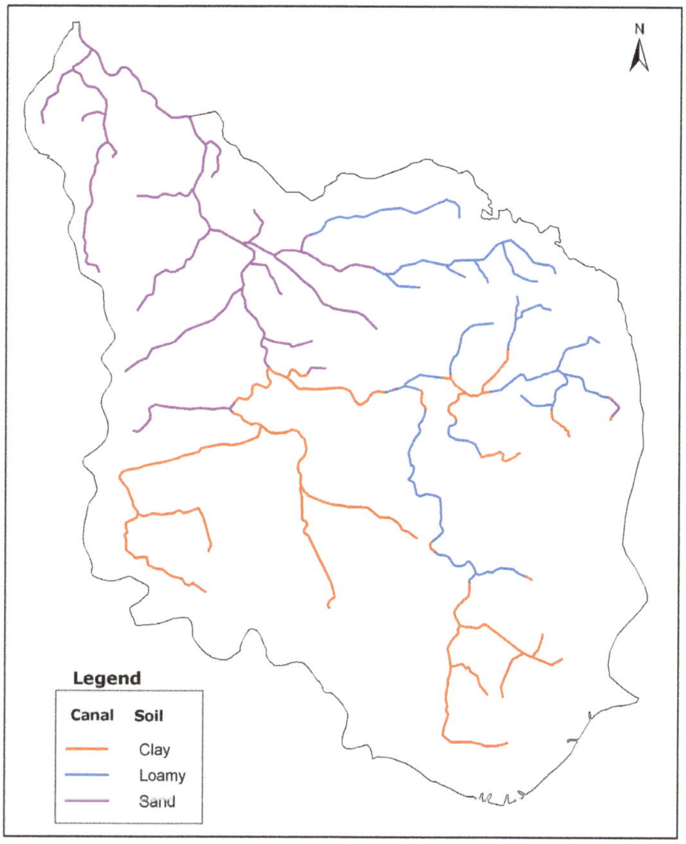

Figure 3. Canals distributed on different soils.

3. Discussion and Results

From the results, it was observed that for WUAs of Chillapuram, Venkatadripalem, Kothagudem, Mirialguda and Chinthapalle, the percolation losses accounts more than 10 million litres per ha. Where as for WUAs of Appalammagudem and Kalleyapalle the percolation losses are 5 million litres per ha. For Kondrapole and Kesavapuram WUAs, the percolation losses accounts 7 to 8 million litres per ha. The conveyance losses calculated from GIS ranging from 1 to 25 MLD (million litres per day) per km. **Figure 4** and **Figure 5** shows WUA wise irrigation demand in the year 2007 and 2009. The percolation losses were calculated using soil, crop, WUA, canal/block themes in the GIS environment using GIS software. Soil type and crop acreage under each command area unit were taken as main parameters for calculation of percolation losses.

The gross irrigation water requirements are compared with the irrigation supplies at WUA level. The study indicates that there is a deficit of water for irrigation in some canal blocks/WUAs, where in water deficiency varies from 5 to 20 percent of total crop water requirement. In some canal blocks and WUAs, water surplus varies from 15 to 40 percent. The data obviously indicates the necessity of additional canal supplies to some blocks and less canal supply to some blocks. **Figure 6** and **Figure 7** shows the WUA wise irrigation demand versus supply for the year 2007 and 2009. It is observed that at block level or WUA level, the water requirements are not only based on crop extent and also depends on the soil and meteorological conditions of the area. The net irrigation water requirements (NIWR) in the years 2007 and 2009 are 54125 and 85208 million litres. The gross water requirements calculated for the years 2007 and 2009 is 4.875 and 7.25 thousand million cubic feet (TMC). The supply to the command area in both the years is 4.35 and 5.49 thousand million cubic feet (TMC). It indicates the supply is lower than the demand for command area. The study found that there is an excess release of water in the year 2007 for Chillapuram, Kothagudem, Kondrapole, Dameracherla and Kesavapuram WUAs and deficiency of water in Venkatadripalem, Chinthapalle, Borrayapalem, Appalammagudem and Kalleyapalle WUAs. Whereas in the year 2009, the excess water is released for Dameracherla and Mirialguda WUAs and remaining WUAs have deficiency in supply of water to fields.

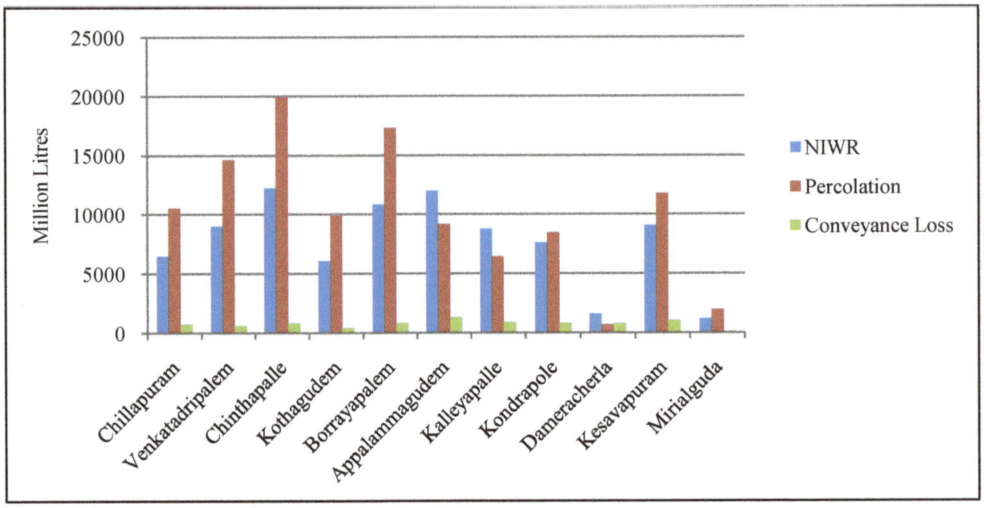

Figure 4. WUA wise irrigation demand in the year 2007.

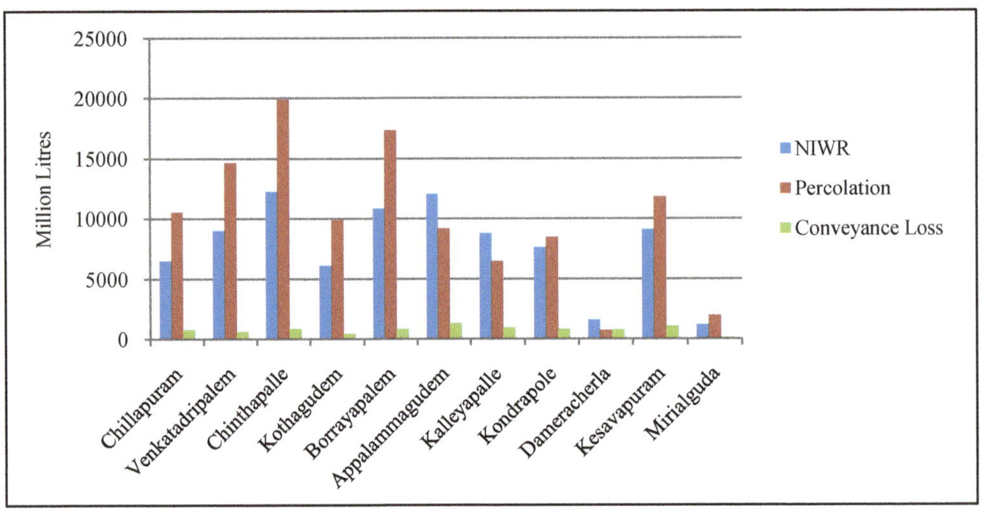

Figure 5. WUA wise irrigation demand in the year 2009.

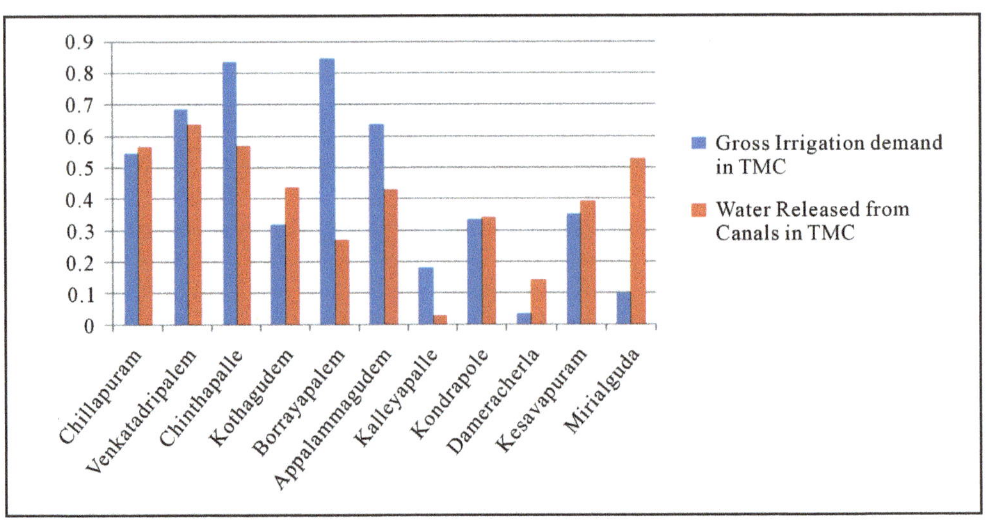

Figure 6. Irrigation demand vs. supply (year 2007).

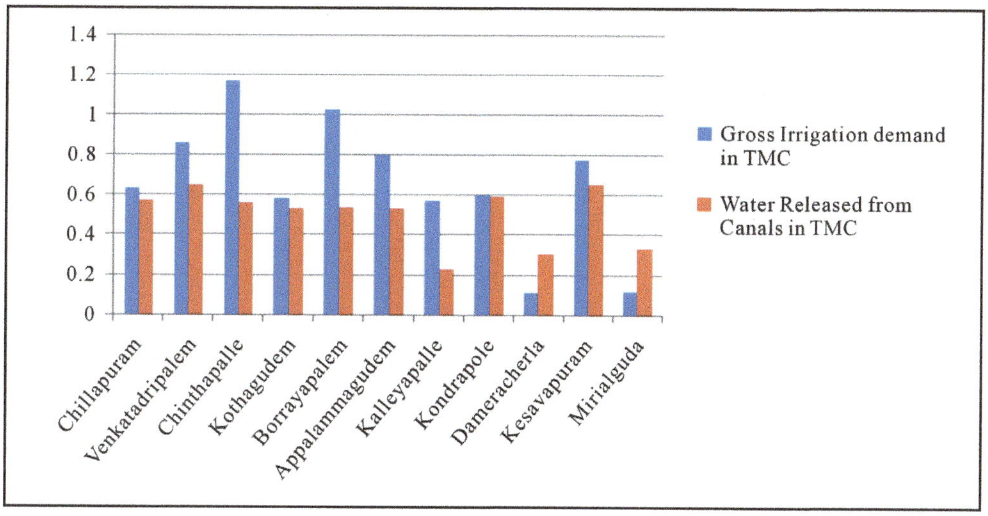

Figure 7. Irrigation demand vs. supply (year 2009).

4. Conclusions

From the study, 11 WUAs, 12 canal blocks and 212 canal sub-blocks were delineated under Wazirabad command area in GIS environment. The total command area was calculated in GIS platform as 26,725 hectares (ha). The gross command area was found maximum in Kalleyapalle WUA with 6482 ha and minimum area was found in Mirialguda WUA with 279 ha. In blocks, the maximum gross command area was found in WL5 canal block with 7721 ha and minimum area was in WL1 canal block with 213 ha. Utilizing GIS network model, the total length of the canal system was identified as 155.846 km. The clay soils were predominant and found 49% in the study area whereas sandy and loamy soils occupied 29% and 22% respectively.

The maximum rice crop was identified in Appalammagudem WUA with 1442 ha in the year 2007 and 1884 ha in the year 2009 in a gross command area of 3254 ha. Whereas in Dameracherla WUA, the rice crop was occupied very less area as 20 ha in the year 2007 and 257 ha in the year 2009 out of 3324 ha of gross command area which was situated in tail end of the command area. The increase of area under rice cultivation demands higher water, as a result greater losses were encountered in the command area.

Due to lengthy canal system coupled with low agriculture area (in Dameracherla WUA), the conveyance losses were higher and this created increase in irrigation demand. Significant difference was observed between the sandy and clayey command areas. From the study, it was observed that for WUAs of Chillapuram, Venkatadripalem, Kothagudem, Mirialguda and Chinthapalle, the percolation losses accounted 10 million litres per ha. Whereas for WUAs of Appalammagudem and Kalleyapalle, the percolation losses were 5 million litres per ha. For Kondrapole and Kesavapuram WUAs, the percolation losses accounted 7 to 8 million litres per ha.

References

[1] Brouwer, C., Hoevenaars, J.P.M. and Van Bosch, B.E., (1992) Scheme Irrigation Water Needs and Supply. FAO, Irrigation Water Management: Training Manual No. 6, 7-43.

[2] El-Magd, I.A. and Tanton, T.W. (2005) Remote Sensing and GIS for Estimation of Irrigation Crop Water Demand. *International Journal of Remote Sensing*, **26**, 2359-2370. http://dx.doi.org/10.1080/01431160420000298261

[3] Palacios, E.V. and Day, J.C. (2012) A New Approach for Estimating Irrigation Conveyance Losses and Their Economic Evaluation. *Journal of the American Water Resources Association*, **13**, 709-720. http://dx.doi.org/10.1080/01431160420000298261

[4] Akkuzu, E., Ünal, H.B. and Karata, B.S. (2002) Determination of Water Conveyance Loss in the Menemen Open Canal Irrigation Network. *Turkish Journal of Agriculture*, **31**, 11-22.

[5] Dadhwal, V.K. (1999) Remote Sensing and GIS for Agricultural Crop Acreage and Yield Estimation. *Proceedings of WG VII/2 Workshop on Application of RS and GIS for Sustainable Development, Inter-national Archives in Photogrammetry and Remote Sensing*, **7**, 58-67.

[6] Jayasekera and Walker (1990) Remotely Sensed Data and Geographic Information System for Management and Ap-

praisal of Large Scale Irrigation Projects in the Developing Countries. Advances in Planning, Design and Management of Irrigation Systems as Related to Sustainable Land Use, 1453-1461.

[7] Navalgund, R.R. and Ray, S.S. (2000) Geomatics in Natural Resources Management. *Proceedings of Geomatics*-2000, Pune, NR-1-14.

[8] EPTRI (2008) ISEA Study Report for the Proposed Nagarjuna Sagar Modernization Project. 5-9.

[9] Krishna Rao, M.V., Hebbar, R. and Venkatartnam, L. (1997) Evaluation of IRS-1C Data for Discrimination and Acreage Estimation of Crops Grown under Multiple Cropping Situation. Remote Sensing for Natural Resources, Joint Publication of ISRS & NNRMS, 205-211.

[10] Dong, Q.H., Herman, E. and Chen, Z.X. (2008) Crop Area Assessment Using Remote Sensing on the North China Plain. *Proceedings in ISPRS*, 957-962.

[11] Allen, E.A. (1998) Crop Evapotranspiration: Guidelines for Computing Crop Water Requirements. FAO Irrigation and Drainage Paper 56, Rome, 300.

[12] Clarke, D. (1998) CropWat for Windows: User Guide. Version 4.2. University of Southampton, UK.

[13] Giridhar, M.V.S.S. and Viswanadh, G.K. (2005) Estimating of Reference Evapotranspiration for an Irrigation Project Site in Andhra Pradesh, India. *Proceedings of International Conference on Environmental Management*, JNTU, Hyderabad, 216-223.

[14] Hassan (2005) Estimation of Rice Evapotranspiration in Paddy Fields Using Remote Sensing and Field Measurements. PhD Thesis, Universiti Putra Malaysia, 3-5.

[15] Konex (1996) Mapping the Spatial Distribution of Irrigation Water Requirement for Main Crop Potatoes Using GIS. *Agricultural Water Management*, **31**, 1-15.

[16] Rao, K.V.G.K., Bower, R.C., Gaur, A. and Visvanatha.(1998) Tertiary Level Irrigation System Management in the Chambal Command by Water User Associations. *Modernization of Irrigation System Operations*: *Proceedings of the 5th ITIS Network International Meeting*, 214-215.

13

Assessment of Physicochemical Characteristics of Sediment from Nwaja Creek, Niger Delta, Nigeria

Adeola Alex Adesuyi[1,2*], Moses Okafor Ngwoke[1], Modupe Olatunde Akinola[2], Kelechi Longinus Njoku[2], Anuoluwapo Omosileola Jolaoso[2]

[1]Environmental Impact Assessment, Environment Department, Shell Petroleum Development Company, Port Harcourt, Nigeria
[2]Environmental Biology Unit, Cell Biology and Genetics Department, University of Lagos, Lagos, Nigeria
Email: *biologistalex@gmail.com

Abstract

Sediments are complex environments, with varying physicochemical characteristics, such as composition and type of organic matter, particle size distribution, and pH. Contaminated sediment is a significant environmental problem affecting many marine, estuarine and freshwater environments throughout the world. Most assessments of water quality have historically focused on water-soluble compounds, with relatively little attention paid to sediment. The aim of this research is to assess the physical and chemical parameters of sediments from Nwaja Creek, Niger Delta, Nigeria. Monitoring of this sediment quality is an important part of preserving and restoring the biological integrity of water bodies as well as protecting aquatic life, wild life and human health. Sediment samples from Nwaja Creek were sampled from seven sampling stations along the creek for over three months May to July, 2015, rainfall peak period, for assessment of their physical and chemical characteristics, such as grain size, organic carbon, pH, conductivity, nitrate and phosphate. These parameters are known to influence the interactions and dynamics of pollutants within sediment matrix. Sediment particle size distribution indicates that they have higher proportion of clay (clay > silt > sand), the mean percentage composition of clay, silt and sand ranged between 64.28% ± 22.04% - 72.36% ± 14.00%, 18.71% ± 12.03% - 27.32% ± 22.17% and 8.40% ± 6.28% - 9.76% ± 4.59% respectively. TOC in the study area is generally above 1% across all stations during the study period with a range between 0.98% and 4.58%. Minimal monthly and spatial variations are observed in particle distribution, pH (3.9 - 8.5) and phosphate (5.5 - 15.5 kg/mg) while significant variations are observed in conductivity (23.0 - 567.0 uS/cm), total organic carbon (0.98% - 4.58%) and nitrate (0.45 - 11.9 mg/kg) concentration. It is concluded that physicochemical characteristics of the sediments from Nwaja Creek are influenced by anthropogenic sources

*Corresponding author.

rather than natural as shown by the elevated phosphate and nitrate levels because the Niger Delta geology is not essentially rich in nitrate and its excess in surface or groundwater is considered as pollutant.

Keywords

Sediments, physicochemical Assessment, Nwaja Creek, Nitrates, Conductivity

1. Introduction

In Nigeria's Niger Delta, the challenges of water (surface and ground) and sediment pollution have been of concern to all stakeholders because of the rate and extent of impairment of the environment and aquatic bodies by human activities, particularly from industrial and domestic sources. Oil production and its other associated activities in the Niger Delta have increased the population growth rate in the region and also the volumes of waste generated [1]. The Niger Delta is the richest part of Nigeria in terms of natural resources with large deposits of hydrocarbon [2] [3].

Sediment is the loose sand, silt and other soil particles that settle at the bottom of a body of water [4]. Sediment strata serve as an important habitat for the benthic macro invertebrates whose metabolic activities contribute to aquatic productivity [5]. Sediment is also the major site for organic matter decomposition which is largely carried out by bacteria. Important macro-nutrients such as nitrogen and phosphorous are continuously being interchanged between sediment and overlying water [5]. Sediment input may impact stream communities through a variety of direct and indirect processes [6], including reduced light penetration, smothering, habitat reduction and introduction of absorbed pollutants (pesticides, metals, nutrients). The structure of the sediments in the intertidal zone plays a major role in the distribution of the organisms that live in or on them [7]. Benthic organisms show habitat preference for specific types of sediment [8]. The physicochemical parameters of the sediments such as salinity, dissolved oxygen, pH, and organic carbon can also influence the occurrence and abundance of species distributed in them [9]. Sediment also serves as reservoir for pollutants and therefore a potential source of pollutants to the water column, organisms, and ultimately human consumers of those organisms. Contaminated sediment can cause lethal and sub-lethal effect in benthic and other sediment associated organisms [10].

Fine grained sediment (silt + clay) is responsible for a significant proportion of the annual transport of metals, phosphorus, chlorinated pesticides and many industrial compounds such as polynuclear aromatic hydrocarbons, polychlorinated biphenyls, dioxins and furans. Of the 128 priority pollutants listed by the United States Environmental Protection Agency [10], 65 percent are found mainly, or exclusively, in association with sediment and biota. Consequently, water quality programmes that focus only on the water phase miss most of the more toxic contaminants. In North America, it has been found that up to 95 percent of the annual phosphorus load in rivers is transported in association with suspended sediment. Organic micropollutants are mainly bound to the organic component of the suspended matter, which is commonly measured as total organic carbon [10].

Monitoring sediment quality is an important part of preserving and restoring the biological integrity of our Nation's water as well as protecting aquatic life, wild life and human health. Sediment is an integral component of aquatic ecosystem providing habitat, feeding, spawning and rearing areas for many aquatic organisms. Results from studying the sediment physical and chemical characteristics from Nwaja River in the Niger Delta area of Nigeria will facilitate the management of the water and similar water bodies. It will also provide base line data for further studies.

2. Materials and Methods

2.1. Study Area

Port Harcourt is located within the Niger Delta Basin of Southern Nigeria. It's located within the eastern lower Niger Delta in the south eastern part of Rivers State of Nigeria (**Figure 1**). It is situated at the right bank of the Bonny River approximately 65 km inland from the Bight of Bonny. Geographically, the area lies between latitudes 4030' and 5000'N and longitudes 6045' and 7030'E. It is bounded on the East and West by meandering Creeks, on the South by first the block-yard creeks, then the Bonny River and finally mangrove swamps and on the north by Abia State. The southern part of the town stands largely on raised levees with silts and clay foundation. These afford permanently dry and firm points within the zone of its fresh water swamps of the Niger Delta. It covers an area of 290 km^2 and the mean annual temperature is about 28°C. Nwaja Creek is in the upper Bonny Estuary of the Niger Delta. Seven sampling stations (S1-S7) were located along the Nwaja Creek, to cover all land-based sources of contaminant inputs into the creek as well as presumably uncontaminated locations (**Figure 2**). The sampling stations and their geographical coordinates were all recorded and documented (**Table 1**).

Figure 1. Showing river state, Niger Delta, Nigeria (Source: Google earth image).

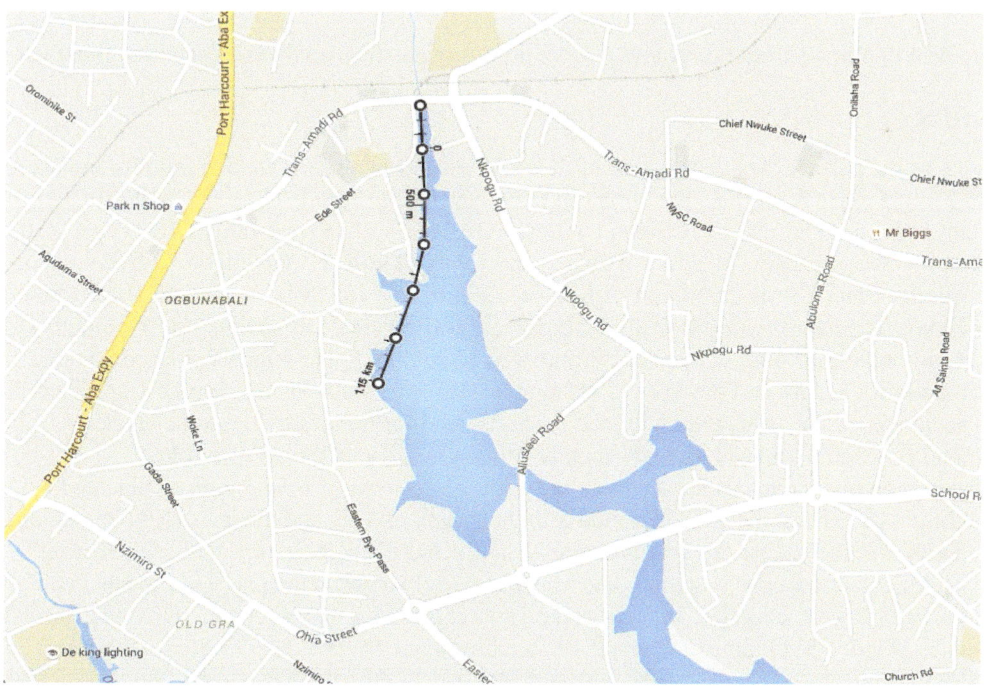

Figure 2. Map showing sampling stations (Source: Google earth map).

Table 1. Nwaja creek sampling stations GPS coordinates.

S/n	Sampling station code	Longitude	Latitude
1	S1	7.015592	4.809313
2	S2	7.015585	4.809011
3	S3	7.015562	4.808353
4	S4	7.015752	4.807685
5	S5	7.015805	4.807083
6	S6	7.015737	4.806236
7	S7	7.015919	4.805435

2.2. Sampling and Analyses

The study adopted both field and laboratory based procedures to generate the data required. Surface water samples were obtained at seven sampling stations (S1 to S7) located at equal distances of 100 km along the stretch of Nwaja creek. Sediment samples were collected monthly with an Ekman grab (15 cm by 15 cm) at each of the stations for a period of 3 months from May to July 2015 (rainfall peak period). Replicate grabs samples were collected for particle size analysis and physicochemical analysis of the sediment. These were transported to the laboratory in ice-cooled boxes.

The sediment samples were air-dried, sieved and used to perform the following physicochemical analysis except pH and conductivity for which wet samples were analyzed. The Bouyoucos hydrometer method was used for the particle size analysis. The pH and conductivity of the sediment were determined using a meter (model H1 8314, membrane HANNA instrument). The sediment samples were mixed in a ratio of 1:1 with distilled water in a beaker before inserting the probes. Readings were taken after allowing the instrument to stabilize. Total Organic Carbon (TOC) in percentage was determined by the wet combustion method of Walkley and Black method of ASTM and APHA [11] [12]. Nitrate (NO_3) levels in sediment were determined following the Brucine Method [13], while available phosphorus in sediment was determined by Bray and Kurt method [14].

2.3. Data Analysis

One-way ANOVA and correlation analyses were used for statistical analysis of each sediment variable.

3. Results

The results of the particle size distribution of the sediments (mean percentage in months and stations) are shown in **Table 2** and **Figure 3** and they are observed to be clay > silt > sand. The mean monthly percentage composition of clay, silt and sand ranged between 64.28% ± 22.04% - 72.36% ±14.00%, 18.71% ± 12.03% - 27.32% ± 22.17% and 8.40% ± 6.28% - 9.76% ± 4.59% respectively (**Table 2**). The highest percentage of clay (90.55%) was recorded at station 3 in the month of May while the least (35.89%) was recorded at station 7 also in the month of May. The lowest value (6.6%) of silt particle was observed at station 3 in the month of May while the highest mean value (60.95%) in the same month was at station 7. The highest value of sand (22.40%) was observed at station 1 in July and the least (2.85%) at station 3 in May. There were no significant differences ($p < 0.05$) in sediment stations and months for clay, silt and sand for this study in Nwaja Creek.

The metal concentration level of heavy metals in plant samples is shown in **Table 3**.

The mean monthly pH values for sediment are 5.39 ± 1.16 in May, 6.54 ± 0.95 in June and 4.65 ± 0.65 in July. The highest pH (8.50) was recorded at station 2 in the month of May and the least (3.90) at station 2 in the month of July (**Table 3**). There were spatial and monthly significant differences ($p < 0.05$) in pH level of the study area (**Figure 4**). Minimal variations in stations and months were observed in the pH of sediments in the study area.

High variations were observed in stations and months for the mean conductivity values of the sediments. Mean values (**Figure 5**) obtained for months ranged between 44.98 ± 20.79 uS/cm (July) and 266.21 ± 151.62 uS/cm (May) while for stations ranged between 79.50 ± 64.97 (Station 6) and 280.83 ± 253.19 uS/cm. The highest conductivity (567.0 uS/cm) was recorded at station 1 in the month of May while the least (23.0 uS/cm)

Table 2. Sediments composition and particle sizes (%).

	May			June			July		
	Clay (%)	Silt (%)	Sand (%)	Clay (%)	Silt (%)	Sand (%)	Clay (%)	Silt (%)	Sand (%)
Station 1	79.3	11.2	9.5	82.5	9.65	7.85	65.2	12.4	22.4
Station 2	85.56	10.8	3.64	79.51	15.85	4.64	72.8	19.45	7.75
Station 3	90.55	6.6	2.85	65.42	18.35	16.23	78.56	14.62	6.82
Station 4	55.65	26	18.35	72.91	16.81	10.28	88.5	7.88	3.62
Station 5	64.52	20.08	15.4	75.6	12.5	11.9	89.05	7.7	3.25
Station 6	38.5	55.61	5.89	40.5	45.9	13.6	52.65	40.64	6.71
Station 7	35.89	60.95	3.16	54.8	41.35	3.85	59.74	28.3	11.96

Table 3. Physicochemical parameters of sediments.

	Month	Statn 1	Statn 2	Statn 3	Statn 4	Statn 5	Statn 6	Statn 7	Mean	SD
	May	5.6	7.56	5.52	5.9	4.54	4.45	4.19	5.394286	1.16017
pH	June	6.1	5.78	6.89	8.5	6.45	5.78	6.25	6.535714	0.94912
	July	4.58	3.9	4.5	5.8	4.99	3.95	4.86	4.654286	0.653449
	Mean	5.42666667	5.74666667	5.63666667	6.73333333	5.32666667	4.72666667	5.1		
	SD	0.77468273	1.83022767	1.19926366	1.530795	0.99851556	0.94585059	1.05076163		
	May	567	340	290.5	148.8	167	150.5	199.7	266.2143	151.6299
Conductivity (uS/cm)	June	189.6	158.9	89	70	90	65	88	107.2143	47.64985
	July	85.9	45	32	56	38	23	35	44.98571	20.79987
	Mean	280.833333	181.3	137.166667	91.6	98.3333333	79.5	107.566667		
	SD	253.19349	73.0897394	135.814518	50.0287917	64.902491	64.9749952	84.0753432		
	May	4.58	3.89	2.5	2.96	3.55	5.45	3.56	3.784286	0.987672
Organic carbon (%)	June	1.7	2.54	1.98	3.09	2.65	1.87	0.98	2.115714	0.701923
	July	2.23	4.4	1.5	1.98	2.9	3.65	1.67	2.618571	1.08215
	Mean	2.83666667	3.61	1.99333333	2.67666667	3.03333333	3.65666667	2.07		
	SD	1.53285137	0.96109313	0.50013332	0.60682232	0.46457866	1.79000931	1.33570206		
	May	10.5	15.5	12.68	11.4	7.89	14.89	16.89	12.82143	3.155828
Phosphate (mg/kg)	June	5.68	12.56	9.5	12.58	6.9	8.55	11.33	9.585714	2.720367
	July	6.54	6.98	9.2	8.56	5.5	9	8.65	7.775714	1.428634
	Mean	7.57333333	11.68	10.46	10.8466667	6.76333333	10.8133333	12.29		
	SD	2.57078458	4.32763215	1.92841904	2.06633331	1.20084692	3.5376593	4.20304651		
	May	3.9	5.05	3.44	6.5	7	11.9	9.95	6.82	3.131075
Nitrate (mg/kg)	June	3.65	3.89	0.45	1.65	1.55	2.56	2.56	2.33	1.217032
	July	1.76	4.06	3.58	3.5	3.25	5.55	6.7	4.057143	1.616567
	Mean	3.1033333	4.3333333	2.49	3.8833333	3.9333333	6.67	6.4033333		
	SD	1.170057	0.626445	1.7680781	2.4476179	2.7885181	4.7696646	3.7039213		

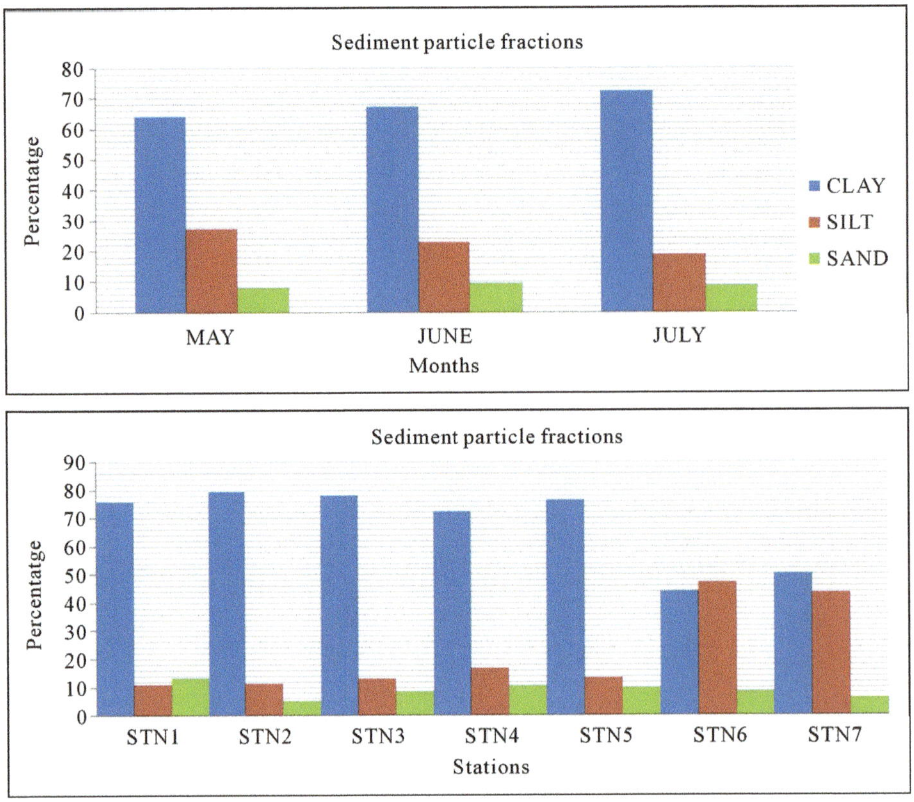

Figure 3. Mean particle size distribution of sediments.

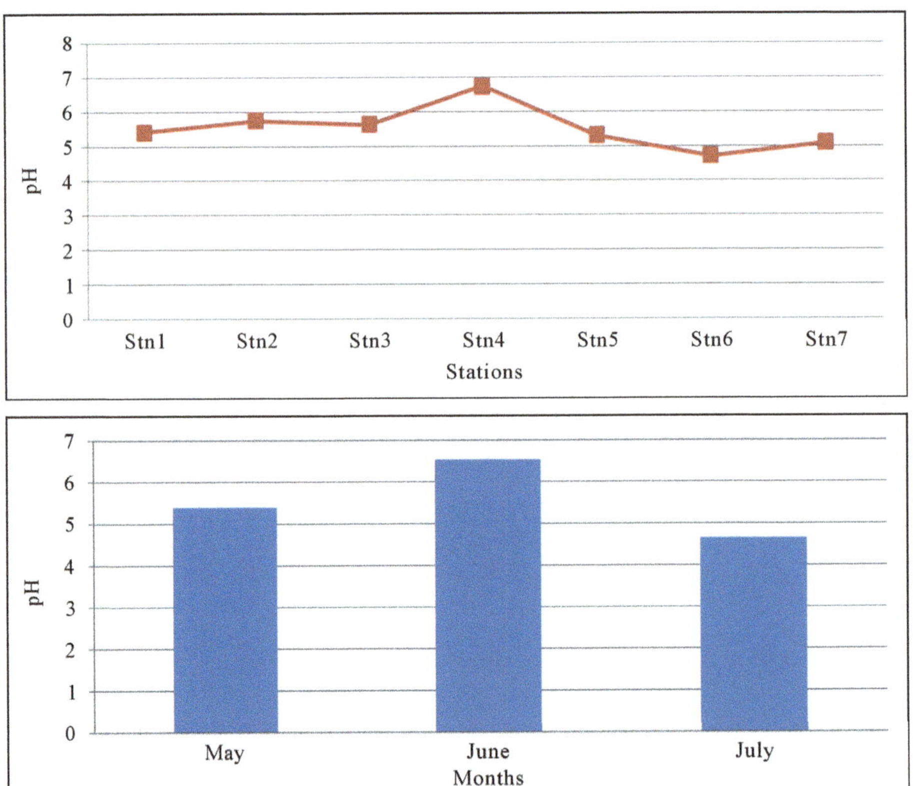

Figure 4. Mean pH values (monthly and station).

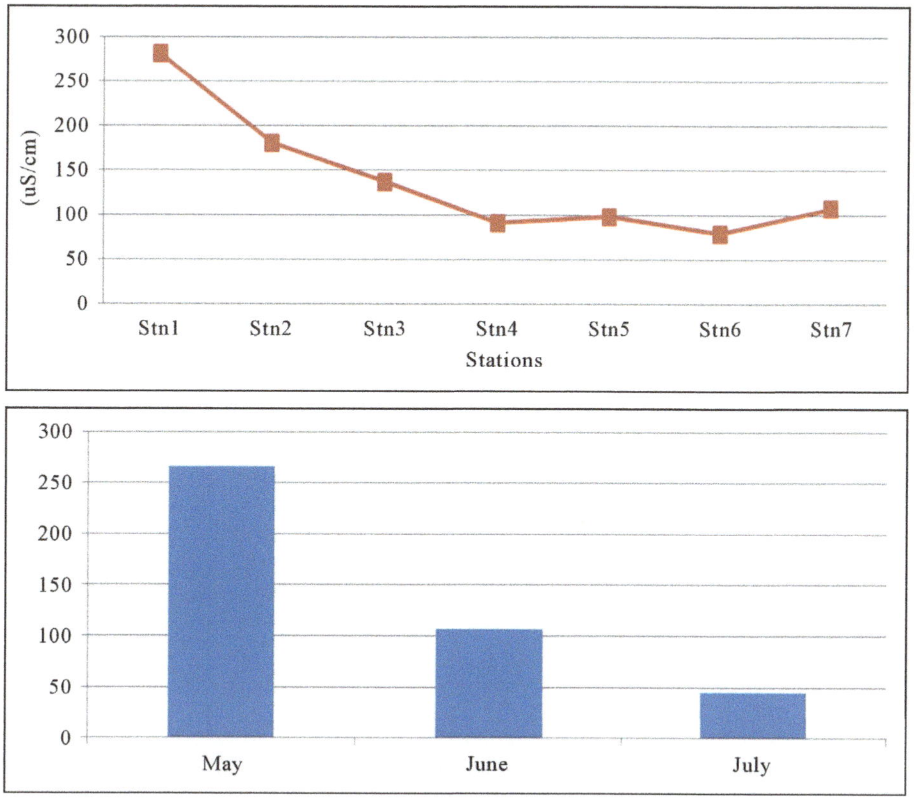

Figure 5. Mean conductivity (monthly and station).

was recorded at station 6. There were significant differences ($p < 0.05$) in conductivity across stations and months in this study of Nwaja Creek sediments.

The values of total organic carbon in the study area varied greatly in months and across stations with higher values observed during the month of May (**Figure 6**). The TOC values in the study area were generally above 1% across all stations during the study period. The lowest mean value recorded was at station 3 (1.99% ± 0.5%) while the highest mean value was at station 6 (3.65% ± 1.79%). In mean monthly variation, the month of May had the highest mean (3.78% ± 0.98%) while the month of June had the lowest mean (2.11% ± 0.70%). Significant differences were observed in the ANOVA results between stations and months ($p < 0.05$).

The mean monthly Phosphate level in Nwaja sediment (**Figure 7**) ranged between 7.77 ± 1.42 mg/kg (July) and 12.82 ± 3.15 mg/kg (May) while the mean phosphate level across stations ranged between 6.76 ± 1.20 mg/kg (station 5) and 12.29 ± 4.20 mg/kg (station 7). The highest value of phosphate was recorded at station 7 (16.89 mg/kg) in the month of May while the least at station 5 (5.50 mg/kg) in the month of July. Phosphate levels in this study were recorded to be higher than nitrate concentration. ANOVA showed significant difference ($p < 0.05$) in mean phosphate concentrations between stations and months, and significant interaction between months and locations.

The nitrate level in Nwaja Creek was quite high and varied across the study stations and sampling months (**Figure 8**). The mean monthly concentration of nitrate in sediment varied between 2.33 ± 1.21 mg/kg in June and 6.82 ± 3.13 mg/kg in May. Mean concentration across the stations varied between 2.49 ± 1.76 mg/kg in station 3 and 6.67 ± 4.76 mg/kg in station 6. Significant differences ($p < 0.05$) were observed in Nitrates between stations and across all the sampling points.

Correlation between variables in sediment samples is shown in the **Table 4** and **Figure 9**. The relationship between pH and conductivity, organic carbon and conductivity, phosphate and pH, Nitrate and organic carbon; and Nitrate and phosphate were all positively correlated. However the correlation between organic carbon and pH, phosphate and conductivity, phosphate and organic carbon, nitrate and pH; and nitrate and conductivity were all negatively correlated. These relations were statistically not significant and it shows weak relationship among the variables. Thecorrelation between soil and plant for all the metals are all positive. These relations

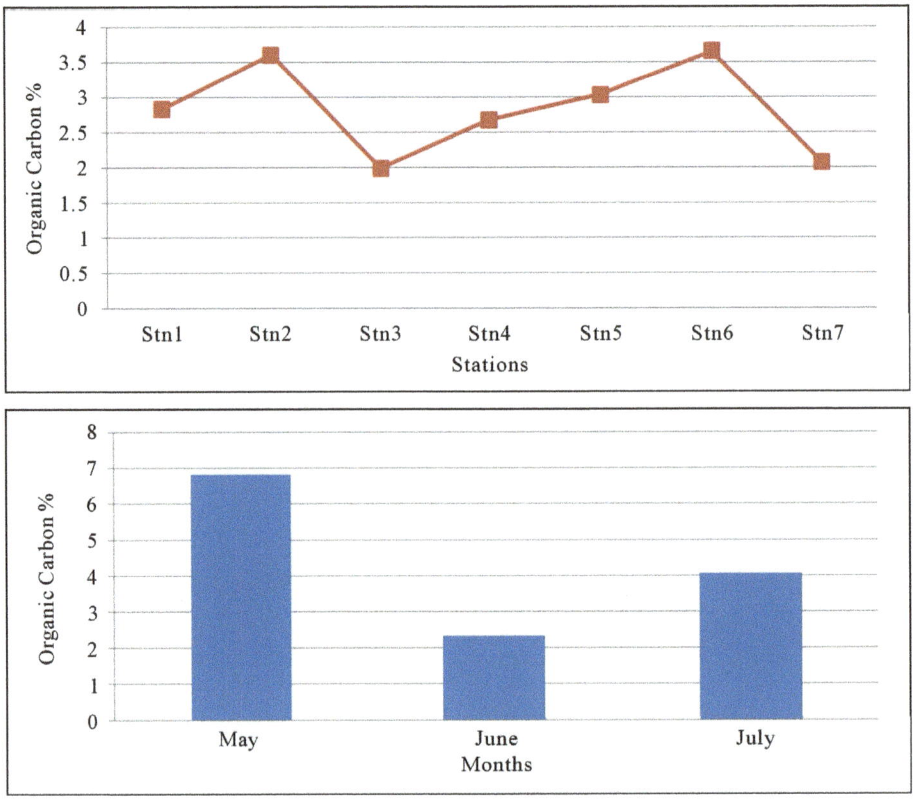

Figure 6. Mean organic carbon (monthly and station).

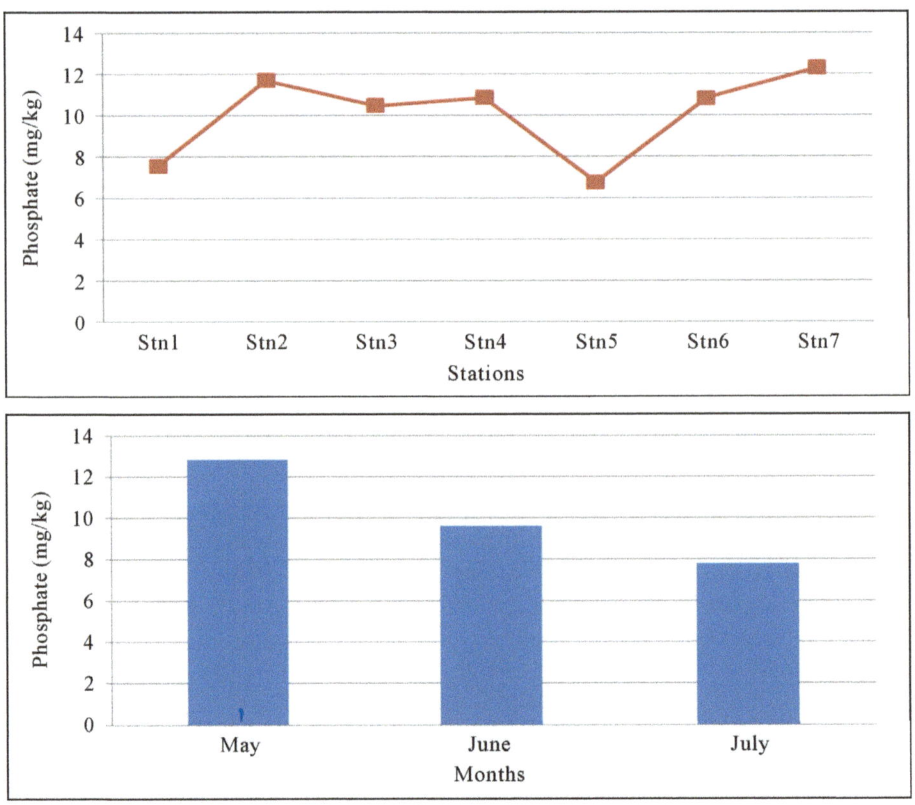

Figure 7. Mean phosphate (mg/kg) in stations and months.

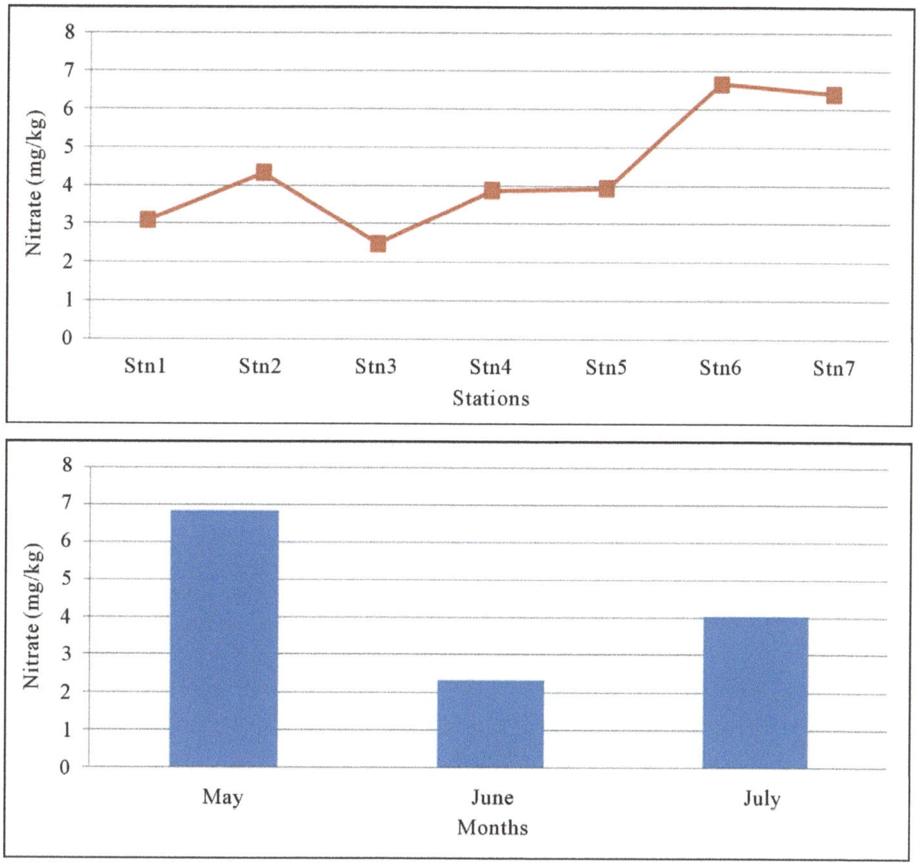

Figure 8. Mean nitrate (mg/kg) in months and station.

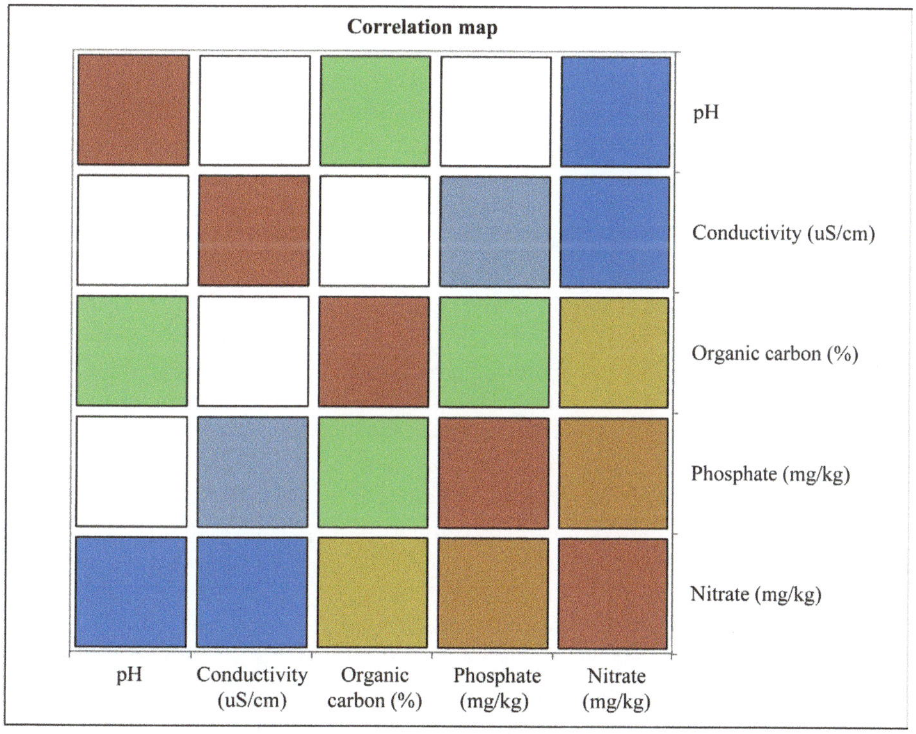

Figure 9. Physicochemical parameters correlation map.

Table 4. Correlation table for studied variables.

	pH	Conductivity (uS/cm)	Organic carbon (%)	Phosphate (mg/kg)	Nitrate (mg/kg)
pH	1				
Conductivity (uS/cm)	0.0255	1			
Organic carbon (%)	−0.1929	0.0323	1		
Phosphate (mg/kg)	0.0892	−0.3412	−0.1149	1	
Nitrate (mg/kg)	−0.5509	−0.5010	0.2986	0.4895	1

were significant for copper, iron, lead and cadmium ($r = 0.718$, $r = 0.644$, $r = 0.705$ and $r = 0.010$ respectively) but not significant for zinc (0.030).

4. Discussion

Sediments with fine particles provide better surface areas for pollutants to adsorb than those with coarse particles. The nature of the sediment and the organic matter composition also determine the benthic community structure found in particular sediments. Grain size analyses show that sediments from the Nwaja Creek were generally clay > silt > sandy in nature with the texture characteristics being fairly constant over the study period. All sampling stations had more clay-silt compositions except in the month of May at station 7. The mean percentage values for clay, silt and sand in this study compared favourably with the observation of Daka and Moslen [15] who recorded mean spatial values ranging from 0.4% ± 0.42% - 45.85% ± 2.16%, 1.57% ± 0.49% - 54.50% ± 6.37% and 5.51% ± 2.68% - 96.70% ± 1.98% for sand, silt and clay respectively. The study revealed that the particle size of the sediments of Nwaja River were generally sandy mud in nature.

There were minimal monthly and spatial variations in the mean pH values of sediment in the study area. pH values were higher during the month of June and lower in the month of July. Station 2 had the highest pH value in the month of May, station 4 in the month of June and station 4 also in the month of July. The values obtained for sediment pH levels during this study were consistent with those of Daka *et al.* [16] who recorded range between of 2.5 - 3.5 in Azuabie creek of the upper Bonny Estuary. These values are acidic and would pose serious challenges to organisms in nature. But this study does not compare favourably well with the report of Braide *et al.* [3] who reported alkaline range of 6.9 - 7.8 from the freshwater stream of Minichida stream also in Niger Delta. Whereas Minichida stream sediment is alkaline that of Azuabie and Nwaja is acidic. This might be due to anthropogenic sources such as oil and gas exploration and production, discharge of industrial effluents, domestic and human wastes etc.

The conductivity of the study area ranged between 23 uS/cm and 567.0 uS/cm. These high variations may be as a result of dilution due to the heavy rainfalls during this period. This compares favorably with the range of 40.0 - 1940.0 uS/cm with a mean of 528.75 uS/cm reported by Ezekiel *et al.* [17] in Sombreiro River, Niger Delta.

The organic carbon content of the study area ranged between 0.98% - 4.58%. This value is similar to those observed in Azuabie Creek (0.82% - 2.16%) by Daka and Moslen [15] and Sombreiro River (2.02% - 4.1%) by Ezekiel *et al.* (2011) [17]. Sediment is a major site for organic matter decomposition which is largely carried out by bacteria. The level of organic matter decomposition may be attributed for the variation in organic carbon content. Fine sediment particles have larger relative surface areas than coarse particles and can absorb colloidal and dissolved organic matter forming sedimentary complexes [15]. The variation in organic carbon content may be attributed to difference in deposition of organic matter at the various stations.

Nitrate and phosphates give an indication of the nutrient level in the study area. High variations were observed monthly and also across stations for the concentrations of these nutrients. Domestic waste input from human settlements near these stations and surface run-offs into the creek could be responsible for the appreciably high nutrient level recorded at these stations. Nitrate concentration ranged between 0.45 - 11.9 mg/kg with a mean of 6.18 mg/kg this is in agreement with the range (2.60 - 4.10 mg/kg) recorded by Ezekiel *et al.* [17] in a study in the Sombreiro River, Niger Delta, Nigeria. This value is quite high especially when it is compared to the low regional nutrient level of the Niger Delta. The Niger Delta geology is not essentially rich in nitrate and its excess in surface or groundwater is considered as pollutant [18]. Similarly, the phosphate values (5.5 - 15.5 mg/kg) observed in this study is in a range similar to values obtained by Daka and Moslen [15] with a mean phosphate concentration of 13.43 mg/kg in a study of Azuabie River sediment. The conditions which affected nitrate concentration also affect phosphate distribution in the stations. Excess phosphate in water is considered a pollutant [19]. As reported by Adesuyi *et al.* [18] about Nwaja Creek surface water becoming highly polluted with nitrates and phosphates, it's obvious that this river need to be monitored and assessed regularly to avoid algal bloom and full scale pollution.

Conclusion

The monitoring of sediment quality is a very important process in the restoration and protection of the biological integrity of our nation's waters as well as our aquatic/wildlife resources. This study presents the results of the physicochemical quality of the bottom sediment of Nwaja Creek in Niger Delta, Nigeria. Minimal monthly and spatial variations are observed in particle distribution, pH and phosphate while significant variations are observed in conductivity, total organic carbon and nitrate concentration. This observation may be due to anthropogenic influences rather than natural as shown by the elevated nitrate level which is known to be minimal in the geology of the Niger Delta. It is concluded that Nwaja Creek should be constantly monitored for trends in surface water and sediments biological and physicochemical parameters.

Acknowledgements

The first author acknowledges research support from Shell Petroleum Development Company in Nigeria especially the Environmental Impact Assessment and the University Relations Teams; Ojesanmi Sola, Prof. Gabriel Umoh, Prof. Valerie Nnodu, Ogbanga Belema, Prof. Emmanuel Ukpebor, Timi Tralagba and others too numerous to mention. However, glory to GOD in the highest.

Adesuyi Adeola Alex is presently a PhD Research Intern in Environmental Impact Assessment, Environmental Department of Shell Petroleum Development Company, Port Harcourt, Nigeria.

References

[1] Adesuyi, A.A., Njoku, K.L. and Akinola, M.O. (2015) Assessment of Heavy Metals Pollution in Soils and Vegetation around Selected Industries in Lagos State, Nigeria. *Journal of Geoscience and Environment Protection*, **3**, 11-19. http://dx.doi.org/10.4236/gep.2015.37002

[2] Moffat, D. and Linden, O. (1995) Perception and Reality: Assessing Priorities for Sustainable Development in the Niger Delta, AMBIO. *Journal of Human Environment*, **24**, 527-538.

[3] Braide, S.A., Izonfuo, W.A.L., Adakwu, P.U., Chinda, A.C. and Obinwo, C.C. (2004) Water Quality of Miniweja Stream, a Swamp Forest Stream Receiving Non-Point Source Waste Discharge in Eastern Niger Delta, Nigeria. *Scientia Africana*, **3**, 1-8.

[4] United States Environmental Protection Agency (USEPA) (2002) Water Quality Monitoring for Coffee Creek (Porter County, Indiana). http://www.usepa/research.htm.modecode=62-28-00-00

[5] Abowei, J.F.N. and Sikoki, F.D. (2005) Water Pollution Management and Control. Double Trust Publications Company, Port Harcourt, 236 p.

[6] Oschwald, W. (1972) Sediment-Water Interactions. *Journal of Environmental Quality*, **1**, 360-366. http://dx.doi.org/10.2134/jeq1972.00472425000100040005x

[7] Ikomi, R.B., Arimoro, F.O. and Odihirin, O.K. (2005) Composition, Distribution and Abundance of Macroinvertebrates of the Upper Reaches of River Ethiope, Delta State, Nigeria. *The Zoologist*, **3**, 68-81.

[8] Atabatele, O.E., Morenike, O.A. and Ugwumba, O.A. (2005) Spatial Variation in Physico-Chemical Parameters and Benthic Invertebrate Fauna of River Ogunpa, Ibadan. *The Zoologist*, **3**, 58-67.

[9] McLusky, D.S. and Elliott, M. (1981) The Feeding and Survival Strategies of Estuarine Molluscs. In: Jones, N.V. and Wolff, W.J., Eds., *Feeding and Survival Strategies of Estuarine Organisms*, Plenum Press, New York, 109-122. http://dx.doi.org/10.1007/978-1-4613-3318-0_9

[10] USEPA (2001) Method for Assessing the Chronic Toxicity of Marine and Estuarine Sediment-Associated Contaminants with the Amphipod *Leptocheirus plumulosus*. US Environmental Protection Agency, Office of Water, Washington DC. http://water.epa.gov/polwaste/sediments/cs/upload/guidancemanual.pdf

[11] Walkley, M. and Black, G.C. (1934) Organic Carbon in Soils and Sediments. Standard Testing Methods for Oil Pollution. ASTM, Philadelphia.

[12] APHA (1995) Standard Methods for the Examination of Waste Water. 14th Edition, APHA, AWIWA-WPCHCF, Washington DC.

[13] APHA (1998) Standard Methods for the Evaluation of Water and Waste Water. 20th Edition, American Public Health Association Inc., New York, Washington DC.

[14] Bray, R.H. and Kurtz, K.T. (1945) Determination of Total Organic and Available Forms of Phosphorus in Soils. *Soil Science*, **59**, 39-46. http://dx.doi.org/10.1097/00010694-194501000-00006

[15] Daka, E.R. and Moslen, M. (2013) Spatial and Temporal Variation of Physico-Chemical Parameters of Sediment from Azuabie Creek of the Upper Bonny Estuary, Niger Delta. *Research Journal of Environmental and Earth Sciences*, **5**, 219-228.

[16] Daka, E.R., Moslen, M., Ekeh, C.A. and Ekweozor, I.K.E. (2007) Sediment Status of Two Creeks in the Upper Bonny Estuary, Niger Delta, in Relation to Urban/Industrial Activities. *Bulletin of Environmental Contamination and Toxicology*, **78**, 515-521. http://dx.doi.org/10.1007/s00128-007-9151-5

[17] Ezekiel, E.N., Hart, A.I. and Abowei, J.F.N. (2011) The Sediment Physical and Chemical Characteristics in Sombreiro River, Niger Delta, Nigeria. *Research Journal of Environmental and Earth Sciences*, **3**, 341-349.

[18] Adesuyi, A.A., Nnodu, V.C., Njoku, K.L. and Jolaoso, A. (2015) Nitrate and Phosphate Pollution in Surface Water of Nwaja Creek, Port Harcourt, Niger Delta, Nigeria. *International Journal of Geology, Agriculture and Environmental Sciences*, **3**, 14-20.

[19] Odiete, W.O. (1999) Environmental Physiology of Animals and Pollution. Diversified Resources Lagos, Nigeria.

14

Simulation of Runoff and Sediment Yield for a Kaneri Watershed Using SWAT Model

Vidula A. Swami, Sushama S. Kulkarni

Rajarambapu Institute of Technology, Sangli, India

Email: vidulaswami@gmail.com, sushma.kulkarni@ritindia.edu

Abstract

Watershed as an entry point acts as a beginning to address the issues of sustainable rainwater management for improving livelihoods. Extraction of watershed parameters using Geographical Information System (GIS) and use of simulation models is the current trend for hydrologic evaluation of watersheds. In the present study, the open Source Tool Quantum GIS 2.2.0 was used for preparation of maps to verify the spatial extent of the area. The Soil and Water Assessment Tool (SWAT) having an interface with Arc-View GIS software (ArcGIS 10.1 with Arc SWAT 2012 extension) was selected for the estimation of runoff and sediment yield from Kaneri watershed, located in Western Maharashtra region. The coefficient of determination (R^2) for the monthly and yearly runoff was obtained as 0.849 and 0.951 respectively for the calibration period 1979 to 2000 and 0.801 and 0.950 respectively for the validation period 2001-2013. The R^2 value in estimating the monthly and yearly sediment yield during calibration period was computed as 0.722 and 0.788 respectively. The R^2 for monthly and yearly sediment yield values for validation period was observed to be 0.565 and 0.684 respectively.

Keywords

Arc SWAT, Calibration, Validation, GIS, Runoff, Sediment Yield, Coefficient of Determination R^2

1. Introduction

Watershed as an entry point acts as a beginning to address the issues of sustainable rainwater management for improving livelihoods. To deal with water management issues, one must analyze and quantify the different elements of hydrologic processes taking place within the area of interest. Obviously, this analysis must be carried out on a watershed basis because all these processes are taking place within individual micro watersheds. Only after understanding the spatial and temporal variation and the interaction of these hydrologic components, one can scientifically formulate strategies for water and soil conservation. To achieve this goal, the choice and use of an appropriate watershed model is a must as stated by Sathian and Shyamala [1]. The Soil and Water Assessment Tool (SWAT) is a physically-based continuous daily step hydrologic model as mentioned by Arnold *et al.* [2] and Neitsch *et al.* [3] and is developed to assess the impact of land management practices on water, sediment, and agricultural chemical yields in large multifaceted watersheds with varying soils, land use and management conditions over long spans of time. Tripathi *et al.* [4] use it to predict water and soil loss in agriculturally dominant small watersheds. For assessing the impact of land management practices and climate on water flow and sediment yield in ungauged Kaneri watershed, Swami and Kulkarni [5] use SWAT (Soil and Water Assessment Tool) model in the study.

1.1. Study Area

Kaneri village is located in south Maharashtra region and is dominated by undulating plateau. It is located south-east to Kolhapur city on Pune-Bangalore highway in Maharashtra state. It is located at 16.6055 and 16.6412N and 74.2535 and 74.2906E and is 11 kms away from Kolhapur. It is surrounded by small hills. **Figure 1** explains the location of the watershed. Geographical area of the village is 975.94 hectares. Kaneri lies in extreme elevation variation from 589 m to 692 m. The village falls in agro-climatic zone IV and is characterized by rainfall ranging from 800 to 1000 mm. The main occupation of the village is agriculture as 68.84% of the total area of village is under cultivation.

1.2. SWAT Model

The water balance is the driving force for the simulation of hydrology. SWAT uses two steps for the simulation

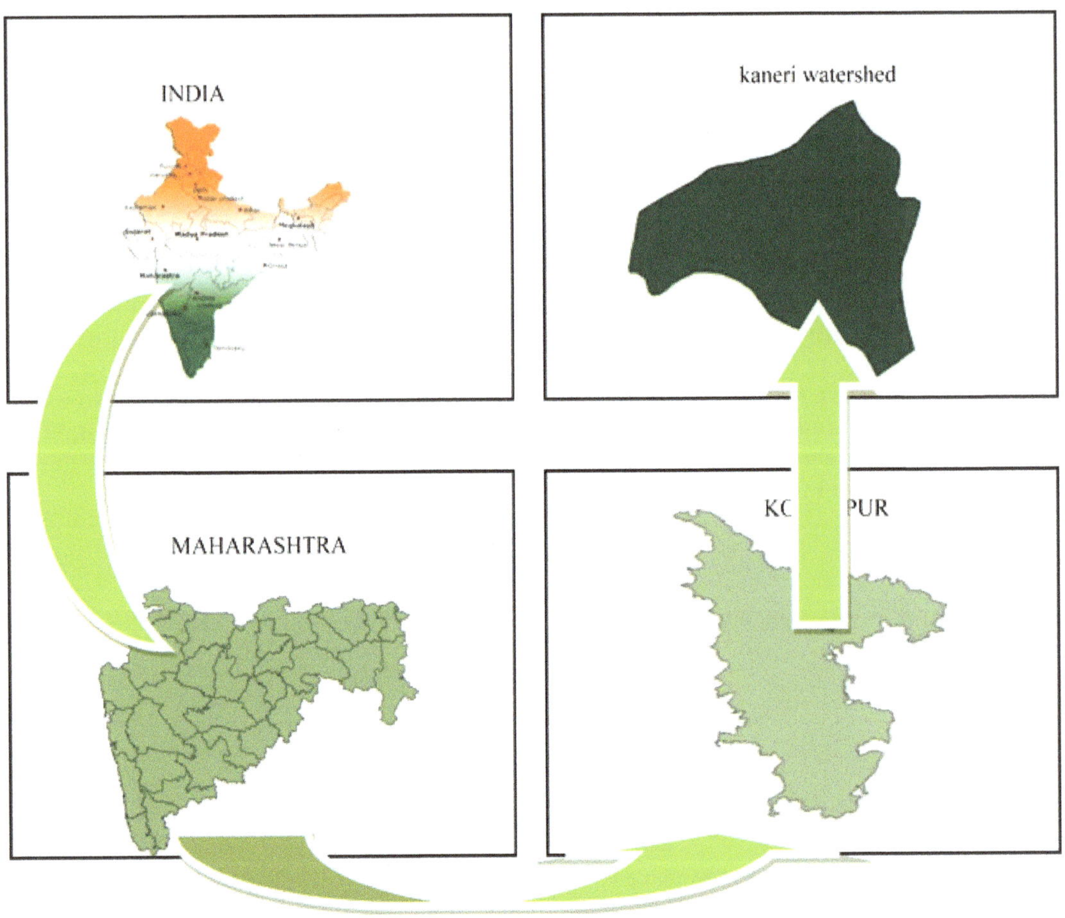

Figure 1. Location of Kaneri.

of hydrology, land phase and routing phase. The land phase is the phase in which the quantity of water, sediment, nutrient and pesticides loadings in the main channel from each subbasin are calculated. Water balance equation for SWAT model is

$$SW_t = SW_o + \Sigma\left(P_{day}\right) - Q_{surt} - AET - Q_{seep} - Q_{gw}$$

where SW_t is the final water content in millimeters, SW_o is the initial soil water content on day *i* (mm), P_{day} is precipitation on day *i* (mm), Q_{surf} is surface runoff on day *i* (mm), AET is the actual evapo-transpiration on day *i* (mm), Q_{seep} is the water flowing into the unsaturated zone form the soil profile on day *i* (mm) and Q_{gw} is the return flow from the shallow aquifer and lateral flow on day *i* and t is time in days [3]. In the present study, SWAT model is used for prediction of runoff and sediment yield in the micro watershed Kaneri. Santosh Kumbhar [6] estimated soil loss for Kaneri watershed as 17.67 Tonnes/Ha/Year, which is beyond the acceptable limits. Permissible soil loss given by Mannering (1981), is from 4.5 to 11.2 tonnes/ha/year.

SWAT model uses the concept that whenever the rate of water application to the ground surface exceeds the rate of infiltration, surface runoff occurs. SWAT uses the Modified Universal Soil Loss Equation (MUSLE) to estimate the soil loss from each HRU. The peak runoff rate is the highest runoff rate that occurs with a given precipitation event. SWAT calculates the peak runoff rate with a modified rational method. The factors K_{USLE}, C_{USLE}, P_{USLE}, LS_{USLE}, and CFRG are taken and used based on previous studies on the watershed and the definition and calculations of the parameters presented in the SWAT documentation. Lateral flow is important in watersheds with soils with high hydraulic conductivities in surface layers and in impermeable or semi-permeable layer at a lesser depth. The water that collects above the impermeable layer is the spring of water for lateral subsurface flow [3].

SWAT calculates percolation for each layer in the profile and this process occurs only when the moisture content of the soil is more than field capacity. Recharge to an unconfined aquifer occurs by percolation to the water table from a major portion of the land surface. Depending on the water table height in the shallow aquifer, there is a base flow contribution to the main channel. This flow occurs only when the water stored in the shallow aquifer is greater than the threshold water level in the shallow aquifer for ground water input to the main channel to occur. This value is defined by the user or in the SWAT interface the variable is presented as GWQMN [3]. Integration of SWAT model with rainfall data available from the WSR 88 D radar network helps us to incorporate the spatial variability of rainfall into the modeling process. Four case studies were presented by Jay Krishnan *et al.* [7]. The study demonstrated the usefulness of radar rainfall data in distributed hydrologic studies and potential of SWAT for application in flood analysis and simulation.

Ashok Mishra *et al.* [8] used SWAT model to assess sediment transport from 17 km^2 Banha watershed located in Northern India. Watershed has mixed land use and check dams are provided for on-stream sediment control. Model is run with and without check dams to test its capability to evaluate their impact on sediment control and proved to be successful. Weruweru catchment of drainage area 101 km^2 with annual precipitation 1500 mm to 3000 mm at the foot slopes of Kilimanjaro in Northern Tanzania is studied for water balance modeling by using SWAT model. Model is performed on annual and monthly basis using spatial and attribute data. The predicted mean daily stream flow is found to be 1.92 m^3/sec. exactly as observed during simulation period. Birhanu *et al.* [9] shows in his study that SWAT model can be a potential monitoring tool for watersheds in hilly terrains. Samira Akhavan *et al.* [10] applied SWAT to simulate the amount and dynamics of nitrate leaching from a crop rotation in Hamadan-Bahar watershed, Iran, presented model and its results showed the potential to provide a strong base for considering different scenarios for reducing nitrate leaching and also for providing Best Management Practices (BMP) in the watersheds. An agricultural watershed of 0.8 Ha area was studied for prediction of runoff flow using SWAT tool with sub daily configuration which assessed flow from the watershed within the range of acceptable accuracy. Ganga Ram Mahajan *et al.* [11] showed that hourly precipitation record for SWAT sub daily with Green Ampt infiltration method was found to be efficient for run off estimation for field sized watershed with better accuracies which could be efficiently used to develop site specific Best Management Practices (BMPs) considering rainfall intensity rather than simply using daily rain fall data.

2. Methodology

2.1. Creation of Database

To delineate the watershed and sub basins and to determine drainage networks SWAT uses the digital representation of the topographic surface. DEM is the digital representation of the topographic surface. A 30 m by 30 m resolution ASTERDEM was derived and re-sampled to 15 m × 15 m for ease in data acquisition. **Figure 2** shows DEM of Kaneri village. Sub basin parameters such as slope gradient, slope length of terrain and the stream network characteristics such as channel length, width and slope were calculated for the base village area and used by the model. **Figure 3** gives drainage map of Kaneri village.

A land use map was created by recording the crop type on each plot in the watershed and by identifying the land cover on areas other than cultivated fields. LULC map was acquired from LISS III (Linear Imaging and Self Scanning Sensors). The digital Google image was geo-referenced by taking control points around and inside the watershed. The shape file representing each plot and other land covers was created using the digitizing tools provided in ArcGIS, ArcMap. The soil map obtained from the NBSS&LUP was geo-metrically registered to the base data to match Landsat & IRS satellite imageries. The geo-referenced soil map was used to assist in visual classification of satellite imagery for obtaining soil categories. The final vector map was stored in a geo-database which is amenable to spatial analyze.

SWAT requires daily or sub-daily meteorological data. For Kaneri watershed the daily climate data from two rain gauge stations was used. Daily rainfall data was used for SWAT Model Run. Multiple Gauges were used for weather data input. The main inputs used for weather data in the model are gridded rainfall and temperature. The curve number method (USDA-SCS, 1986) was chosen for calculating runoff. Penman-Monteith method (Monteith, 1965) was chosen for calculating potential evapotranspiration. The variable storage method was chosen as the channel routing mechanism with the assumption that channel dimensions remain constant. The model was set to run from 1st January 1979 to 31st July 2014 with a monthly printout interval. By considering the drainage lines the stream network was prepared. The watershed outlet was manually added and selected for finalizing the watershed delineation (**Figure 4**). With this information the model automatically delineated a watershed of 535.48 ha and 28 sub basins were produced. Multiple HRUs were defined within a sub basin by ignoring land uses less than 2% of the subbasin and also ignoring soil types in a subbasin covering less than 5% of the subbasin. **Figure 5** shows land use land cover map of Kaneri watershed.

The watershed of total area 535.48 Ha has been classified in following land uses (**Table 1**).

Land use map is a critical input for SWAT model. Land use/land cover map was prepared using remote sensing data of Landsat ETM+. The classification of satellite data mainly follows two approaches *i.e.* supervised and unsupervised classification. The intent of the classification process is to categorize all pixels in a digital image into one of several land cover classes, or themes [12]. The detailed Sub Basin wise land use distribution is as in

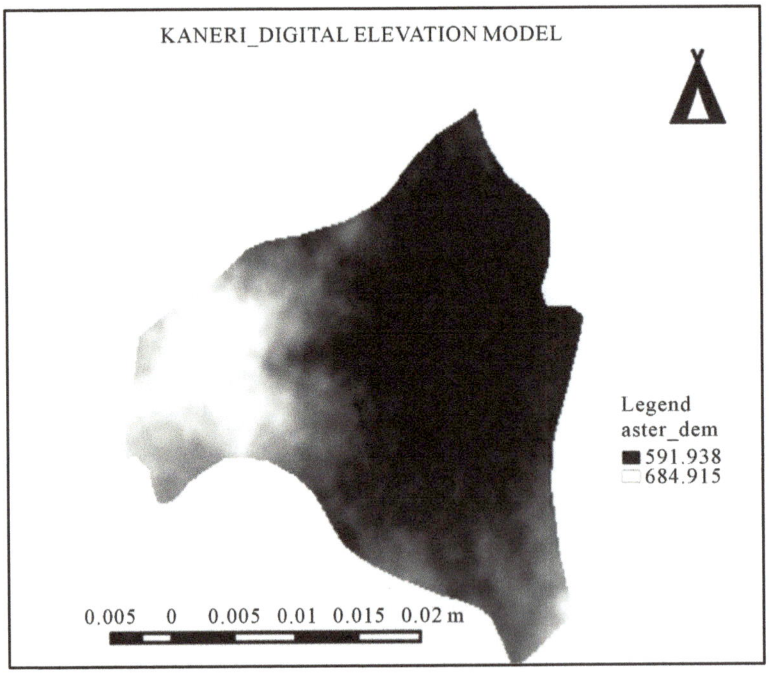

Figure 2. Digital elevation model of Kaneri.

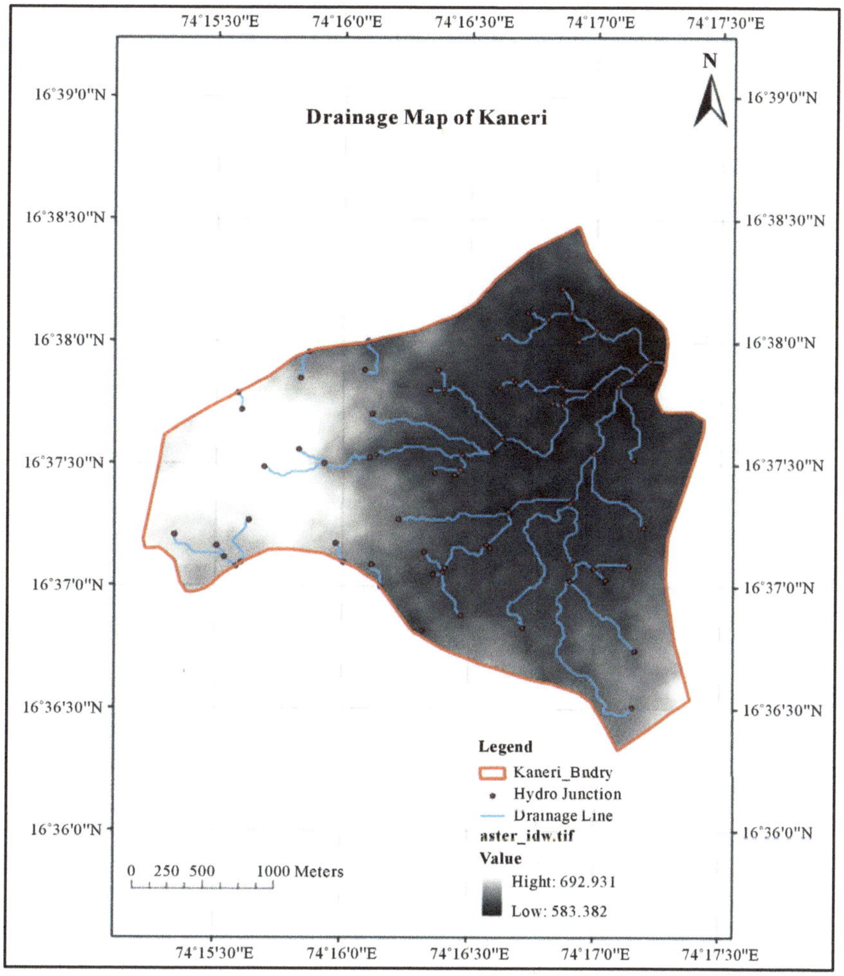

Figure 3. Drainage map of Kaneri Village.

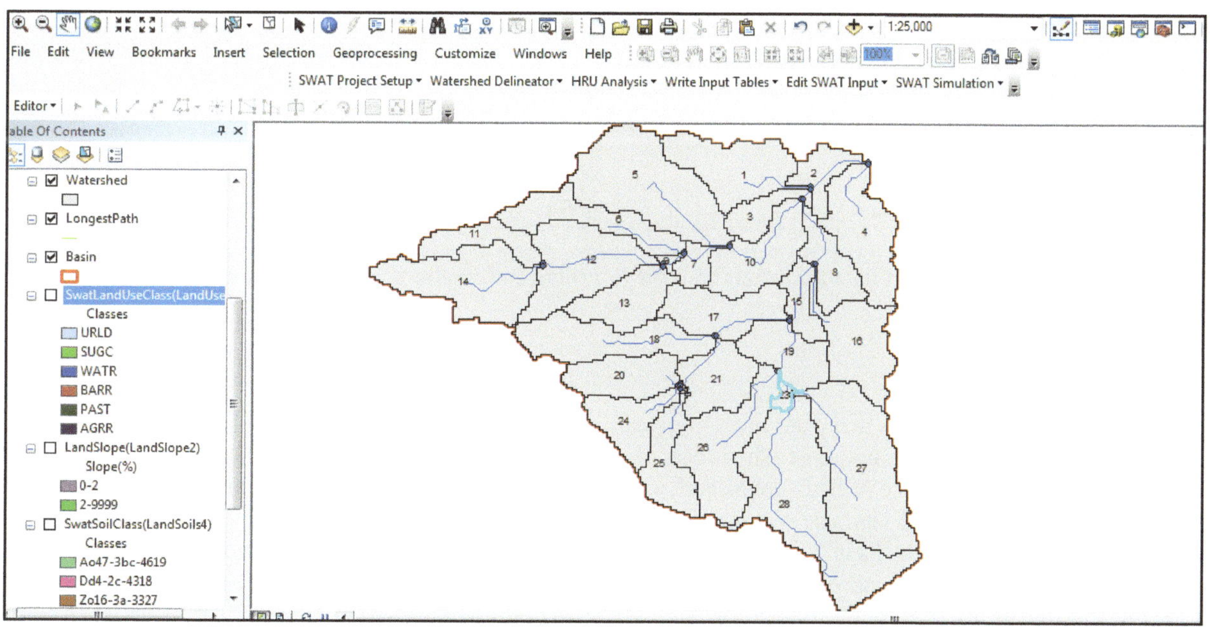

Figure 4. Watershed delineation map.

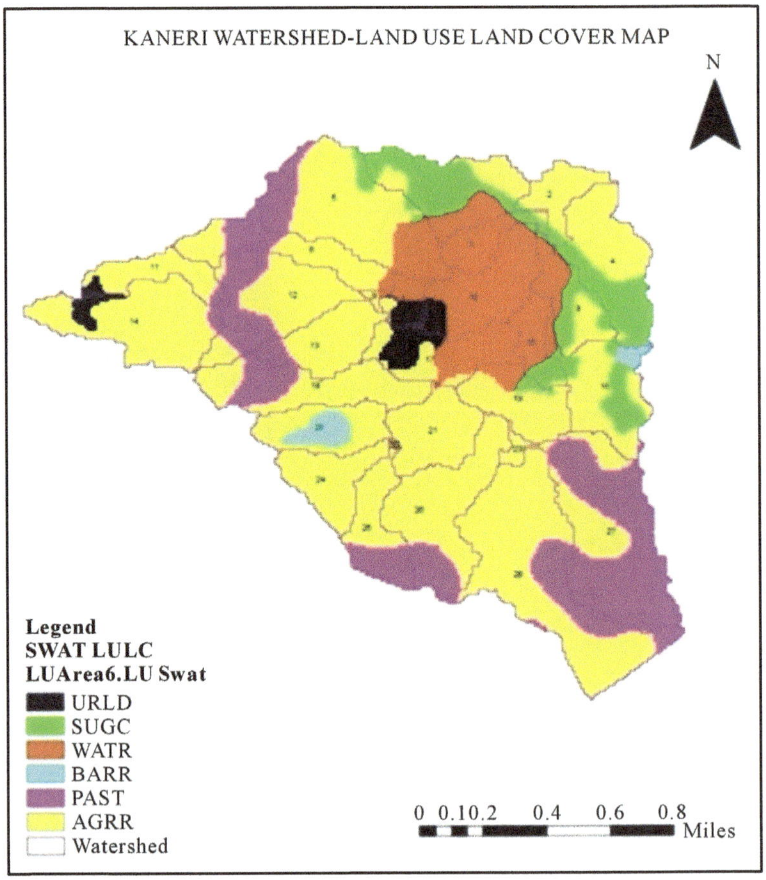

Figure 5. LULC of Kaneri Watershed.

Table 1. Land use classification with total area distribution.

Sr. No.	Land Use Category	Code	Area (Ha)	% of Watershed Area
1	Urban Residential Low density	URLD	11.93	2.228
2	Sugarcane	SUGC	37.84	7.069
3	Water bodies	WATR	64.20	11.99
4	Barren	BARR	85.25	15.919
5	Pasture	PAST	5.72	1.069
6	Agriculture Land	AGRR	330.54	61.724

Table 2. This categorized data may then be used to produce thematic maps of the land cover present in an image.

2.2. Model Setup

SWAT allows the user to delineate the watershed and sub basins using the Digital Elevation Model (DEM). Drainage network is also prepared which can be useful for delineation. **Figure 3** shows the drainage map of Kaneri Watershed. **Figure 4** gives delineated watershed through Arc-SWAT. The delineated watershed by Arc SWAT and the prepared land use map overlapped 100%. **Figure 6** shows Land Use Land Cover map overlay. For runoff to produce, the basic necessity of the model is to have the total rainfall more than field capacity of the soil. If it is less, then the subsurface flow is modeled. SWAT runs for both the possibilities and computes peak rate, transmission losses and sediment yield in first case and computes soil water routing, evapo-transpiration, crop growth, pond, wetland balances, ground water flow in the second case. While running the model land use

Table 2. Detailed sub basin wise land use distribution of Kaneri Watershed.

OID	SUBBASIN	URLD	SUGC	WATR	PAST	BARR	AGRR
1	1		12.27188	5.145494			6.739872
2	2		0.724717				8.261779
3	3		0.917975	7.053916			0.338201
4	4		9.711213	0.096629	0.096629		14.97749
5	5		1.594378	7.005602		6.329199	26.25893
6	6			0.507302		4.56572	12.92413
7	7	2.19831		2.391567			0.70056
8	8		4.082575	4.855607			4.29999
9	9			0.26573			0.748875
10	10	2.149995	0.144943	18.45614			
11	11	0.773032				0.628088	9.590427
12	12					9.759528	18.72187
13	13					2.174152	13.72132
14	14	2.294939				0.869661	27.75668
15	15		1.956737	5.870211			0.26573
16	16		5.580324	1.256177	1.521907	0.144943	13.45559
17	17	4.517405		8.817395			5.459538
18	18					6.280884	18.38367
19	19		0.869661	2.439882			12.19941
20	20				4.106732		11.76458
21	21			0.048314			16.30614
22	22						0.169101
23	23						1.811794
24	24						14.32525
25	25					5.483695	9.276383
26	26					5.169651	24.49545
27	27					33.02296	12.75503
28	28					10.82245	44.83585

land cover map, soil map, slope map and weather data is provided as an input on which the processing is done through SWAT model.

Modeling process once over, gives the output in the form of surface runoff, PET (Potential Evapo Transpiration), Evapo Transpiration, Percolation, Ground water flow, Soil Moisture, Water Yield and Sediment Yield. In this study by Ashok Mishra *et al.* [8], the estimated runoff and sediment yield from each sub-watershed were compared with measured values and the effect of LU/LC, soil and topographic characteristics was then evaluated. **Figure 7** shows soil map overlay for SWAT model setup.

Multiple HRUs were defined within a sub basin by ignoring land uses less than 2% of the subbasin and also ignoring soil types in a subbasin covering less than 5% of the subbasin. A total of 76 HRUs for 28 sub basins were created. **Figure 8** shows slope map overlay of Kaneri Watershed.

SWAT tool having an interface with Arc View GIS software (AVSWAT 2000) was used by Jain S. K. *et al.* [12] for runoff and sediment yield from area of Suni to Kasol, watersheds of satluj river, western Himalaya, performed well for runoff prediction. Coefficient of determination (R^2) for daily and monthly runoff was 0.53 and 0.90 respectively for calibration period (1993-1994) and 0.33 and 0.62 respectively for validation period (1995 to 1997) R^2 values in estimating daily and monthly sediment yield during calibration period was found as 0.33 and 0.38 respectively, and for validation period it was observed 0.26 and 0.47 respectively showing a little low performance for sediment yield estimation.

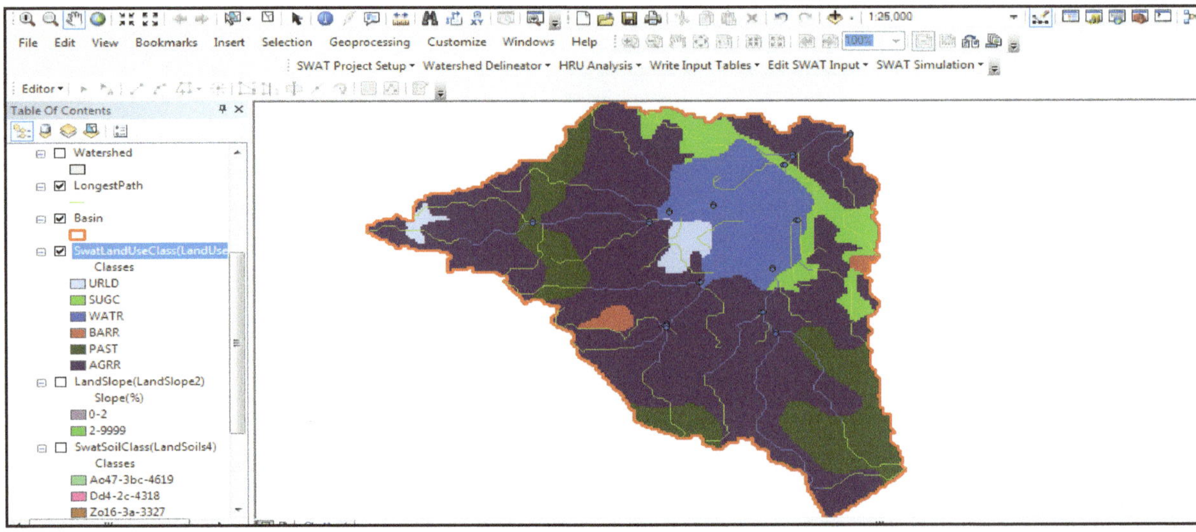

Figure 6. Land use land cover map overlay for SWAT model setup.

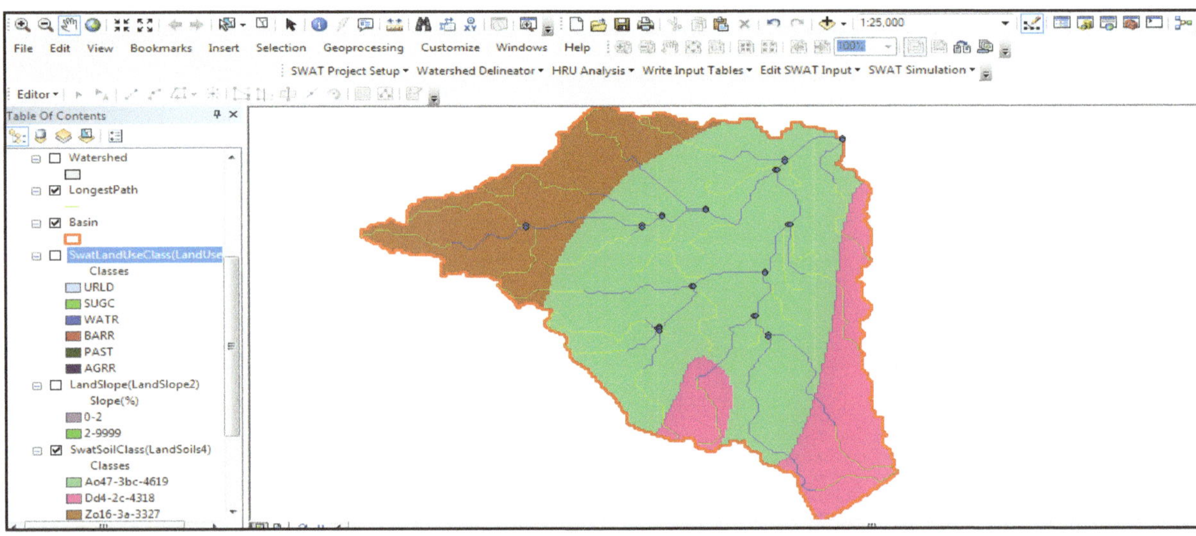

Figure 7. Soil map overlay for SWAT model setup.

Narayan K. Shreshtha *et al.* [13] examined SWAT model performance of Kilene Nete Watershed Belgium with area 581 sq. km. Seven SWAT model parameters were calibrated (1994-1998) and validated (1999-2002). The Parameter Ch_K_2 (Channel effective hydraulic conductivity) was found to be the most sensitive. R^2 and NSE values for calibration and validation were found to be 74 and 67 percent respectively, showing satisfactory goodness of fit statistics. Puwadon Phomcha *et al.* [14], while simulating monthly stream flow in Lam Sonthi watershed in Thailand found R^2 and NSE values greater than 0.7 and deviation of runoff volumes with acceptably accurate values. Some months of simulated flows showed over estimated values but most simulated flows were close to observed flows by both visual and statistical approaches.

Calibration and validation of a SWAT model applied by Arun Babu *et al.* [15] for Krishnagiri Reservoir Watershed, TamilNadu were carried for testing model performance. Model Calibration (2000-2005) and validation (2006-2010) were carried by using the measured discharge and sediment load from a gauge station. It showed a satisfactory agreement between the measured monthly flow with simulated flow values with R^2 value 0.9 and NSE value 0.8. For a small agricultural watershed in Tamilnadu, India, SWAT model was applied by Kaviya K. *et al.* [16] for prediction of water balance parameters. The investigation was carried out using a 10 year rainfall record from Jan. 88-Jan. 98 for calibration and validation. Model performed well with R^2 value greater than 0.98 and correlation coefficient value greater than 0.9.

2.3. Performance Evaluation of the Model

For calibration and validation, different techniques are being used. In the present study, conventional method along Arc SWAT tool was used. The main function of an interface is to provide a link between the input/output of a calibration program and the model. The simplest way of handling the file exchange is through text file formats. The model was run for thirty six years 1979 to 2014. Surface runoff and sediment calibration for the Kaneri watershed was conducted for the years 1979 to 2000. Similarly, surface runoff and sediment validation for the Kaneri watershed was carried out for the years 2001 to 2013. The most widely used criteria, for testing performance of a model is coefficient of determination R^2

$$R^2 = \frac{\left[\sum (Q_{sim\,i} - Q_{asim\,i})(Q_{obs\,i} - Q_{aobs\,i})\right]^2}{\sum (Q_{sim\,i} - Q_{asim\,i})^2 \sum (Q_{obs\,i} - Q_{aobs\,i})^2}$$

where,

R^2 = coefficient of determination;

$Q_{obs\,i}$ = Observed value at time step *i*;

$Q_{sim\,i}$ = Simulated value at time step *i*;

Q_{aobsi} = Average of observed value at time step *i*;

$Q_{asim\,i}$ = Average of simulated value at time step *i*.

R^2 describes the percentage of the variance in calculated data experienced by the model. According to the criteria developed by Sameh *et al.* [17] R^2 ranges from 0 to 1, with greater values indicating low error variance, and typically values greater than 0.5 are considered acceptable. Krause *et al.* [18] stated that R^2 statistic provided an estimate of how well the variance of measured values are replicated by model predictions. The general performance rating criteria developed by Sameh *et al.* [17] for calibration and validation of SWAT model are given in **Table 3**.

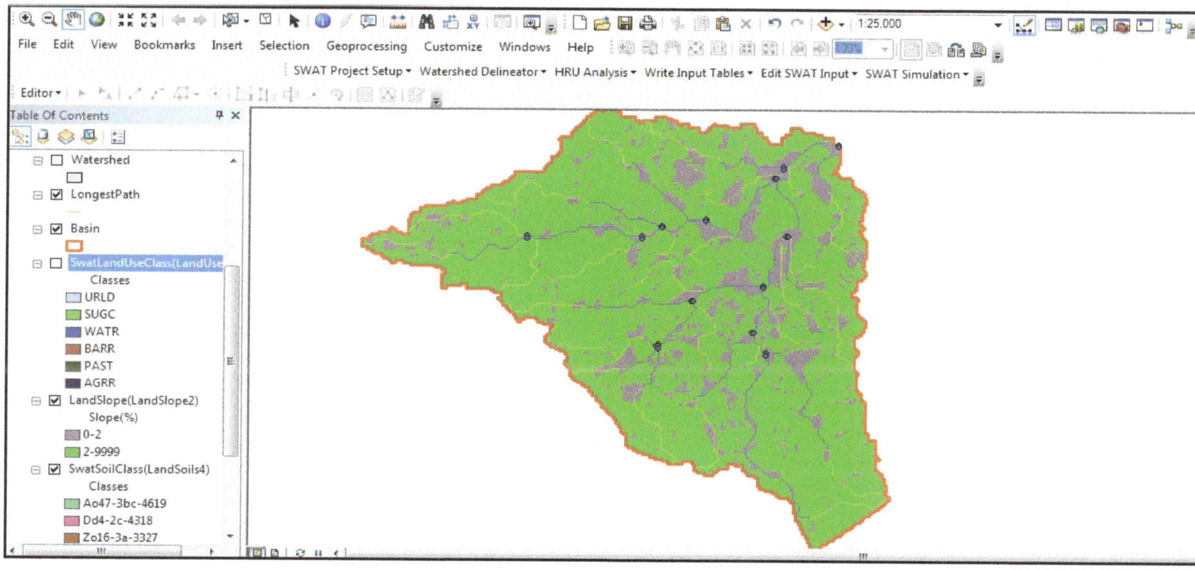

Figure 8. Slope map overlay for SWAT model setup.

Table 3. Performance rating for calibration and validation of SWAT model.

Performance rating	R^2
Very Good	$R^2 > 0.70$
Good	$0.60 < R^2 \le 0.70$
Satisfactory	$0.50 < R^2 \le 0.60$
Unsatisfactory	$R^2 < 0.50$

The model goodness-of-fit was evaluated on a yearly and monthly basis. The linear graphs for the measured and simulated values both for flow and sediment on yearly basis for calibration and validation are presented. The linear graphs for the measured and simulated values both for flow and sediment on monthly basis for calibration and validation are also produced.

3. Results and Discussion

3.1. Model Calibration

The SWAT model was calibrated using the yearly data of runoff and sediment yield recorded at the outlet of the study watershed for the years 1979 to 2000. Several simulation runs were applied to achieve the model calibration. The time series of the observed and simulated monthly sediment yield for the calibration period were plotted for visual comparison (**Figure 9** and **Figure 10**). The total monthly runoff computed by the model was, found to be 6792.11 mm against the observed runoff of 6992.77 mm during 1979 to 2000. The sediment yield computed by the model during respective months was obtained as 124.61 t/ha against the observed sediment yield of 118.26 t/ha.

The observed and predicted values were plotted against each other in order to determine the goodness-of fit criterion of coefficient of determination (R^2) both for runoff and sediment yield. The R^2 for yearly and monthly values was obtained as 0.951 and 0.849 respectively for runoff (**Figure 11(a)**, **Figure 11(b)**) and 0.788 and 0.722 respectively for sediment yield (**Figure 11(c)** and **Figure 11(d)**). It was reported that SWAT's yearly flow predictions, in general, were not as good as monthly flow predictions.

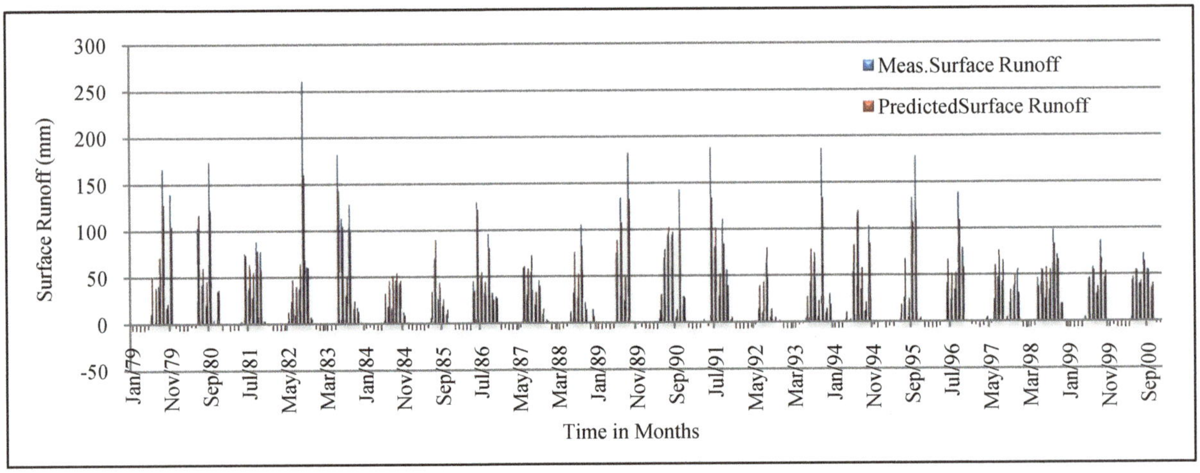

Figure 9. Measured and predicted surface runoff for calibration period 1979-2000.

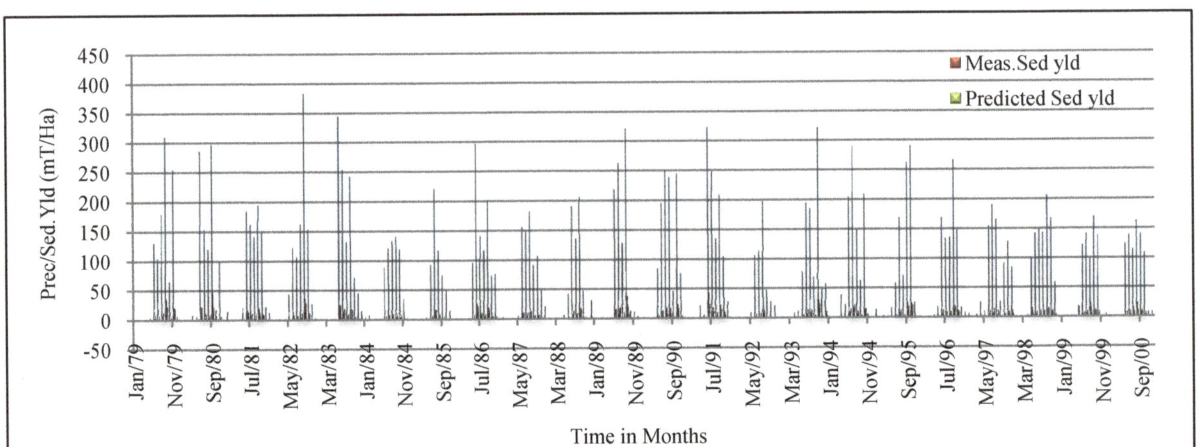

Figure 10. Measured and predicted sediment yield for calibration period 1979-2000.

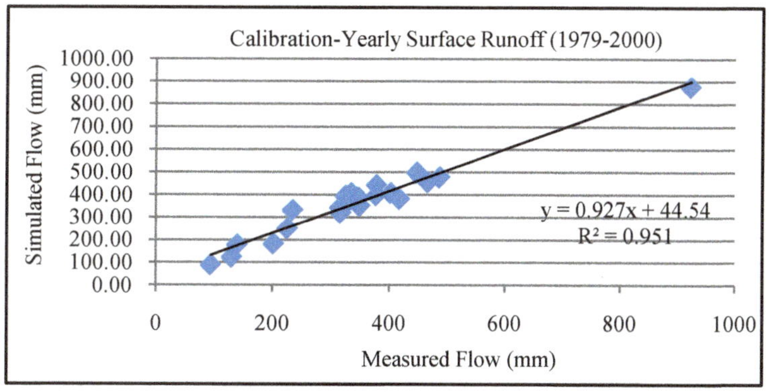

(a)

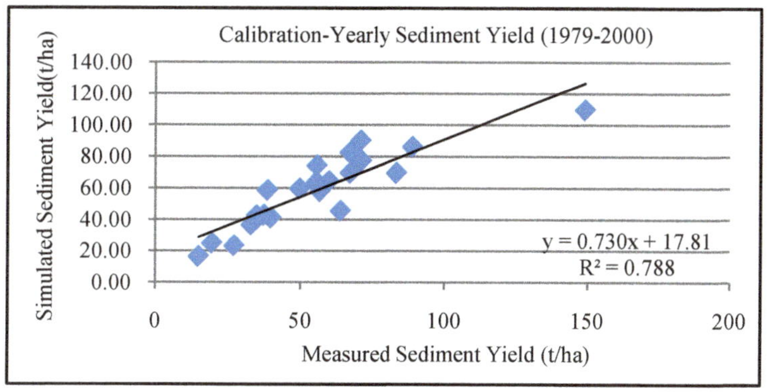

(b)

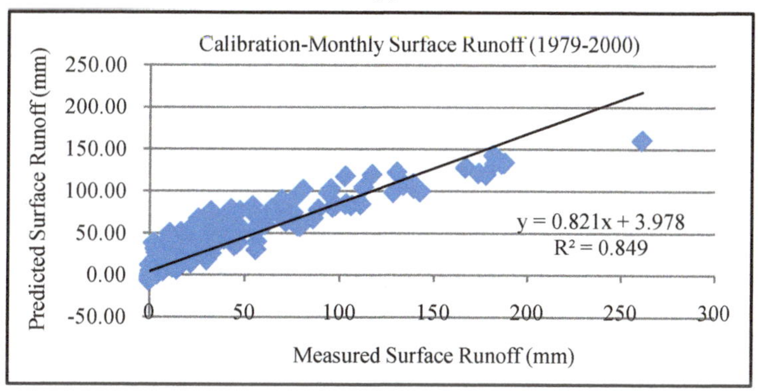

(c)

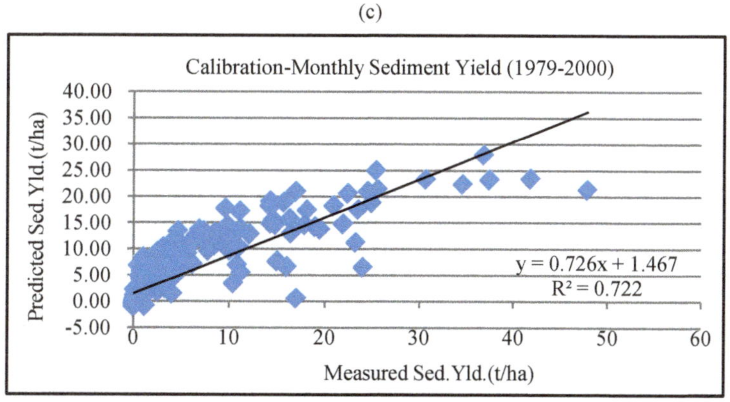

(d)

Figure 11. (a) Determination of coefficient of determination for yearly surface runoff; (b) Determination of coefficient of determination for yearly sediment yield; (c) Determination of coefficient of determination for monthly surface runoff; (d) Determination of coefficient of determination for yearly sediment yield.

3.2. Model Validation

The model validation was carried out for yearly and monthly surface runoff and sediment yield for the years 2001 to 2013. A graphical comparison of the observed and simulated yearly and monthly flows and sediment yield are shown in **Figure 12** and **Figure 13**.

The total monthly surface runoff computed by the model was, found to be 3228.04 mm against the observed runoff of 3027.38 mm during 2001 to 2013. The sediment yield computed by the model during respective months was obtained as 60.06 t/ha against the observed sediment yield of 66.40 t/ha. The observed and predicted values were plotted against each other in order to determine the goodness-of fit criterion of coefficient of determination (R^2) both for runoff and sediment yield. The R^2 value for monthly and yearly surface runoff was obtained as 0.801 and 0.950 respectively (**Figure 14(a)**, **Figure 14(b)**) and 0.565 and 0.684 respectively for sediment yield (**Figure 14(c)** and **Figure 14(d)**). It was reported that SWAT's yearly flow predictions, in general, were better than monthly flow predictions.

4. Conclusions

In the present study, The Soil and Water Assessment Tool (SWAT) having an interface with Arc-View GIS software (ArcGIS 10.1 with Arc SWAT 2012 extension), was applied to the hilly watershed for modeling runoff and sediment yield. After preparing all the thematic maps and database as per the format of AVSWAT model,

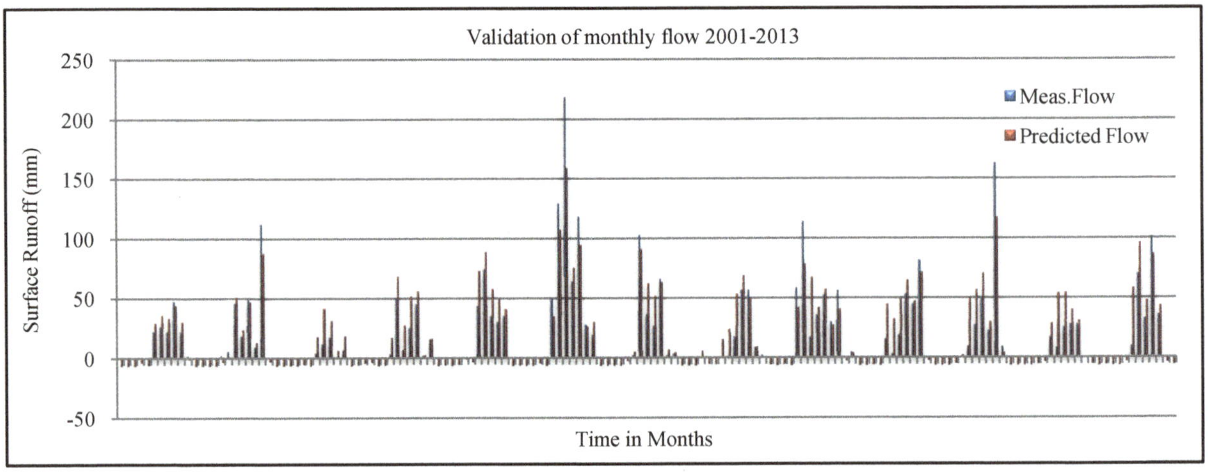

Figure 12. Measured and predicted surface runoff for validation period 2001-2013.

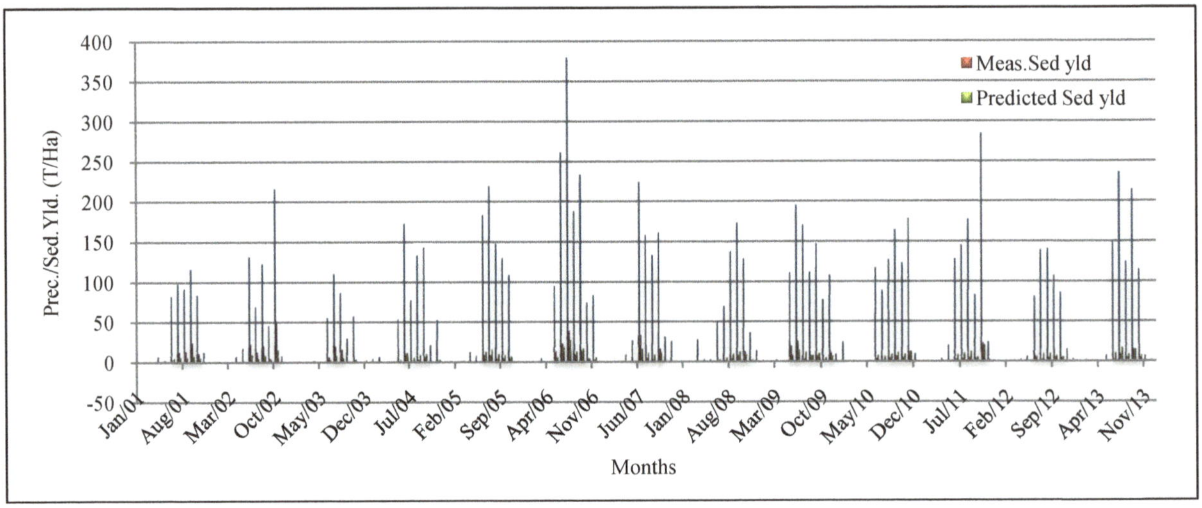

Figure 13. Measured and predicted sediment yield for validation period 2001-2013.

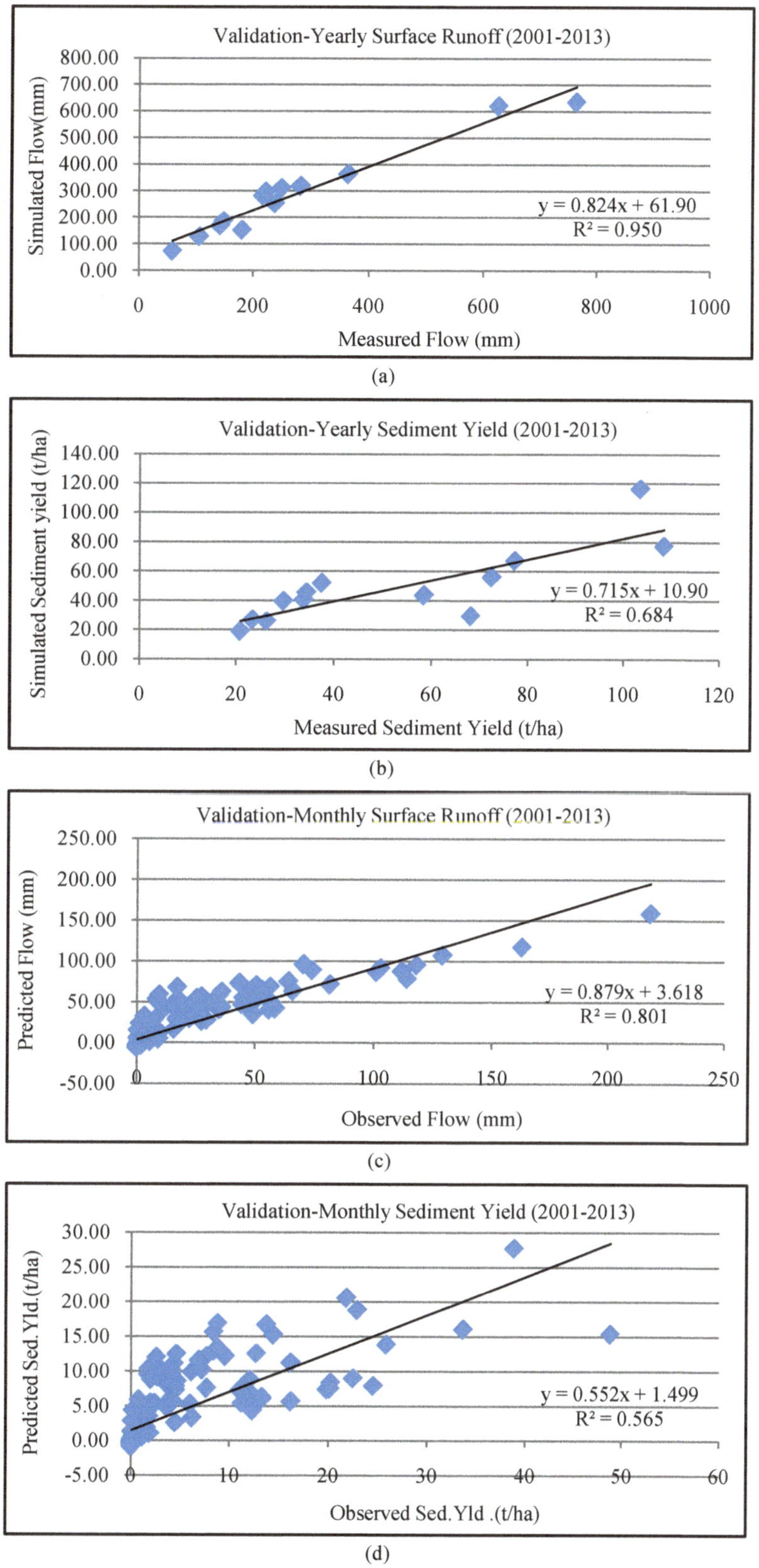

Figure 14. (a) Determination of coefficient of determination for yearly surface runoff; (b) Determination of coefficient of determination for yearly sediment yield; (c) Determination of coefficient of determination for monthly surface runoff; (d) Determination of coefficient of determination for monthly sediment yield.

the model was calibrated for the yearly and monthly surface runoff and sediment yield using the observed data of 1979 to 2000. The model validation was carried out for a data set of thirteen years of 2001 to 2013. The simulation performance of the model for calibration and validation was evaluated using graphical and statistical methods.

The coefficient of determination (R^2) for the yearly and monthly runoff was obtained as 0.849 and 0.951 respectively for the calibration period and 0.801 and 0.950 respectively for the validation period. The R^2 value in estimating the yearly and monthly sediment yield during calibration was computed as 0.722 and 0.788 respectively. The R^2 for yearly and monthly sediment yield values during validation period was observed to be 0.565 and 0.684. Thus, the values of R^2 can be considered satisfactory for estimating runoff and sediment yield from a hilly watershed with available data.

References

[1] Sathian, K.K. and Shyamala, P. (2009) Application of GIS Integrated SWAT Model for Basin Level Water Balance. *Indian Journal of Soil Conservation*, **37**, 100-105.

[2] Arnold, J.G., Moriasi, D.N., Gassman, P.W., Abbaspour, K.C., White, M.J., Srinivasan, R., Santhi, C., Harmel, R.D., Van Griensven, A., Van Liew, M.W., Kannan, N. and Jha, M.K. (2012) SWAT Model Use: Calibration and Validation. *Transactions of the ASABE*, **55**, 1491-1508. http://dx.doi.org/10.13031/2013.42256

[3] Neitsch, J.R., Arnold, J.G., Kiniry, J.R. and Williams, J.R. (2005) Input/Output File Documentation Version 2005. User Manual Grassland, Soil and Water Research Laboratory, Agricultural Research Service, Blackland Research Center, Texas Agricultural Experiment Station, Temple, Texas.

[4] Tripathi, C.N. and Goasin, A.K. (2003) Micro Watershed Modeling in India Using GIS Technologies and Agricultural Policy Environmental Extender (APEX) Model—A Case Study. *International Journal of Engineering Research and Application*, **3**, 1640-1648.

[5] Swami Vidula, A. and Kulkarni Sushma, S. (2014) Evaluation of the Best Management Practices Using SWAT Model for the Kaneri Micro-Watershed, Southern Maharashtra, India. *International Journal of Advanced Technology in Engineering and Science*, **2**, 9-16.

[6] Kumbhar, S. (2011) Watershed Development Plan for Kaneri Watershed. M.E. Thesis, Rajarambapu Institute of Technology, Sakharale.

[7] Jay Krishnan, R., Sriniwasan, R., Santhi C. and Amold, J.G. (2005) Advances in the Application of the SWAT Model for Water Resources Management. *Hydrological Processes*, **19**, 749-762. http://dx.doi.org/10.1002/hyp.5624

[8] Mishra, A., Kar, S. and Singh, V.P. (2007) Prioritizing Structural Management by Quantifying the Effect of Land Use and Land Cover on Watershed Runoff and Sediment Yield. *Water Resources Management*, **21**, 1899-1913. http://dx.doi.org/10.1007/s11269-006-9136-x

[9] Birhanu, B.Z., Nodomba, P.M. and Mtalo, F.W. (2007) Application of SWAT Model for Mountainous Catchment. *LARS Conference at Catchment at Arba Minch University*, 2007, 182-187.

[10] Akhavan, S., Koupai, J.A., Mousavi, S., Afyuni, M., Eslamian, S. and Abbaspour, K.C. (2010) Application of SWAT Model to Investigate Nitrate Leaching in Hamadan Bahar Watershed, Lran. *Agriculture, Ecosystems and Environment*, **139**, 675-688. http://dx.doi.org/10.1016/j.agee.2010.10.015

[11] Maharajan, G.R., Park, Y.S., Kim, N.W., Shin, D.S., Choi, J.W., Hyun, G.W., Jeon, J.H., Ok, Y.S. and Lim, K.J. (2012) Evaluation of SWAT Sub Daily Runoff Estimation at Small Agricultural Watershed in Korea. *Frontiers of Environmental Sciences and Engineering*, **7**, 109-119. http://dx.doi.org/10.1007/s11783-012-0418-7

[12] Jain, S.K., Tyagi, J.V. and Singh, V. (2012) Simulation of Runoff and Sediment Yield for a Himalayan Watershed Using SWAT Model. *Journal of Water Resources and Protection*, **2**, 267-281. http://dx.doi.org/10.4236/jwarp.2010.23031

[13] Shreshtha, N.K., Shakti, P.C. and Gurung, P. (2010) Calibration and Validation of SWAT Model for Low Lying Watersheds : A Case Study on the Kliene Nete Watershed, Belgium. *Hydro Nepal*, **6**,47-51.

[14] Phomcha, P., Wirojanagud, P., Vangpaisal, T. and Thaveevouthti, T. (2011) Suitability of SWAT Model for Simulating of Monthly Stream Flow in Lam Sonthi Watershed. *Journal of Industrial Technology*, **7**, 49-56.

[15] Arun Babu, E., Suresh, B. and Ravichandran, S. (2013) Application of SWAT Model to an Agricultural Watershed in Tamil Nadu, India. *Proceeding of SWAT Conference* 2012, IIT, Delhi, 16-20 July 2012, 120-132.

[16] Kaviya, K. and Ramalingam, M. (2012) Hydrological Cycle Simulation of Kodanavar River (Athur Block) Watershed Using Soil and Water Assessment Tool (SWAT). *Proceeding of SWAT Conference* 2012, IIT, Delhi, 16-20 July 2012, 109-117.

[17] Al-Muqdadi, S.W. and Merkel, B.J. (2011) Automated Watershed Evaluation of Flat Terrain. *Journal of Water Re-*

sources and Protection, **3**, 892-903. http://dx.doi.org/10.4236/jwarp.2011.312099

[18] Krause, P.D., Boyle, P. and Base, F. (2005) Comparison of Different Efficiency Criteria for Hydrological Model Assessment. *Advanced Geo Science*, **5**, 89-97. http://dx.doi.org/10.5194/adgeo-5-89-2005

15

Kinetics of Sorption of Pendimethalin on Soil Samples Obtained from the Banks of Rivers Katsina-Ala and Benue, Central Nigeria

Justin Kpagh, Rufus Sha'Ato*, Raymond A. Wuana, Terrumun A. Tor-Anyiin

Department of Chemistry, University of Agriculture, Makurdi, Nigeria
Email: *rshaato@gmail.com

Abstract

Kinetics of the sorption of the herbicide, pendimethalin (3,4-Dimethyl-2,6-dinitro-*N*-pentan-3-yl-aniline) on four soil samples obtained from the banks of Rivers Benue and Katsina-Ala, in Central Nigeria was studied by contacting 5.0 g soil samples with 100 mL 7×10^{-6} M herbicide solution. The sorption was monitored by following the herbicide concentration in the aqueous phase spectrophotometrically at one hour intervals for 5 hours (at 25°C, 35°C, 45°C, 55°C and 65°C, respectively). Pseudo-first order kinetics treatment of the data showed that sorption rates as well as the accompanying activation energies were low, indicating that physical adsorption is the operative mechanism for its retention in the soils and that the herbicide tended to reside more in the aqueous phase, which raised its potential to contaminate natural water systems, when applied on the field.

Keywords

Kinetics, Sorption, Pendimethalin, Soil, Nigeria

1. Introduction

The use of herbicides in the Lower Benue Valley (6°15' - 8°00'N; 6°15' - 11°00'E) in Central Nigeria is not only on the increase, but also indiscriminate. This is because the area has a predominantly agrarian economy and one of the major problems farmers face is the aggressive growth of weeds which negatively affects farm produce. Among other herbicides [1], pendimethalin (3,4-Dimethyl-2,6-dinitro-*N*-pentan-3-yl aniline, **Figure 1**), a pre-emergence herbicide effective against many annual grasses and broadleaf weeds enjoys widespread use in the area.

*Corresponding author.

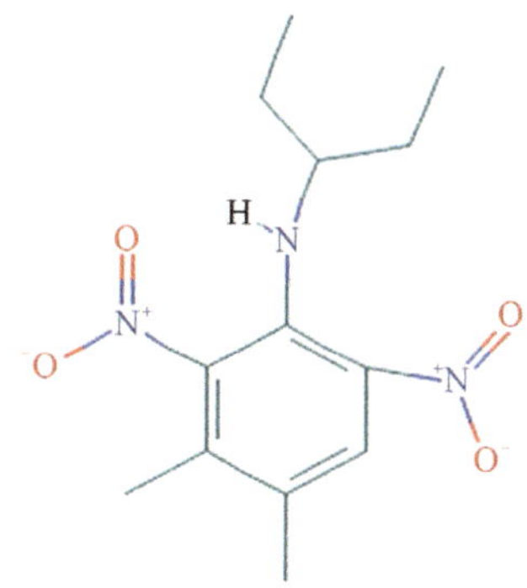

Figure 1. Structure of pendimethalin (3,4-dimethyl-2,6-dinitro-*N*-pentan-3-yl aniline).

Being a soil-applied herbicide, once applied, it is taken up by the roots and/or hypocotyls of the target weeds, interfering with their growth, and eventually killing them. It has been noted, for a long time, that the efficacy of soil-applied herbicides depends, in part, on soil type [2]. Performance is further influenced by herbicide adsorption. According to Wauchope and Koskinen [3], sorption is a major factor controlling herbicide activity, mobility and persistence in soil. It also determines the quantity of herbicide in soil solution and thus the amount that is available to be leached [4]. Thus, for effective assessment of the mobility and pollution potentials of herbicides to be made and for better understanding of the fate of these chemicals in soil, a good knowledge of their sorption behaviour, including kinetics, in different soil types is desired. While pendimethalin is known to adsorb strongly to soil organic matter and clay, soils in the Lower Benue Valley are generally low in organic matter and clay [5]. It is, therefore, expected that this herbicide effective against many annual grasses and broadleaf weeds may be highly mobile in them and so may pose a surface and groundwater contamination hazard in the area. Considering that it is a suspected carcinogen [6], it is not subject to microbial degradation and only slight losses of the herbicide may be expected from photodecomposition and volatilization on the field (its soil half-life is found to be 90 days [7], it is imperative that its fate in the Lower Benue Valley environment, where it is widely used, is investigated. The aim of this study is, therefore, to investigate the sorptive behaviour of pendimethalin in selected soils in the area, since there is little or no information on this in the literature.

2. Materials and Methods

2.1. Soil Sampling and Preparation

Soil samples were taken at the surface horizon (0 - 25 cm) at four different locations along the banks of Rivers Benue and Katsina-Ala, where both rain-fed and irrigation agriculture (vegetable cropping) is very active. The samples were identified as follows:

1) RB1 Benue River bank (7.746756N 8.61755E)
2) RB2 Benue River bank (7.748797N 861755E)
3) RKA1 Katsina-Ala River bank (7.194838N 9.287592E)
4) RKA2 Katsina-Ala River bank (7.154097N, 9.285760E)

The whole soil samples, each weighing 1.5 kg, were taken to the Chemistry Laboratory, University of Agriculture, Makurdi, in black polythene bags and subsequently spread in a dust-free wooden cupboard to air-dry. After air-drying (7 days), the samples were gently crushed and sieved using a 2 mm sieve/shaker. The <2 mm soil fraction was retained for sorption studies.

2.2. Pendimethalin

The commercial formulation of the herbicide (500EC *i.e.* 500 g/L) as used by farmers, was purchased from Lado Agrochemicals Company, Katsina-Ala. This was used without purification to simulate the condition in which it reaches the soil in practice.

2.3. Determination of Some Physico-Chemical Properties of Soil Samples

Laboratory soil samples moisture content was determined using the method employed by Charturvediand Sankar

[8]; pH was determined using a portable pH meter (Jenway Digital Meter model 3505) after calibration at pH 4 and 7. Electrical conductivity was measured directly using a portable conductivity meter (HACH C0150 model). Organic matter was determined using the dichromate method, while Cation Exchange Capacity (CEC) was determined at pH 7 using 1M ammonium acetate. In each case, three replicate readings were taken and the mean calculated.

2.4. Sorption Kineticsstudies

A wavelength scan of the herbicide solution was done using chloroform from 370 nm to 480 nm to determine the wavelength of maximum absorption λ_{max} for use in the sorption experiments; this was found to be 416 nm compared to 437 nm in the literature [9]. The experimental λ_{max} was used in this study. For each soil sample, 5.0 grams were placed in a 250-mL glass-stoppered conical flask. Hundred (100) mL of standard pendimethalin solution (7×10^{-6} M) was added and the volume made up to 250 mL with distilled water. Five sets of experiments were conducted at 25˚C, 35˚C, 45˚C, 55˚C and 65˚C using an incubator. At one-hour intervals, aliquots of the solutions were withdrawn, cooled, centrifuged and the herbicide concentration in the supernatant determined spectrophotometrically using the UV-Visible spectrophotometer at 416 nm. At each stage, the difference between the concentration of pendimethalin added and the one found in the supernatant was used to calculate the amount of the herbicide adsorbed after definite time intervals. All readings were triplicated and mean values used.

3. Results

Table 1 gives the physicochemical properties of the soil samples: the low moisture content in storage showed that the soils were well dried as at the point of analysis; the pH values indicate the soils are fairly acidic under field conditions, as their CECs are typical of soils in the Lower Benue Valley [10]. While the soils from the banks of the River Benue (RB1 and 2) have high clay contents, the soils from the banks of River Katsina-Ala (RKA1 and 2) are largely sandy loam in nature. Overall, the soils have organic matter content typical of their alluvial origins. **Table 2** shows kinetic parameters (in the temperature range of 25˚C - 65˚C) which were calculated from slopes of pseudo-first order kinetic plots (of natural logarithm of amount of pendimethalin adsorbed (Q) versus time (t, in hours), in accordance with Equation (1).

$$\ln Q = k_r \qquad (1)$$

where, k_r is the pseudo-first order rate constant. The linearity of the plots (with negative slopes), gave support to the primary assumption that the sorption considered in this study followed first order kinetics [11]. From the rate constants, the sorption half-life (a measure of the persistence in the aqueous phase) of pendimethalin was calculated appropriately (Equation (2)).

$$t_{1/2} = \frac{0.693}{k_r} \qquad (2)$$

Application of the linearized Arrhenius Equation (3) (plots of $\ln k_r$ vs. $\frac{1}{T}$, where, T = absolute temperature)

Table 1. Physicochemical properties of soil samples taken from the banks of Rivers Benue and Katsina-Ala, Central Nigeria.

Soil sample	Moisture content (%)	pH	Electrical conductivity µs/cm	% Clay	Organic matter (%)	Soil CEC at pH7 (cmol(+)kg)
RB1	1.03	5.8	59.60	45.7	8.29	3.63
RB2	1.06	5.6	37.50	50.6	8.17	3.80
RKA1	0.44	6.1	46.03	16.2	4.58	0.82
RKA2	0.72	6.3	33.47	12.2	4.47	3.41

Table 2. Parameters derived from pseudo-first order kinetic plots for pendimethalin sorption on soil samples taken from the banks of Rivers Benue and Katsina-Ala, Central Nigeria.

1/Temp	Slope	k_r (hour^{-1})	lnk_r	$t_{1/2}$ (hours)	$E^{\ddagger}$ (kJ·mol^{-1})
		Pendimethalin sorption on soil sample RB1			
3.35E-03	−0.0598	0.137	−1.98	5.05	
3.25E-03	−0.0627	0.144	−1.93	4.81	
3.14E-03	−0.0265	0.061	−2.79	11.4	10.6
3.05E-03	−0.0577	0.132	−2.01	5.25	
2.96E-03	−0.0837	0.192	−1.64	3.61	
		Pendimethalin sorption on soil sample RB2			
3.35E-03	−0.0749	0.172	−1.75	4.03	
3.25E-03	−0.0540	0.124	−2.08	5.59	
3.14E-03	−0.0604	0.139	−1.97	4.98	10.6
3.05E-03	−0.0450	0.103	−2.26	6.73	
2.96E-03	−0.0633	0.145	−1.92	4.78	
		Pendimethalin sorption on soil sample RKA1			
3.35E-03	−0.0762	0.175	−1.74	3.96	
3.25E-03	−0.0556	0.128	−2.05	5.41	
3.14E-03	−0.0615	0.141	−1.95	4.91	10.3
3.05E-03	−0.0590	0.135	−1.99	5.13	
2.96E-03	−0.0602	0.138	−1.97	5.02	
		Pendimethalin sorption on soil sample RKA2			
3.35E-03	−0.0696	0.160	−1.83	4.33	
3.25E-03	−0.0624	0.143	−1.93	4.85	
3.14E-03	−0.0621	0.143	−1.94	4.85	10.9
3.05E-03	−0.0476	0.109	−2.21	6.36	
2.96E-03	−0.0500	0.115	−2.16	6.03	

$$\ln k_r = \ln A - \frac{E^{\ddagger}}{RT} \quad (3)$$

gave access to the activation energy ($E^{\ddagger}$) for the sorption process in the soils.

4. Discussion

Of the four soil samples studied, percentage clay was highest in sample RB2 followed by sample RB1, RKA1 and RKA2 in that order (**Table 1**). Soil samples RB1 and RB2 with the highest percentage clay also had the highest percentage organic matter. For example, it was earlier noted that the more clay or organic matter present in a soil, the higher the CEC [12]. This was the case for soil samples RB1 and RB2, which also showed high water – holding capacities (moisture content in storage of circa 1% compared to <1% for the other soil samples). The pH values of all the soil samples were less than 7, and generally within the same acidic range (**Table 1**). On the whole, the soils have properties that are expected of soils in this area of Nigeria.

Sorption rate constants obtained from the pseudo-first order kinetics treatment of the data were low (**Table 2**) and half-lives relatively long, although they showed no clear trend for samples RKA1 and RKA2. However, adsorption rates on samplesRB1 and RB2 showed an increase in adsorption rate with temperature, similar to observations by [11] who studied dazomet adsorption on natural adsorbents. This trend was in agreement with the fact that the amount of a pesticide adsorbed in soils is related to the organic carbon contents of the soils [13]. Torrents and Jayasundera [14] also conducted the sorption study of non-ionic pesticides and found that the intensity of adsorption was a function of herbicide and clay contents. Nevertheless, Shimizu and Liljestrand [15] reported that there was no relationship between sorption and organic carbon in soils in the case of low organic carbon soils. Therefore, soil organic carbon and clay contents may not have been the only factors relevant to pendimethalin adsorption in this study. This fact may have been responsible for the unclear adsorption trend in soil samples RKA1 and RKA2.

The activation energies ($E^{\ddagger}$) obtained in the study ranged from 10.3 kJ·mol^{-1} to 10.9 kJ·mol^{-1}. According to Bansal [16], activation energies of 10 - 28 kJ·mol^{-1} represent physical adsorption. Since the calculated values of $E^{\ddagger}$ were within this range, it implied that only physical adsorption of the herbicide on soil surfaces must have taken place. This was expected, looking at the molecular structure of the herbicide. Low sorption ensured a longer residency time of the herbicide in the aqueous phase (half-lives of about 6 hours generally), suggesting that pendimethalin could be fairly mobile (susceptible to leaching transport) in the soils under study and hence could easily contaminate ground- and surface water resources in the area.

5. Conclusion

In conclusion, rate constants, half-lives and activation energies of adsorption of pendimethalin on four soil samples obtained from the Lower Benue Valley in Central Nigeria were determined in this study, along with some physicochemical parameters of the soils. Physicochemical parameters showed significant variations in line with what is known about tropical soils. Values of sorption rate constants, half-lives and activation energies indicated low adsorption rates on soil samples. Activation energies indicated physical adsorption of the herbicide. On the whole, results in the study showed that the herbicide has potential to cause contamination of surface and ground waters by its persistence through leaching and runoff because of its low adsorption.

References

[1] Kughur, P.G. (2012) The Effects of Herbicides on Crop Production and Environment in Makurdi Local Government Area of Benue State, Nigeria. *Journal of Sustainable Development in Africa*, **14**, 206-216.

[2] Stevenson, F.J.J. (1972) Organic Matter Interactions Involving Herbicides in Soil. *Journal of Environmental Quality*, **1**, 333-343. http://dx.doi.org/10.2134/jeq1972.00472425000100040001x

[3] Wauchope, R.D. and Koskinen, W.C. (1983) Adsorption Desorption Equilibrium of Herbicides in Soil: Thermodynamic Perspective. *Weed Science*, **31**, 504-512.

[4] Bailey, G.W. and White, J.L. (1970) Factors Influencing the Adsorption, Desorption and Movement of Pesticides in Soils. *Residue Review*, **32**, 29-92. http://dx.doi.org/10.1007/978-1-4615-8464-3_4

[5] Sha'Ato, R., Ajayi, S.O. and Ojanuga, A.G. (2012) Total and Extractable Copper, Iron, Manganese and Zinc in Major Agricultural Soils in the Lower Benue Valley, Central Nigeria and the Concept of Extractant Efficiency. *Nigerian Journal of Chemical Research*, **17**, 59-62.

[6] Coats, J.R. and Yamamoto, A. (2003) Environmental Fate and Effect of Pesticides. Symposium 853, ACS, Washington. http://dx.doi.org/10.1021/bk-2003-0853.fw001

[7] United States Department of Agriculture (1990) Soil Conservation Service. Pesticide Properties Database, New York.

[8] Charturvedi, R.K. and Sankar, K. (2006) Laboratory Manual for Physiochemical Analysis of Soil, Water and Plants. Wildlife Institute of India, Dehradun, 6-9.

[9] Pest Management Regulatory Agency (2007) A. L 6605C, Ontario, Canada pp 7.

[10] Sha'Ato, R., Ojanuga, A.G. and Ajayi, S.O. (2000) Forms and Profile Distribution of Potassium in Typical Soils of the Lower Benue Valley of Nigeria. *Communications in Soil Science and Plant Analysis*, **31**, 437-453.

[11] Sismanoglu, T., Ercag, A., Pura, S. and Ercag, E. (2004) Kinetics and Isotherms of Dazomet Adsorption on National Adsorbents. *Journal of the Brazilian Chemical Society*, **15**, 669-675. http://dx.doi.org/10.1590/S0103-50532004000500010

[12] Agronomy Fact Sheet No. 22 (2007) Dept. of Crop and Soil Sciences. Cornell University Co-Operative Extension, 1-2.

[13] Kanazawa, J. (1989) Relationship between the Soil Sorption Constants Pesticide and Their Physicochemical Properties.

Environmental Toxicology and Chemistry, **8**, 477-484. http://dx.doi.org/10.1002/etc.5620080604

[14] Torrents, A. and Jayasundera, S. (1997) The Sorption of Nonionic Pesticides onto Clays and the Influence of Natural Organic Carbon. *Chemosphere*, **35**, 1549-1565. http://dx.doi.org/10.1016/S0045-6535(97)00206-3

[15] Shimizu, Y. and Liljestrand, H.M. (1990) Sorption of Polycyclic Aromatic Hydrocarbons onto Natural Solids: Determination by Fluorescence Quenching Method. *Water Science & Technology*, **23**, 427-436.

[16] Bansal, O.P. (2004) Kinetics of İnteraction of Three Carbamate Pesticides with Indian Soils: Aligrah District. *Pest Management Science*, **60**, 1149-1155. http://dx.doi.org/10.1002/ps.915

16

The Spatial and Temporal Variation Characteristics of CH_4 and CO_2 Emission Flux under Different Land Use Types in the Yellow River Delta Wetland

Qingfeng Chen, Junjian Ma, Changsheng Zhao, Rongbin Li

Key Laboratory for Applied Technology of Sophisticated Analytical Instruments of Shandong Province, Shandong Provincial Analysis Test Center, Jinan, China

Email: chensdcn@163.com

Abstract

The Yellow River Delta Wetland is one of the youngest wetlands, and also the most complete, extensive wetlands in China. The wetland in this delta is ecologically important due to their hydrologic attributes and their roles as ecotones between terrestrial and aquatic ecosystems. In the study, the spatial and temporal variation characteristics of CH_4 and CO_2 emission flux under five kinds of land use types in the wetland were investigated. The results indicated that the greenhouse gas emission flux, especially the CO_2 and CH_4, showed distinctly spatial and temporal variation under different land use types in the wetland. In the spring, the emission flux of CO_2 was higher than that of CO_2 in the autumn, and appeared negative in HW3 and HW4 in the autumn. CH_4 emission flux of HW4 and HW5 was negative in the spring and autumn, which indicated that the CH4 emission process was net absorption. Among the five kinds of land use types, the CO_2 emission flux of HW4 discharged the largest emission flux reaching 29.3 $mg \cdot m^{-2} \cdot h^{-1}$, but the CH_4 emission flux of HW2 discharged the largest emission flux reaching 0.15 $mg \cdot m^{-2} \cdot h^{-1}$. From the estuary to the inland, the emission flux of CO_2 was decreased at first and then appeared increasing trend, but the emission flux of CH_4 was contrary to CO_2.

Keywords

Wetland, CH_4 and CO_2, Emission Flux, Land Use, Spatial and Temporal Variation

1. Introduction

Global warming has attracted wide attention and advanced research hotspot of global environmental problems, which is caused by increased greenhouse gas (GHG) emissions and the change of land use. Both CO_2 and CH_4 are considered as the most important greenhouse gases, accounting for 70% and 23% of the contribution to the temperature rising efficiency respectively [1].

Wetlands account for 6% of the world's land surface [2] and play an important role in the global carbon cycle by acting as natural carbon sinks [3]. Wetlands contain about 12% of the global carbon pool, and are very close related to climate change [4]. Wetlands provide a productive ecosystem and favorable habitat for a wide variety of plants and animal species in the world. However, wetlands ecological systems are also ecologically sensitive and adaptive systems, and show enormous diversity according to their genesis, geographical location, water regime and chemistry, dominant species, and soil and sediment characteristics [5].

The Yellow River Delta, one of the largest deltas in China, is situated in the northeast of Shandong Province on the southern bank of the Bohai Sea [6]. The delta covers an area of 7870 km^2 and is composed of large wetland areas, where the total area of the wetlands amounts to 4167 km^2 [7]. Among the total wetlands, natural wetlands cover 3131 km^2 (or 75.1% of the whole delta), and artificial wetlands cover 1036 km^2 (or 24.9% of the study area) [8]. The Yellow River Delta Wetland is one of the youngest wetlands, and also the most complete, extensive wetlands in warm temperate area in China. The wetlands in this delta are ecologically important due to their hydrologic attributes and their role as ecotones between terrestrial and marine ecosystems [9].

In this study, the spatial and temporal variation characteristics of CH_4 and CO_2 emission flux under different land use types in the Yellow River Delta Wetland were investigated, including: 1) The variation characteristics of CH_4 and CO_2 emission flux under different seasons; 2) The variation characteristics of CH_4 and CO_2 emission flux under different years; 3) The variation characteristics of CH_4 and CO_2 emission flux under different land use types. This study may have a large contribution to the protection of new-born frangibility, typical habitat and biodiversity in the wetland ecological system. It will also be beneficial for investigating the influence of the wetland carbon storage change on the terrestrial ecosystem carbon cycle and the global climate change.

2. Materials and Methods

2.1. Site Description

The study was conducted at the Yellow River Delta Wetland (N36°55' - N38°16', E117°31' - E119°18'), which is located in the southern bank of the Bohai bay and western bank of the Bohai Sea (**Figure 1**). It belongs to the warm temperate and semi-humid monsoon climate zone, with 594.3 mm of mean annual precipitation, 2049.4 mm of average annual evaporation, 12.4°C of mean annual temperature and 217.8 days of mean annual frost-free period. The soil types of this zone have high salinity, including tidal soil, saline tidal soil and coastal tidal soil. Tidal soil is neutral or alkalescence, and is mainly distributed along the river and south central plains. Salt soil distributes in the coastal areas, with a small amount of salt cultivated [10].

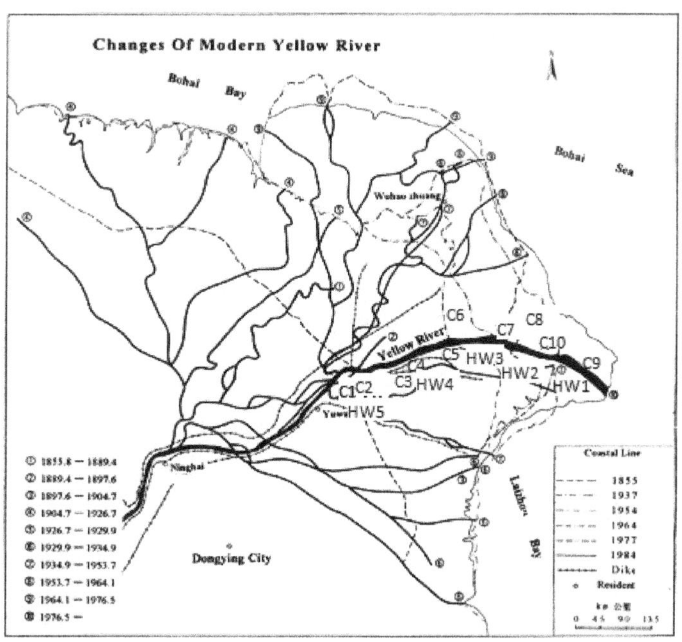

Figure 1. Location of the Yellow River Delta Wetland and sampling.

2.2. Sampling Sites Selection

The monitoring sites and Lland use characteristics of the Yellow River Delta Wetland were shown in **Figure2**, **Table 1** and **Table 2** [10]. There were 10 sites of soil samples and 5 kinds of typical salt marsh plant communities as carbon emissions monitoring site, including beaches bare land, *Suaeda salsa* community, mixed community of *Phragmites australis* and *Suaeda salsa*, *Phragmites australis* community, *Tamnrix chinesi* community and farmland community. The five types of vegetation communities in Yellow River Delta Wetland are the most typical and representative, and have a zonal distributing phenomenon from the coastal to the inland [11] [12].

2.3. Experimental Methods

The emissions concentration and fluxes of CH_4 and CO_2 were measured by using the static opaque chamber-GC technique, an eddy covariance technique. Five sampling sites were selected to collect 0 - 20 cm of soil samples in every typical salt marsh plant community. The samples of soil, plants and water were stored at 4˚C and analyzed in 48 h after sampling. The other parameters, such as TN, TP, pH, and OM, were measured according to the Standard Methods of APHA [13] [14].

The frequency of samples was taken every quarter of one year. The method of vegetation coverage degree is quadrat sampling method. The size of quadrat is 100 cm × 100 cm. In the quadrat, every vegetation coverage degree can be obtained.

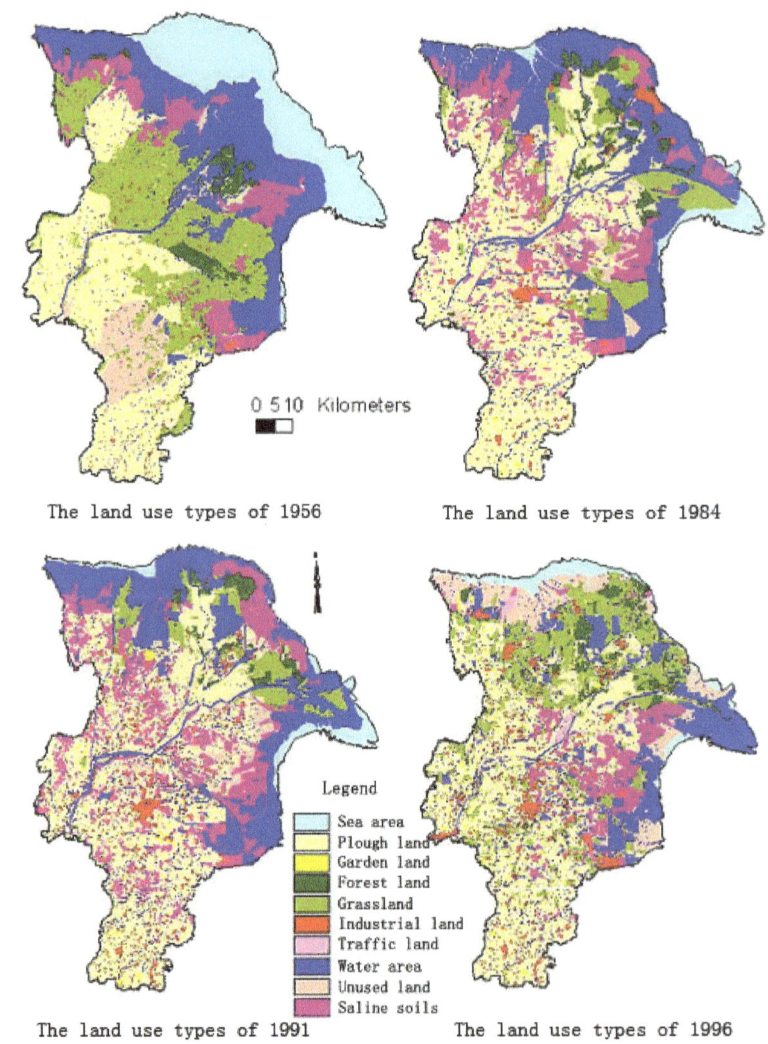

Figure 2. Land use characteristics of the Yellow River Delta Wetland.

Table 1. Soil sampling sites and description of ecosystem situation.

Number	Sampling site	Longitude and latitude	Description of ecosystem situation
C1	Woodland	E118°55'32" N37°45'96"	Woodland ecosystem, the vegetation types are mainly poplars.
C2	Cotton field	E118°55'39" N37°46'11"	Farmland ecosystem, the vegetation types are mainly cotton.
C3	*Imperata cylindrica* community	E118°58'21" N37°46'4"	The vegetation types are mainly *Imperata cylindrical* and *Phragmites australis*, with 0.5 - 1.2 m of plant height and about 80% of cover degree.
C4	*Tamnrix chinesi* community	E118°58'21" N37°46'9"	The vegetation types are mainly *Tamnrix chinesi*, with 0.5 - 2.5 m of plant height and about 60% of cover degree.
C5	*Tamnrix chinesi* community	E119°1'1" N37°45'51"	The vegetation type is *Phragmites australis*, with 0.5 - 1.5 m of plant height and about 40% of cover degree.
C6	*Phragmites australi* community	E119°04'07" N37°45'90"	The vegetation type is *Phragmites australis*, with 0.5 - 1.8 m of plant height and about 85% of cover degree.
C7	Mixed community of *Phragmites australi* and *Suaeda salsa*	E119°9'20" N37°44'48"	The vegetation types are mainly *Phragmites australis* and *Suaeda salsa*, with 0.5 - 1.2 m of plant height and about 65% of cover degree.
C8	*Suaeda salsa* community	E119°11'22" N37°44'68"	The vegetation types are mainly *Suaeda salsa*, with 0.5 - 1.0 m of plant height and about 45% of cover degree.
C9	Beaches bare land	E119°13'44" N37°43'04"	The vegetation types are mainly *Suaeda salsa*, with 0.2 - 0.6 m of plant height and about 15% of cover degree.
C10	*Suaeda salsa* community	E119°12'76" N37°43'46"	The vegetation types are mainly *Suaeda*, with 0.2 - 0.5 m of plant height and about 25% of cover degree.

Table 2. Typical salt marsh plant community and description of ecosystem situation.

Number	Community type	Longitude and latitude	Description of ecosystem
HW1	Beaches bare land	N37°43'4" E119°13'45"	The major land use is tidal flats, and scattered vegetation such as *Phragmites australi* and willow, height of 0.5 - 1 m.
HW2	*Suaeda salsa*	N37°45'55" E119°08'50"	The vegetation types are *Suaeda salsa* and *Phragmites australi*.
HW3	*Phragmites australis*	N37°45'2" E119°7'43"	The vegetation type is *phragmites australis* community, mainly including *Phragmites australis, Suaeda salsa, Tamnrix chinesi* and *wild chrysanthemum*, with 2 cm layer of litter at the surface.
HW4	*Tamnrix chinesi*	N37°46'04.6" E119°09'27.1"	The vegetation type is community of *Tamnrix chinesi-Phragmites australi*, and 80% of cover degree. There are oilfield pipelines and vehicles and other human activities around.
HW5	Farmland	N37°46'2" E118°55'38"	The vegetation type is cotton.

2.4. Date Analysis

The size of the static opaque chamber is 100 cm × 100 cm × 60 cm. The static opaque chamber method was used to measure CH_4 and CO_2 flux. The concentrations of CH_4 and CO_2 were determined with infrared carbon dioxide analyzer or G-C. The sampling time was 0, 20, 40, 60, 90, 120 min in 120 min sample period. At the same time, the temperature, air pressure and the concentration of CO_2 were measured in the static opaque chamber. CH_4 and CO_2 flux was calculated by using the following formula [15].

$$J = \frac{\mathrm{d}c}{\mathrm{d}t} \cdot \frac{M}{V_0} \cdot \frac{P}{P_0} \cdot \frac{T_0}{T} \cdot H$$

where J represents the gas flux ($\mathrm{mg \cdot m^{-2} \cdot h^{-1}}$); $\mathrm{d}c/\mathrm{d}t$ is the straightslope for the gas concentration at the time change of sampling; M is molar mass of gas to be measured; P is the pressure in sampling site; T is the absolute temperature; V_0, P_0, T_0 are molar volume of gas, air pressure and absolute temperature under the standard state condition; H is the height of sampling box above the water surface.

The load of annual emissions was calculated by using the following estimation formulas:

$$L = J \cdot S \cdot 24\,\text{h} \cdot 365\,\text{d} \cdot 10^{-6}$$

where L represents the load of annual emissions ($t \cdot a^{-1}$); J is the mean gas flux ($mg \cdot m^{-2} \cdot h^{-1}$); S is the zone area (m^2).

3. Results and Discussion

3.1. The Variation Characteristics of CH_4 and CO_2 Emission Flux under Different Seasons

Five different plant communities were selected to monitor the carbon emissions on-site under different seasons. The emissions flux of CH_4 and CO_2 in different kinds of salt marsh plant communities was calculated. The results were shown in **Figure 3.**

The results of CH_4 and CO_2 emission flux presented distinct season diversity in the spring and autumn. In the spring, CO_2 emission flux was higher than that in the autumn, and appeared negative in HW3 and HW4 in the autumn. CH_4 emission flux of HW4 and HW5 was negative in the spring and autumn, which indicated that the CH_4 emission process was net absorption.

3.2. The Variation Characteristics of CH_4 and CO_2 Emission Flux under Different Years

The emissions flux of CH_4 and CO_2 in different kinds of salt marsh plant communities was calculated under different years. The results were shown in **Figure 4**.

From the **Figure 4**, emission fluxes of CO_2 were all positive in 2011, performance for carbon emissions. But emission flux of CH_4 was all negative in 2011, showing the net carbon absorption. Except for HW2 and HW5, the emission flux of CH_4 was contrary to that of CO_2 in 2012. The emission flux of CH_4 was contrary to that of CO_2 for HW4 and HW5 in 2013.

3.3. The Variation Characteristics of CH_4 and CO_2 Emission Flux under Different Land Use Types

CO_2 emission flux of HW3 and HW4 was opposite in the spring and autumn (**Figure 5**). The performance of HW3 and HW4 for CO_2 emission was released in the spring, and performance for carbon sequestration in the autumn. While other land use types, the CO_2 emission flux was characterized by carbon emissions.

CH_4 emission flux of HW4 and HW5 was all negative in the spring and autumn. While for other land use types, emission flux of CH_4 was characterized by net carbon emissions.

From the **Figure 6**, the results of CH_4 and CO_2 annual emission flux presented distinct space diversity under different land types. Among the 5 kinds of land use types, the HW4 discharged the largest emission flux of CO_2 reaching 29.3 $mg \cdot m^{-2} \cdot h^{-1}$. It can be concluded that the emission flux of CO_2 was increased by the human activities. The emission flux of CO_2 was distinct because of the large hydrological change of Yellow River's water level, which made the soil condition of oxidation and reduction alternately changed frequently. The order of CO_2 emission flux: HW4 > HW5 > HW1 > HW2 > HW3. Except for CO_2 emission flux of HW1 and HW3 was

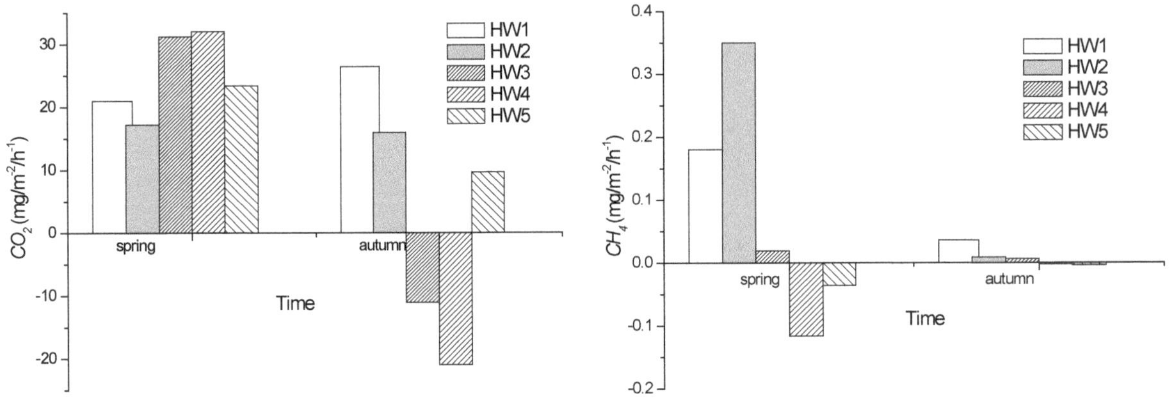

Figure 3. The variation characteristics of CH_4 and CO_2 emission flux under different seasons.

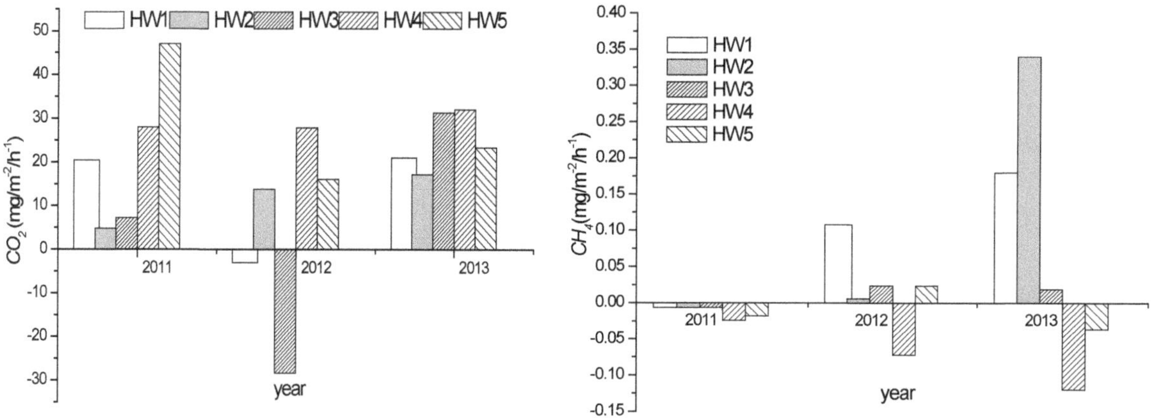

Figure 4. The variation characteristics of CH_4 and CO_2 emission flux under different years.

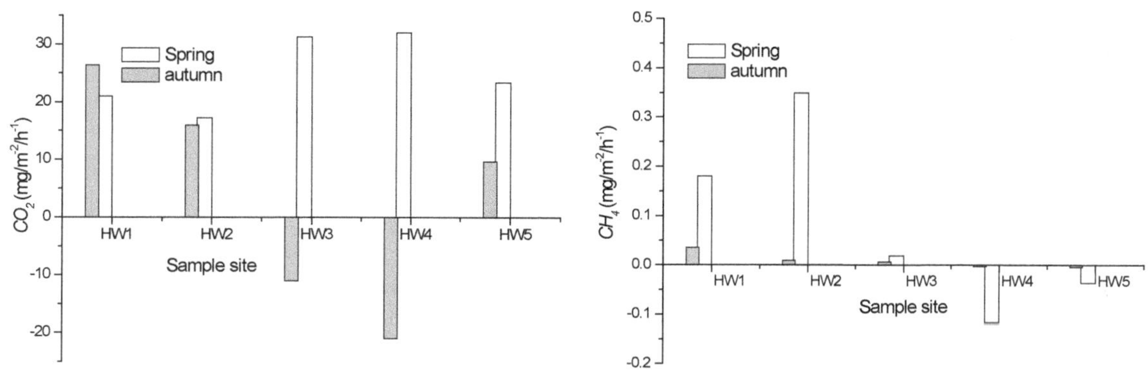

Figure 5. The seasonal variation characteristics of CH_4 and CO_2 emission flux under different land use types.

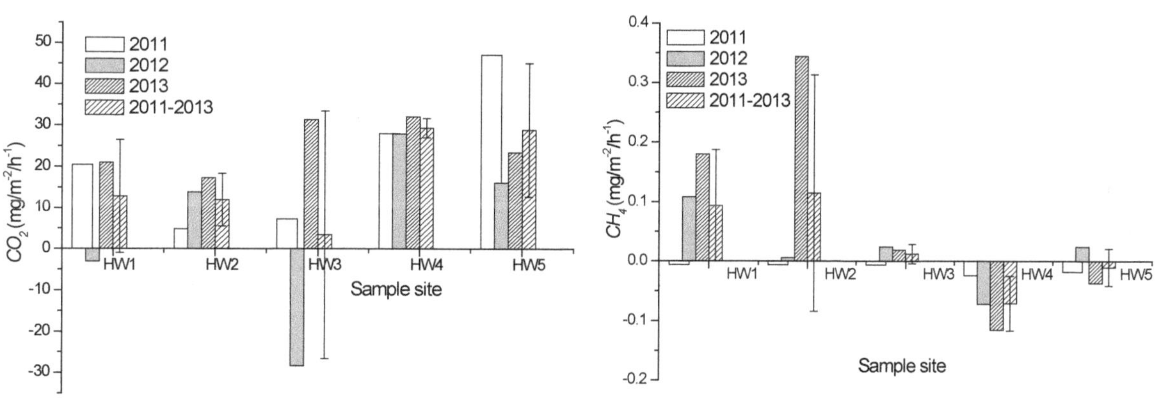

Figure 6. The annual variation characteristics of CH_4 and CO_2 emission flux under different land use types.

negative in 2012, the others were all positive.

Among the 5 kinds of land use types, the HW2 discharged the largest emission flux of CH_4, reaching 0.15 $mg\cdot m^{-2}\cdot h^{-1}$. From the estuary to the inland, the emission flux of CH_4 was increased at first and then showed decreasing trend. The order of CH_4 emission flux: HW2 > HW1 > HW3 > HW4 > HW5. CH_4 emission flux of HW4 and HW5 was negative, and showed the net carbon absorption.

4. Conclusions

The greenhouse gas emission flux, especially the CO_2 and CH_4, showed distinctly spatial and temporal variation under different land use types in the Yellow River Delta Wetland. In the spring, the emission flux of CO_2 was higher than that of CO_2 in the autumn, and appeared negative in HW3 and HW4 in the autumn. CH_4 emission flux of HW4 and HW5 was negative in the spring and autumn, which indicated that the CH_4 emission process was net absorption.

Among the 5 kinds of land use types, the HW4 discharged the largest emission flux of CO_2, reaching 29.3 $mg \cdot m^{-2} \cdot h^{-1}$, but the HW2 discharged the largest emission flux of CH_4, reaching 0.15 $mg \cdot m^{-2} \cdot h^{-1}$. From the estuary to the inland, the emission flux of CO_2 was decreased at first and then showed decreasing trend, but the emission flux of CH_4 was contrary to CO_2. Among the 5 kinds of land use types, the order of CO_2 emission flux: HW4 > HW5 > HW1 > HW2 > HW3. Except for CO_2 emission flux of HW1 and HW3 was negative in 2012, the others were all positive. The order of CH_4 emission flux: HW2 > HW1 > HW3 > HW4 > HW5. CH_4 emission flux of HW4 and HW5 was negative and showed the net carbon absorption.

Acknowledgements

This study was jointly sponsored by National Natural Science Foundation of China (No. 41003033), and Major Science and Technology Program for Water Pollution Control and Treatment (2015ZX07203-005, 2015-ZX07203-007).

References

[1] Nnoby, R. (1997) Carbon cycle: Inside the Black Box. *Nature*, **388**, 522-523. http://dx.doi.org/10.1038/41441

[2] Sahagian, D. and Melack, J. (1998) Global Wetland Distribution and Functional Characterization: Trace Gases and the hydrologic Cycle. IGBP Report 46.

[3] Han, G., Xing, Q., Yu, J., *et al.* (2014) Agricultural Reclamation Effects on Ecosystem CO_2 Exchange of a Coastal Wetland in the Yellow River Delta. *Agriculture, Ecosystems & Environment*, **196**, 187-198. http://dx.doi.org/10.1016/j.agee.2013.09.012

[4] IPCC (International Panel on Climate Change) (1996) Climate Change 1996—Impacts, Adaptations and Mitigation of Climate Change: Scientific Technical Analysis. Contribution of Working Group II to the Second Assessment Report of the IPCC. Cambridge University Press, Cambridge.

[5] Bassi, N., Kumar, M.D., Sharma, A., *et al.* (2014) Status of Wetlands in India: A Review of Extent, Ecosystem Benefits, Threats and Management Strategies. *Journal of Hydrology: Regional Studies*, **2**, 1-19. http://dx.doi.org/10.1016/j.ejrh.2014.07.001

[6] Qi, S. and Fang, L. (2007) Environmental Degradation in the Yellow River Delta, Shandong Province, China. *AMBIO: A Journal of the Human Environment*, **36**, 610-611. http://dx.doi.org/10.1579/0044-7447(2007)36[610:EDITYR]2.0.CO;2

[7] Xu, X., Lin, H. and Fu, Z. (2004) Probe into the Method of Regional Ecological Risk Assessment—A Case Study of Wetland in the Yellow River Delta in China. *Journal of Environmental Management*, **70**, 253-262. http://dx.doi.org/10.1016/j.jenvman.2003.12.001

[8] Liu, X.Z. and Qi, S.Z. (2011) Wetlands Environmental Degradation in the Yellow River Delta, Shandong Province of China. *Procedia Environmental Sciences*, **11**, 701-705. http://dx.doi.org/10.1016/j.proenv.2011.12.109

[9] Qin, Y., Yang, Z. and Yang, W. (2010) A Novel Index System for Assessing Ecological Risk under Water Stress in the Yellow River Delta Wetland. *Procedia Environmental Sciences*, **2**, 535-541. http://dx.doi.org/10.1016/j.proenv.2010.10.058

[10] Chen, Q.F., Ma, J.J., Liu, J.H., *et al.* (2013) Characteristics of Greenhouse Gas Emission in the Yellow River Delta Wetland. *International Biodeterioration & Biodegradation*, **85**, 646-651. http://dx.doi.org/10.1016/j.ibiod.2013.04.009

[11] Funk, D.W., Noel, L.E. and Freedman, A.H. (2004) Environmental Gradients, Plant Distribution, and Species Richness in Arctic Salt Marsh near Prudhoe Bay, Alaska. *Wetlands Ecology and Management*, **12**, 215-233. http://dx.doi.org/10.1023/B:WETL.0000034074.81373.65

[12] He, Q., Cui, B.S., Zhao, X.S., Fu, H.L. and Liao, X.L. (2009) Relationships between Salt Marsh Vegetation Distribution/Diversity and Soil Chemical Factors in the Yellow River Estuary. *Acta Ecologica Sinica*, **29**, 676-686.

[13] Tessier, A., Campbell, P.G.C. and Bisson, M. (1979) Sequential Extraction Procedure for the Speciation of Particulate Trace Metals. *Analytical Chemistry*, **51**, 844-851. http://dx.doi.org/10.1021/ac50043a017

[14] APHA (2005) Standard Methods for the Examination of Water and Wastewater. 21st Edition, American Public Health Association, Washington DC.

[15] Li, M. and Li, W. (2009) Review on Carbon Cycle of Wetland Ecosystem. *Huazhong Agri. Univ.*, **28**, 116-123.

17

Fueling Water-Intensive Economic Growth: What Hope for Water Conservation?

Ola Busari, Jeremiah Mutamba

Trans-Caledon Tunnel Authority, Centurion, South Africa
Email: obusari@tcta.co.za, jmutamba@tcta.co.za

Abstract

In virtually every country has strategies for pursuing greater broad-based economic growth and fighting poverty and inequality that underscore the pivotal role of water, including its availability in reasonable quantity and desired quality. While GDP growth in South Africa slowed down from 2.2% in 2013 to 1.5% in 2014, and while there are suggestions that current growth drivers are weak and inflation drivers strong, hopes remain for moderate economic growth of upwards of 2.5% going forward. Aside from stabilizing power supply and securing an upward movement in consumer confidence, such optimism rests on the return of mining and manufacturing activities to their previously high levels, with implications for water requirements. Growing demand for water-intensive growth will be occurring side-by-side with the expanding needs of urbanization and social advancement, in the face of an increasing threat of climate change, recurring droughts, environmental pollution and limited accessible water resources. Effective reconciliation of the demand and supply of water would require the strengthening of water conservation and demand management beyond their currently low to moderate levels, calling for a paradigmatic shift in approaches to water management, sound appreciation of the potential benefits vis-à-vis the allocation of requisite resources and firm political leadership and support.

Keywords

Demand Management, Economic Development, South Africa, Water Conservation

1. Introduction

From its currently suppressed GDP growth of 1.5%, South Africa projects to progress into renewed periods of economic performance in the short to medium term. In the frame of the country's long-term development perspective, the 2030 National Development Plan [1], efforts are being intensified to explore alternative paths for inclusive growth and broad-based participation in the economy. Such planning recognizes that water is pivotal to growing the economy, building strong communities and ensuring the well-being of citizens, considering its multi-faceted use in driving agriculture, energy, mining, tourism, urban growth and rural development.

In particular, South Africa's National Infrastructure Plan [2] makes the bold move to prioritize significant resources for the expansion of critical economic and social infrastructure, including that for augmenting water resources to address the strategic requirements of the industrial sector and the troubling electricity sector, as well as for developments in mining and agriculture.

The energy sector contributes about 15% to the country's GDP, and supporting the economic growth targets of the country is linked to closing the gaps in energy and fuel requirements, with the operation of additional plants dependent on huge quantities of water. Mining, a critical destination of the largest foreign investment flows and accounting for about 9% of GDP, promises to continue to significantly contribute to the economy. But its growing water requirements occur simultaneously with the increasing need to ensure resource protection from pollution from mining activities, and to deal with the complex environmental threat presented by acid mine drainage. Agriculture, while witnessing growing competition for water from industrial and domestic uses, still accounts for the largest water use, receiving nearly 60% of the country's water resources. Over the full decade leading to 2013, domestic water use only increased from 22% to 27% of the total available water resource [3]. Although the agricultural sector contributes a relatively low 3% to GDP, the sector provides employment and livelihoods for millions of South African households, especially in rural communities.

Against the backdrop of strategic water infrastructure initiatives, the country is water-scarce and already faces several challenges as a result of the limited and variable nature of its water resources, including the impacts of climate change and environmental pollution [4]. The situation of water scarcity is exacerbated by the geographic mismatch between its relative availability in certain catchments and the centres of strategic industrial activities with the potential to propel economic growth. But also, as [5] argues, the water supply system loses large amounts of water, worsening the pressure on the resource and inflicting supply challenges. Indeed, [6] tracked the level of non-revenue water for urban supply systems across the country during a six-year period, indicating an average of 37%, but pointing to figures as high as 90% in some municipal water supply schemes.

In a bid to augment and improve water supplies to meet the anticipated demand for socio-economic advancement, water resources planners and managers will more easily opt to continue to turn to supply-driven approaches. Experiences elsewhere, for example [7], suggest that, in several situations, the traditional solution to water problems has been to quickly increase water availability through developing additional sources of water. However, [8] finds that the continued application of traditional water management concepts in contemporary times will not deliver the required results, especially as such approaches would not meet the combined tests of efficiency and cost-effectiveness.

Along similar lines, South Africa's latest National Water Resource Strategy recognizes pointedly that efforts to increase water availability must move beyond conventional engineering approaches to infrastructure development, to emphasize a better mix of reuse of groundwater, rainwater harvesting, desalination and treated acid mine drainage [3]. Also, the genuine pursuit of a sustainable water balance is now strongly acknowledged as requiring renewed and greater attention to water conservation and demand management.

First, this paper provides a global overview of the concept and principles of water conservation and demand management, tracking their application in different settings and summarizing the risks and benefits of efforts to improve application. Second, a review of the experiences in South Africa is presented, including the policy framework and level of actual implementation, so as to identify key constraints and critical success factors, in view of enhancing the hope for better influencing demand and conserving water into the future.

2. Water Conservation and Demand Management: An Overview

2.1. Core Principles

The principles governing water conservation and demand management (WCDM) are well researched and documented [6] [9] [10]. The principles are predicated on the notion of improving water use efficiency and promoting sustainable use, and their effective application is meant to ensure optimum, long-term, environmentally sustainable social and economic benefit for society [11]. Also, [11] identifies three fundamental principles that underpin WCDM, as follows:

- Water institutions should supply water in an efficient and effective manner by minimizing water losses.
- Consumers should endeavour to use water efficiently.
- WCDM should be considered as part of the planning process for water resources, water supply and sustainable water services.

The foregoing principles emphasize a focus on the reduction of the quantity of water required to accomplish a specific task; adjusting the nature of the task so that it can be accomplished with less water or lower quality water; reduction of losses in movement from source, through use, to disposal; shifting time of use to off-peak periods; and increasing the ability of the system to operate during droughts. It is important to note that WCDM principles involve the active participation of all parties in the water value-chain, from the water institutions mandated to manage water resources right down to the end user. Also, as [12] advise, for the principles to make positive impact, there is need for strong support from sound policies and governance.

Universally, conserving water and managing demand are motivated by a number of inter-linked factors. Some of the important drivers are the general lack of new water resources to develop; a growing demand for water supplies to feed economic development, industrial expansion and rising population; diminishing water resources due to continued pollution of the limited resource; and unacceptably high water losses experienced in various uses, including agriculture, communities and municipal centres. Other contributing factors include rapid urbanization and increasing over-exploitation of the groundwater resource [8].

2.2. Key Measures

Efforts to conserve water and manage demand at various points in the water value-chain commonly adopt a number of measures. For ease of understanding and application, water practitioners frequently cluster those WCDM measures into certain categories. The four groups summarized by [13] are economic instruments, structural methods, operational procedures and socio-political interventions.

The economic instruments include the use of appropriate pricing policies, block tariff structures, long run marginal costing, cost-benefit analysis techniques and penalties. These approaches are normally deployed to ensure marginal water use is given its real marginal value. Block tariffs are usually applied to ensure that those who use above-average amounts of water pay for the privilege, with the approach of increasing block tariffs a common feature in several countries, including South Africa. The increasing block tariffs achieve two objectives: that of dis-incentivizing high levels of water consumption and that of cross-subsidizing access to minimum levels of use. Economic approaches need to be applied with caution, as they are considered to be the least equitable.

Structural methods emphasize the fitting of on-site pressure reduction devices (retrofitting), the installation of efficient irrigation systems, the adoption of recycling and reclamation systems, segregation of wastewater flows into greywater and sewage, and the use of water efficient devices. Operational procedures focus particularly on pressure reduction measures, leak detection and repairs. Socio-political interventions revolve around the establishment of supportive laws and regulations, including code modifications, as well as the mounting of public awareness programmes on water conservation.

Water utilities are encouraged to use the outlined WCDM measures in combination, as it has been shown that such combination results in the achievement of maximum utility [14]. An example of a possible matrix of generic interventions at different points along the water supply value-chain is presented in **Figure 1**.

2.3. Application of the Principles

Globally, there is a heightened awareness among organizations to improve their water use efficiency and reduce water footprint, as part of sustainability strategies. This consciousness is growing at all levels, regional and national, as well as among manufacturing companies, mining firms and municipal entities. The region of Southern Africa has witnessed the same trend, with governments and organizations demonstrating progress in embracing the principles of WCDM. For example, Zimbabwe and South Africa since introduced new policies and legislation underscoring demand-side management and improved water conservation. Some of the ongoing interventions in Botswana include water tariff reviews, water restrictions and education campaigns seeking to manage water use and water demand. Namibia, in particular, has made significant strides in applying a range of WCDM measures.

As the driest country in sub-Saharan Africa, it is understandable that Namibia would demonstrate the early promotion of an integrated approach to water resource management, combining both supply-oriented and demand-side efficiency measures, especially in its capital city, Windhoek. Windhoek uses treated effluent that applies a dual pipe system for municipal irrigation; water reclamation for potable use was introduced in 1968, with a new plant commissioned in 2002, accounting for 26% of daily potable supply [15] [16]. In 1994, the city approved an integrated water demand management programme consisting of policy improvements, block tariffs, retrofitting and awareness campaigns, seeking to stimulate efficient water use. The success of integrated WCDM measures resulted in a zero overall growth in water consumption over the period 1990-1997, despite an increase of 35% in the city's population [15].

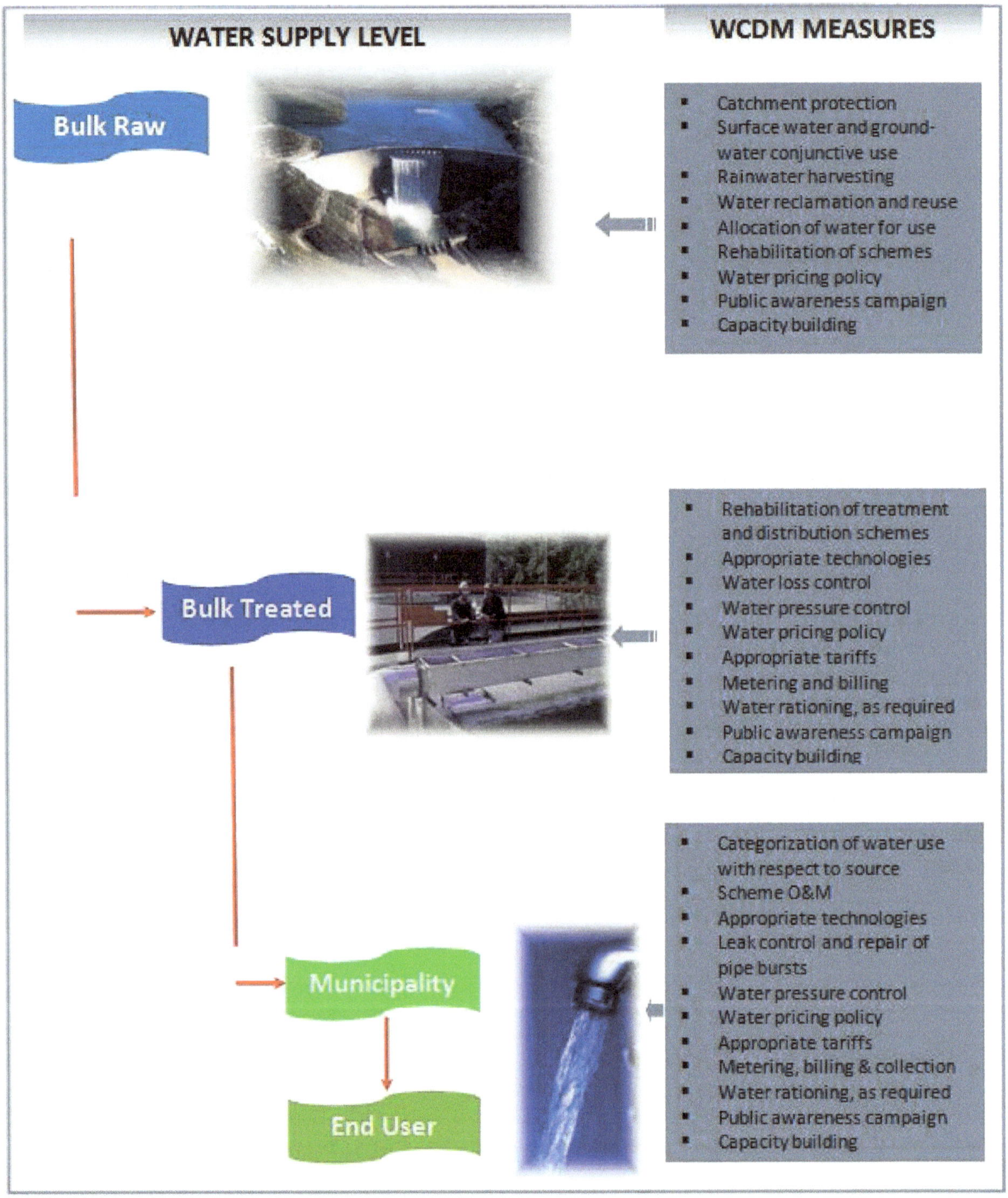

Figure 1. Combination of WCDM measures along the water value-chain.

Across the globe, several cities are known for their innovative programmes on water conservation and demand management. For example, the city of Seattle, USA is reported to have reduced its water consumption by 1% each year over 23 years, despite an aggregate increase in population of more than 20% [17]. Also, the city of Athens, Greece aggressively introduced water demand management measures in the 1990s after suffering from severe drought: the mix of strategies implemented included carrying out information campaigns, banning the use of drinking water for other purposes, and adjusting the price of water. Reference [18] reported that, based on the level of water consumption in 1989, Athen's initiatives resulted in a 25% decrease in consumption in 1993 and 35% reduction in 1995. In Australia, [19] documents the efforts of the Sydney Water Corporation to manage demand as centering on expanded water reuse and rainwater harvesting, as well as retrofitting, pricing reform and awareness campaign.

There are also many applications of WCDM measures at the industrial level. Reference [20] captures the interventions of a German fruit packaging company, Dole Europe, focusing on four areas of water efficiency: maximization of rainwater use, implementation of advanced irrigation techniques, optimization of the water used during the packing process, and special attention to the quality of return water. The results suggest that Dole managed to reduce its water use per unit of production. For example, it is reported that Dole used around 100 litres of water to produce one kilogram of Dole pineapples, and about 250 litres for a kilogram of Dole bananas. Reference [20] juxtaposes its water use with a global average of around 1600 litres of water used for the production of one kilogram of plums, 3000 litres of water for one kilogram of rice and about 15,000 litres of water for one kilogram of beef.

Examples of businesses aggressively looking for opportunities to improve their water use efficiency through WCDM measures are commonplace, with some launching ambitious global water conservation programmes and many other setting targets of reducing water use in corporate buildings through the installation of low-flow toilet systems, urinals and sinks. It is now common knowledge that conserving and saving water is critical for continued global business development and sustainability.

2.4. Risks and Benefits

Several advantages and benefits can be derived from the dedicated and successful application of the principles underpinning water conservation and demand management. Reference [15] identifies a number of such benefits, as follows:

- More efficient use of existing water supplies.
- Reduction in water leakage and promotion of the equitable use of water.
- Deferment of additional spending on water and wastewater infrastructure.
- Cost savings to water users, as a result of paying only for water received rather than for water lost in the distribution system, as well as cost savings for bulk water suppliers due to a reduction in the amount of water that would be purchased.
- Promotion of environmental quality improvements and integrity through the lowering of water losses and unwarranted development of large infrastructure.
- Energy savings through the deferment of huge pumping facilities associated with large water transfer schemes.
- Reduction in supply disruptions to water users, arising from frequent bursts and repairs.
- Improvements in water quality through the lessening of pollution of water resources.
- Increased revenues from previously undercharged customers and formerly illegal and previously un-paying customers.
- Better management of future water uses and certainty in water demand forecasts.
- Lowering of the financial risk of miscalculating demand far into the future.
- Water saving technologies are not usually capital-intensive, thus they can be implemented at affordable cost.
- Opportunity to build water management databases and to develop water balances, as appropriate implementation requires the measurement of all components of the water cycle.
- Expanded knowledge of the water distribution system, as well as of the water system in general.
- High benefit-to-cost ratios in excess of 10:1 in several urban situations.

In spite of the numerous advantages, carrying out demand-side management and conservation faces certain risks and challenges that water practitioners should take cognizance of and plan mitigation measures for. In particular, once significant efforts have been expended implementing the applicable measures, it may be difficult to sustain the levels of water loss reduction achieved. It needs to be understood that WCDM is not a once-off initiative, but rather an ongoing exercise that should form part of the water management process for water institutions. In order to sustain the achieved levels of gains, it is important to continuously dedicate a competent team and appropriate budget to the program.

Further, WCDM should be considered as a participatory process requiring the interwoven and complex involvement of an array of stakeholder groups. At times, the stakeholders are not willing or available to participate in the process-which jeopardizes the chances of success. There is also a risk of implementing sub-optimal strategies, a frequent occurrence where limited budgets are made available to tackle huge challenges.

3. The South African Setting

3.1. Legal and Policy Framework

Water management in South Africa is regulated primarily through two pieces of legislation – the National Water Act of 1998 and the Water Services Act of 1997. Both Acts prescribe several requirements and provisions for the implementation of strategies for water conservation and demand management. Importantly, the Constitution of South Africa affirms the right of all citizens to a healthy living environment, including access to adequate and affordable water services. In this vein, government introduced a Free Basic Water policy in 2000, to facilitate sustainable access to a basic water supply by subsidizing the ongoing operation and maintenance costs of a basic service.

Direct responsibility for the provision of water services was vested in local government, with anticipated backstopping by provincial and national government. In response to new municipal legislation, the Strategic Framework for Water Services was approved by Cabinet in 2003, updating national policy on water services and, notably, effectively revising the provisions of the 1997 Water Services Act [21].

The Water Services Act requires water services institutions to establish modalities for the supply of water that prioritize the promotion of demand-side management and conservation. Also, the Act makes provision for the Minister of Water and Sanitation to prescribe compulsory national standards relating to the effective and sustainable use of water resources. On its part, the Strategic Framework for Water Services points to the scarcity of water resources in South Africa and underscores that due attention is paid to water conservation and demand management. In particular, the framework requires water services institutions to develop appropriate water conservation and demand management strategies, which should be reflected in their water services development plans.

The National Water Act provides for the development of a National Water Resources Strategy, first published in 2004 and revised in 2013, as a comprehensive tool for managing water resources throughout the country. More specifically, South Africa also adopted a National Water Conservation and Demand Management Strategy, emphasizing the critical elements and requirements for effective implementation across several sectors, including industry, mining and power generation, agriculture, forestry and water services.

3.2. Implementation Initiatives

Until the mid-1990s, the approach to water resources planning and management in South Africa was predominantly traditional, focusing on the supply-side assessment and development of the resource [15], with extensive investment in medium to large-scale storage dams and water transfer infrastructure feeding growth regions experiencing water deficits. As part of the broad sector reform of the post-1994 democratic dispensation, the new government launched a National Water Conservation Campaign, signifying a shift towards more holistic approaches to water management.

The national initiative was designed and driven against the background of the international perspective that efforts to conserve water and manage demand were more likely to succeed when the measures introduced were multi-pronged and simultaneous, incorporating technological, economic and socio-political dimensions. Several projects, 56 in all, were identified for implementation across all major water use sectors, spanning agricultural, municipal, industrial, mining, power generation and environmental. The scale of the program left little room for prioritization and effective coordination, affecting the achievement of meaningful results. Commitment to the roadmap already charted in the 1994 national conservation campaign was reinforced with the reform of the policy and legislative setting for the fast-tracking of water services and the broader management of water resources, as described in the foregoing section.

Yet, as [22] argues, in spite of the mainstreaming of measures to conserve water and manage demand into national policies and regulations, actual implementation progress remains inadequate. There is agreement that South Africa still loses plenty of water through theft, leakages and frequent pipe bursts as a result of ageing infrastructure, and a recent study by [6] documents an average proportion of non-revenue water of 37%, reaching a very high 90% in some municipal areas. Reference [6] particularly reports that, notwithstanding the setting of clear targets to reduce water loss by 50% by 2014, the carry-through of the appropriate interventions at the municipal level has been rather limited. Reference [10] believes that the insufficient attention to implementation in municipal water supplies may be related to lack of appreciation of the implications and potential benefits of WCDM.

3.3. Constraints to Implementation

It is worthy of note that the foregoing trend of muted implementation of demand-side management and conservation is not unique to South Africa: Several workers report that, while the concept and principles are well understood and embraced, limited application in practice is widespread [12] [23]-[26]. The main constraints to wider application, beyond the progress in policy reform, include the following:

- Political support and leadership: In juxtaposition with the more visible development of new water supply sources, such as the construction of dams and large-scale transfer schemes, demand-side programs and conservation initiatives are not very popular with political heavyweights, as they are generally viewed as *software* and, therefore, low-impact, interventions.
- Attitude and awareness: There is a low awareness of the value of water, partly because of the perception of its natural availability and/or accessibility in some form for some use, and partly due to the consideration of water as a public good, with government-subsidized supply free of charge to rural households and indigent residents in urban centres. In several places, the interest of stakeholders in actively participating in mounted initiatives is weak, and encouraging willingness by stakeholders is often a slow process.
- Institutional capacity: Implementing agents frequently point to weaknesses in their capacity to roll-out well-designed programs, including the requisite technical and managerial skills, as well as financial resources, for sustaining and tracking the interventions.

Moving water conservation and demand management into greater adoption in programmatic improvements, from its current broad-based theoretical appeal, would require that more concerted efforts be devoted to tackling the aforementioned impediments.

4. Conclusions

Now and into the future, managing water smartly is critical to sustainably fueling economic growth and driving social development across South Africa. Conventional supply-oriented management approaches have proved inadequate for addressing the yawning disparity between available resources and the expanding, competing water requirements. While the need for demand-side management and conservation is well understood and widely embraced with fruitful application growing in some countries and industries, the level of implementation in several developing countries, including South Africa, remains low to moderate.

Critical elements required for the successful implementation of water conservation and demand management are two-fold. The first relates to the need for the high-level political buy-in and stewardship that would lead to an effective paradigmatic shift for taking initiatives into implementation, as well as, ultimately, for going to scale. The second calls for the allocation of sufficient resources on an ongoing basis, including born-again water managers with the capacity to ensure that planning is not only integrated but inclusive, and that it is followed by careful implementation.

References

[1] National Planning Commission (2013) National Development Plan 2030: Our Future, Make it Work. National Planning Commission, Government of South Africa, Pretoria.

[2] PICC (2012) A Summary of the South African National Infrastructure Plan. Presidential Infrastructure Coordinating Commission, Government of South Africa, Pretoria.

[3] DWA (2013) National Water Resource Strategy: Water for an Equitable and Sustainable Future. Department of Water Affairs, Government of South Africa, Pretoria.

[4] Busari, O. and Ndlovu, J. (2012) Leveraging Water Infrastructure for Transformative Socio-Economic Development in South Africa. *WIT Transactions on Ecology and the Environment*, **162**, 435-446. http://dx.doi.org/10.2495/EID120381

[5] Mutamba, J. and Busari, O. (2014) Strategic Coordination for Sustainable Investment in Critical Infrastructure. *Journal of Geoscience and Environment Protection*, **2**, 79-86. http://dx.doi.org/10.4236/gep.2014.25011

[6] McKenzie, R.S., Siqalaba, Z.N. and Wegelin, W.A. (2012) The State of Non-Revenue Water in South Africa. Research Report WRC-TT-522/12, Water Research Commission, Pretoria.

[7] Gumbo, B. and van der Zaag, P. (2002) Water Losses and Political Constraints to Demand Management: The Case of the City of Mutare. *Physics and Chemistry of the Earth*, **27**, 805-813. http://dx.doi.org/10.1016/S1474-7065(02)00069-4

[8] Kayaga, S. and Smout, I. (2008) Water Demand Management: A Key Building Block for Integrated Resource Planning for the City of the Future. *Annual Conference of Allied Social Sciences Association*, New Orleans, 4-6 January.

[9] McKenzie, R.S. and Bhagwan, J.N. (2000) Managing Unaccounted for Water in Potable Water Distribution Systems: Recent Software Developments through the WRC. *Journal of the Institution of Municipal Engineers*, **January**, 53-58.

[10] Wegelin, W.A. and Jacobs, H.E. (2013) The Development of a Municipal Water Conservation and Demand Management Strategy and Business Plan as required by the Water Services Act, South Africa. *Water SA*, **39**, 415-422. http://dx.doi.org/10.4314/wsa.v39i3.11

[11] DWAF (2004) National Water Conservation and Demand Management. Department of Water Affairs and Forestry. Government of South Africa, Pretoria.

[12] Mohapatra, S.P. and Mitchell, A. (2009) Groundwater Demand Management in the Great Lakes Basin—Directions for New Policies. *Water Resources Management*, **23**, 457-475. http://dx.doi.org/10.1007/s11269-008-9283-3

[13] UNCHS (1989) The Conservation of Drinking Water Supplies: Techniques for Low-income Settlements. United Nations Centre for Human Settlements, Habitat, Nairobi.

[14] Pansegrouw, J.P. (2005) Water Demand Management—Current Water Productivity Methodology and Water Management Tool in South Africa. Conference Series Papers, International Water Management Institute.

[15] Herbertson, P.W. and Tate, E.L. (2001) Tools for Water Use and Demand Management in South Africa. Technical Reports in Hydrology and Water Resources #73, World Meteorological Organization, Geneva.

[16] Lahnsteiner, J. and Lempert, G. (2007) Water Management in Windhoek, Namibia. *Water Science & Technology*, **55**, 441-448. http://dx.doi.org/10.2166/wst.2007.022

[17] US-EPA (2002) Cases in Water Conservation: How Efficiency Programs Help Water Utilities Save Water and Avoid Costs. United States Environmental protection Agency, Washington DC.

[18] Kampragou, E., Lekkasb, D.F. and Assimacopoulos, D (2009) Water Demand Management: Implementation Principles and Indicative Success Stories. http://www.samos.aegean.gr/actuar/dlekkas/cv/

[19] White, S. (2001) Demand Management and Integrated Resource Planning in Australia. Paper Presented at the Conference on Efficient Use and Management of Water for Urban Supply, Madrid.

[20] Dole (2011) Corporate Responsibility and Sustainability. *Dole Food Company Newsletter*, **1**.

[21] Busari, O. and Jackson, B. (2006) Reinforcing Water and Sanitation Sector Reform in South Africa. *Water Policy*, **8**, 303-312. http://dx.doi.org/10.2166/wp.2006.038

[22] Mwiinga, G., Gumbo, T. and Mkoka, I. (2008) Water Conservation and Demand Management in Practice: South Africa Case Studies. Paper Presented at the WISA Conference, Sun City.

[23] Stiles, G. (1996) Demand-Side Management, Conservation and Efficiency in Use of Africa's Water Resources. In: Rached, E., Rathgeber, E. and Brooks, D., Eds., *Water Management in Africa and the Middle East: Challenges and Opportunities*, IDRC Books, Ottawa.

[24] Arlosoroff, S. (1998) Water Demand Management. *UNCHS (Habitat) Regional Conference on Sustainable Consumption Patterns in Asian Cities*, Fukuoka, 29 June-1 July 1998. http://www.fukuoka.unchs.org/english/information/Occassional/water-e.html

[25] Mwendera, E.J., Hazelton, D., Nkhuwa, D., Robinson, P., Tjijenda, K. and Chavula, G. (2004) Overcoming Constraints to the Implementation of Water Demand Management in Southern Africa. *Physics and Chemistry of the Earth*, **28**, 761-778. http://dx.doi.org/10.1016/j.pce.2003.08.002

[26] Mutamba, J. and Busari, O. (2012) Municipal Lessons in Water Demand Management: Case Study from South Africa. *Proceedings of* 34*th Hydrology and Water Resources Symposium*, Sydney, 1385-1392.

18

Clay Minerals Channels Identification in the Tindikala-Boutou Area (Eastern-Cameroon) along the Kadey River Using Direct Current (DC) Method

Daniel Hervé Gouet[1*], Arsène Meying[2], Stéphane Patrick Assembe[3], Théophile Ndougsa-Mbarga[4]

[1]Department of Petroleum and Gas Explorations, Institute of Mines and Petroleum Industries, University of Maroua, Maroua, Cameroun
[2]Department of Applied Geophysics, Geology and Mining Exploitation College, University of Ngaoundéré, Ngaoundéré, Cameroun
[3]Department of Physics, Faculty of Science, University of Yaounde I, Yaoundé, Cameroon
[4]Department of Physics, Advanced Teacher's Training College, University of Yaounde I, Yaoundé, Cameroon
Email: [*]gouetdanyl@yahoo.fr

Abstract

To achieve the current study, geoelectrical surveys along six (06) profiles of 4 km long in a 100 m × 200 m grid defined according to the triangulation principle in the Tindikala-Boutou (Eastern-Cameroon) area along the Kadey River have been made through electrical sounding and profiling following Schlumberger array. The instrument is the resistive meter Syscal Junior 48 (IRIS Instrument). The data have been processed and modelled with Res2Dinv and Winsev softwares, and then interpolated with Surfer software. Investigation method used is the Direct Current (DC) method. Interpretations and analyses of results from the investigation method highlight weak zones or conductive discontinuities. The latter has been identified as shear zones within granitic structures of the Precambrian basement, according to the geologic and tectonic background of the area. The structural trend of these shear zones is E-W approximately. The mineralization characterized by conductive zones proves the presence of clay minerals disseminated in weathered quartz vein, which cross the shear zones. The intense activities of gold washers encountered in the studied area are able to attest the presence of clay minerals concentrations.

[*]Corresponding author.

Keywords

Geoelectrical Surveys, Direct Current Method, Resistivity, Shear Zones, Clay Minerals

1. Introduction

The reliability of any geophysical exploration result is supported by the geological background assessment of a given area, no matter its purpose. In accordance with this, a geoelectrical investigation has been made in the Tindikala-Boutou village of the Ngoura subdivision (Eastern-Cameroon). The approach consists in collecting apparent resistivity data using the Direct Current Schlumberger's geoelectrical investigation (sounding and profiling). Data will be processed with Res2Dinv, Winsev and Surfer softwares. Results, given as pseudo sections of resistivity, resistivity maps will be able to plot tectonic unevenness of the area under study and also to characterize the associated mineralization. The geological sections realized through the interpretation of electrical soundings will bring out the correlation between the resistivity data and the geological background of the studied area.

2. Geological and Tectonic Setting

The study area (Tindikala-Boutou village) is located in the Ngoura subdivision in the heart of the East Cameroon's region. Its easting stretches from 426,000 m to 437,800 m and the northing from 548,750 m to 557,750 m following UTM33 WGS84 system (**Figure 1** and **Figure 2**).

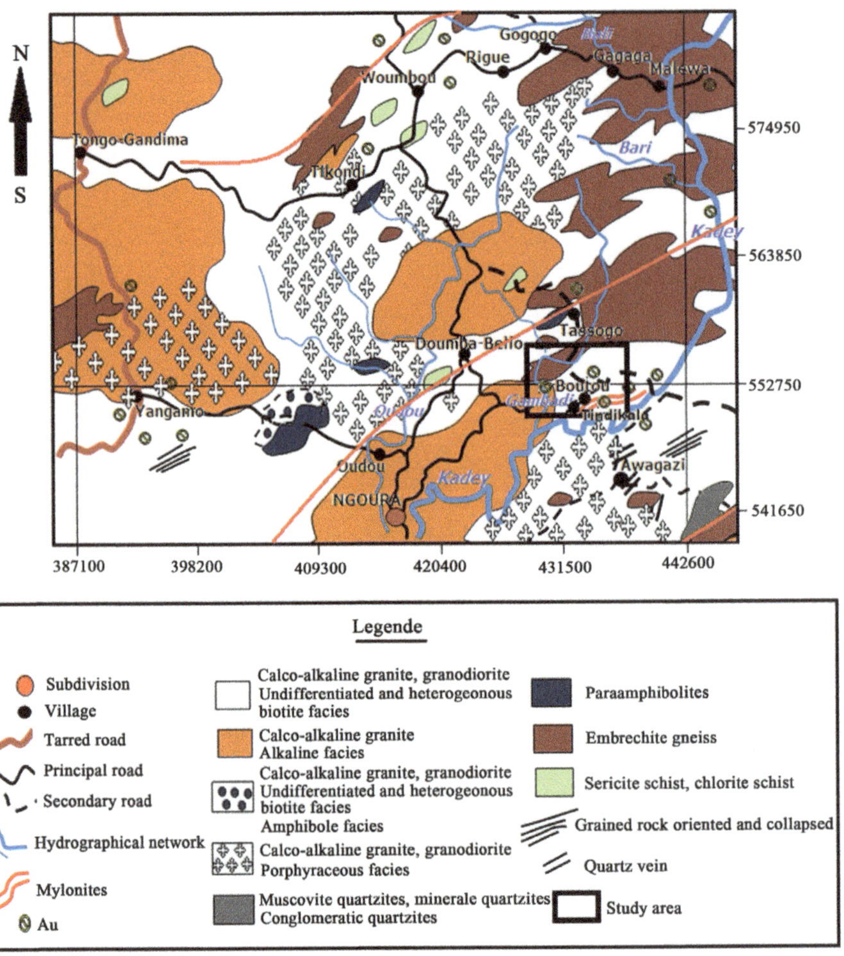

Figure 1. Geological map of the study area.

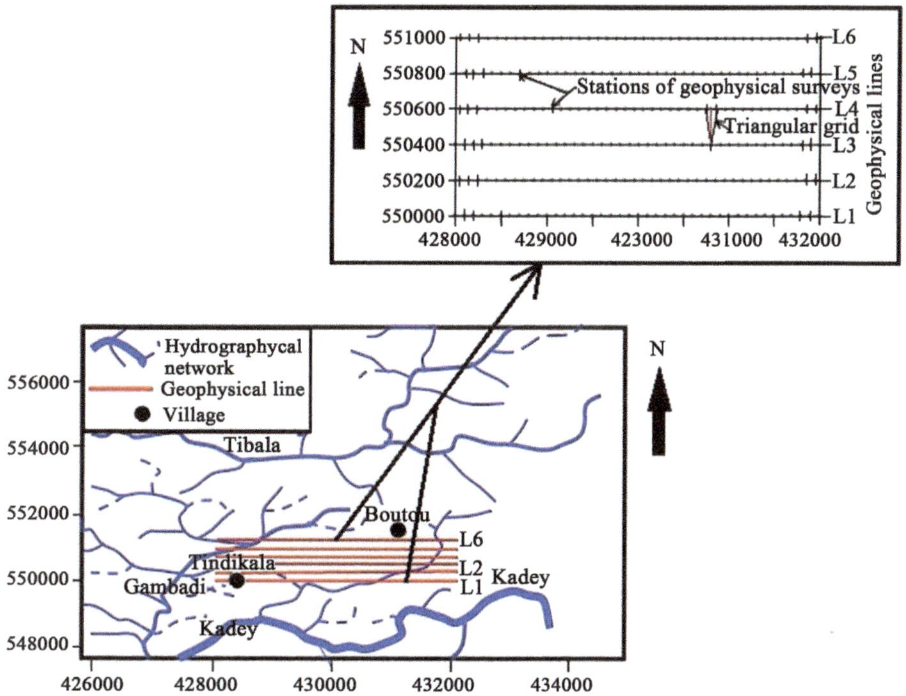

Figure 2. Profiles localization map in the study area.

The Ngoura area is located in the transition zone between the Pan African domain and the Congo Craton [1] [2], in the northern edge of the Cameroon faults' zone. This region of Cameroon is essentially made up of a Precambrian basement comprising metamorphic and magmatic rocks [3]-[5]. These are Precambrian rocks which sometimes outcrop. These are mainly granites and migmatites rejuvenated during the panafrican event. The area is mainly made up of (**Figure 1**) [6] [7]:

- Biotite and muscovite quartzites, sericitic muscovite and conglomeratic quartzites, chloritic and sericitic schists, paraamphibolites, orthogneisses, biotite gneisses and, pegmatites and quartz veins constituting the precambrian basal complex aged between 2.5 and 1.8 billion years.
- Plutonic and metamorphic formations made up of calc-alkaline granites with a porphyroic and alkali facies; granodiorites and syenites containing heterogeneous and undifferentiated biotite facies; quartz-diorites; micaschists, migmatites and embrechite gneisses.
- The main sedimentary rocks encountered are sandstones, sand, marl, limestones and, Paleozoic and Mesozoic conglomerates.

The tectonic facts revealed that, the study area is characterized by four deformations phases D1-D4 [8]. The observed tectonic lines are directed SW-NE below, and turned to be SE-NW above the 4 N parallel (**Figure 1**). Previous geophysical studies [9] [10] have shown evidence of some buried faults directed W-E and have confirmed tectonic nappes with a southern vergency. According to Olinga *et al.* (2010), the Pan-African deformation affecting the study area which appertains to the southern segment of the Neoproterozoic fold belt of Central Africa in Cameroon, is controlled by thrust tectonics and late strike-slip shear zones: the thrusting of the Pan-African Nappe over the Congo Craton (D2 deformation phase) is followed by a strike-slip shearing trending ENE-WSW (D3 deformation phase). During these stages deforming conditions were ductile to brittle-ductile. The dominant structural features of the D3 phase are penetrative foliation steeply dipping N or S, an associate ENE-WSW stretching lineation, and an N-S to NE-SW folding. Deformation criteria in the distinguished rock units indicate dextral sense of shear. A dextral trans-pressional model is assumed by Olinga *et al.* (2010) to explain the observed thrust and shear movements.

3. Method

The resistivity is a suitable parameter in characterizing the nature and the weathering of materials [11]. Hence electrical prospecting methods have been used for a long time in geological and geotechnical engineering. These both qualitative and quantitative methods are based on the Ohm law [11]. They consist in the injection of a direct current in the ground and the measurement of the electrical potential which enables to obtain the true resistivity of encountered formations. In our study area, while considering the geological setting, Schlumberger sounding and profiling methods have been used to determine: the thickness, the lateral extension and the nature of formations encountered along a profile; and to highlight the geometry of geological bodies related to contrasted electrical characteristics [12].

Field electrical methods (D.C.) consist in injecting an electrical current in the ground between two electrodes A and B (**Figure 3**), and then, measuring the induced potential drop between two so-called potential electrodes M and N [12]. For the current intensity is known and the potential drop measured, it is therefore possible to determine the ground apparent resistivity. This apparent resistivity depends on the current and potential electrodes array. The apparent resistivity ρ_a (Rho) can be expressed function of the potential drop and the current intensity [13]:

$$\rho_a = K\left(V_{MN}/I_{AB}\right) \tag{1}$$

where ρ_a is in Ohm·m, and K (in m) is the geometric factor depending on the electrodes' array which is given by the Formulae (2) below.

$$K = 2\pi\left(AM^{-1} - AN^{-1} - BM^{-1} + BN^{-1}\right) \tag{2}$$

-V_{MN}: potential drop between electrodes M and N, in mV;

-I_{AB}: electric current injected between electrodes A and B, in mA.

For the Schlumberger symmetrical configuration (**Figure 3**), the apparent resistivity ρ_a is given by Equation (3) as follow [14]:

$$\rho_a = \pi/4\left(AB^2 V_{MN}/MN \cdot I_{AB}\right) \tag{3}$$

This resistivity value enables to characterize a formation in the point (o) or station (**Figure 3**). The resistivity of an earth's material depends essentially on the humidity and the clay proportion in a given volume of that material [15]. While clay and water fill in any vacuum in a rock, one assumes that the resistivity is function of parameters such as fracturing, fractures and fissures clay filling in, porosity, the clayey clogging of alluvium [11] [15].

4. Material and Data Acquisition

For a suitable coverage of the study area, data have been collected through 236 VES (vertical electrical sounding) along six (06) profiles of 4 km long in a 100 m × 200 m grid, defined according to the triangulation principle (**Figure 2**). This grid has been designed according to the topographic data of the area whose coordinates have been expressed in UTM33 WGS84. The acquisition data is made by combining electrical sounding and profiling following the Schlumberger configuration along a profile (**Figure 3**). The AB maximum length was held at 600 m to detect desired lithological formations at an approximate depth between 114 and 130 meters [16]. To avoid miscellaneous due to formations' anisotropies, sounding and profiling surveys were E-W oriented (**Figure 2**).

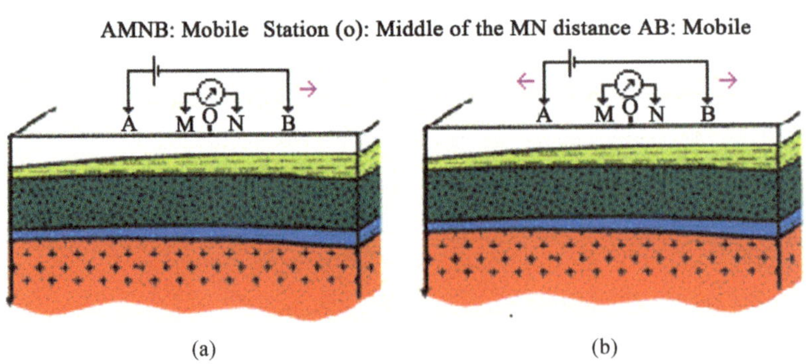

Figure 3. (a) Schlumberger electrical profiling; (b) Schlumberger electrical sounding.

We used the direct current resistivimeter, Syscal Junior 48 (IRIS Instrument) system. This unit runs under the Rho mode which enables to measure the resistivity (Rho) of ground structures. Acquisition of electrical data (sounding and profiling) is made using the Schlumberger direct method where the current is injected through A and B electrodes (**Figure 3**) and the potential difference is measured through receiving electrodes M and N [17].

The apparent resistivity data from Schlumberger electrical profiles were processed and modelled using the Geotomo Res2Dinv software [14] [18] to obtain pseudosections or inverse pseudosections that reflect the true resistivity values of local subsurface structures. The electrical soundings were interpreted using WinSev from Geosoft [19] which permits to obtain the depth distribution of layers at each station, hence enabling to plot geological sections. Resistivity maps were plotted using Surfer software [20] which determines the spatial distribution of resistivity by interpolating their values in the area under study.

5. Results

5.1. Pseudo-Sections

Electrical sections of profiles L1 to L6 (**Figures 4(a)-(c)** and **Figures 5(a)-(c)**) were plotted with Res2Dinv [21]. These figures reveal an approximate investigation depth of 128 m for each profile. This depth corresponds to the maximum cable length (AB = 600 m) according to $\mathbf{x}_l$ position along a profile [16] [22].

For the L1 profile (**Figure 4(a)**), the resistivity is slightly low (Rho < 1500 Ω·m) from near subsurface to about 40 m depth in the eastern part and less than 20 m in the western part. Downward along E-W, one notices a bedding of layers and an increase of resistivity values above 10,000 Ω·m. These resistive and bedded layers suggest unweathered granitic structures [23]. Along the profile (1100 < $\mathbf{x}_l$ < 1900 m), the resistivity is fairly low (Rho < 2500 Ω·m) and iso-resistivity lines are nearly subvertical. Low resistivity values and the shape of iso-re-

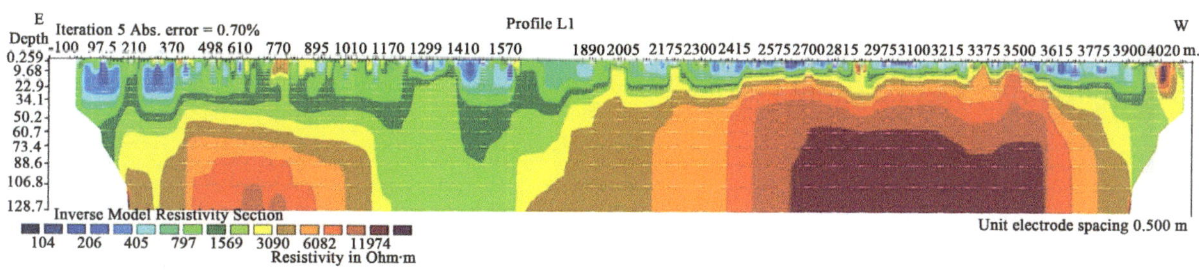

(a)

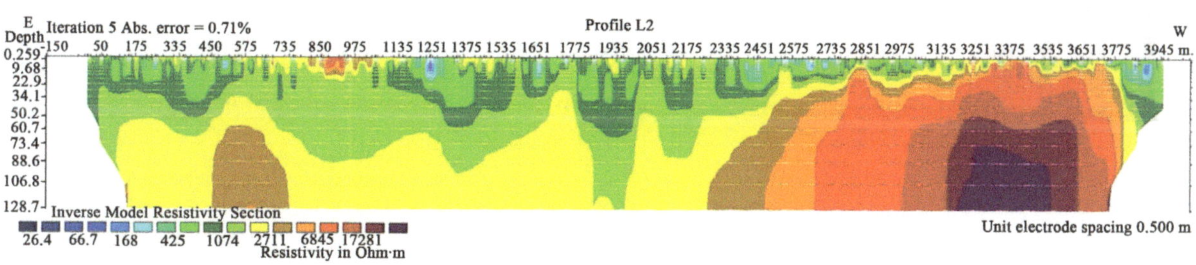

(b)

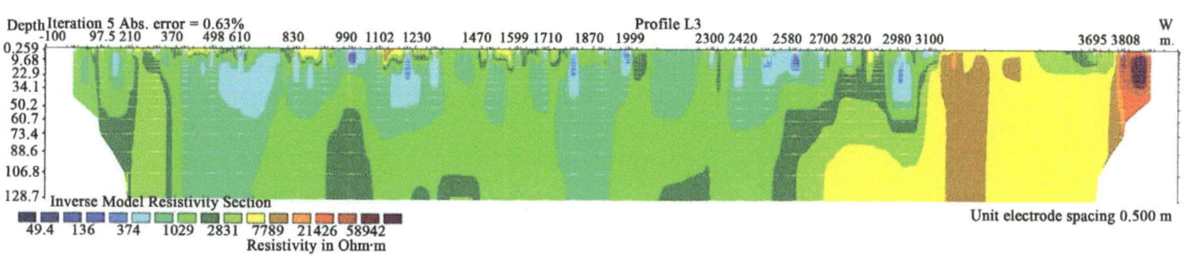

(c)

Figure 4. (a) Pseudo-section of apparent resistivity of Profile L1; (b) Pseudo-section of apparent resistivity of Profile L2; (c) Pseudo-section of apparent resistivity of Profile L3.

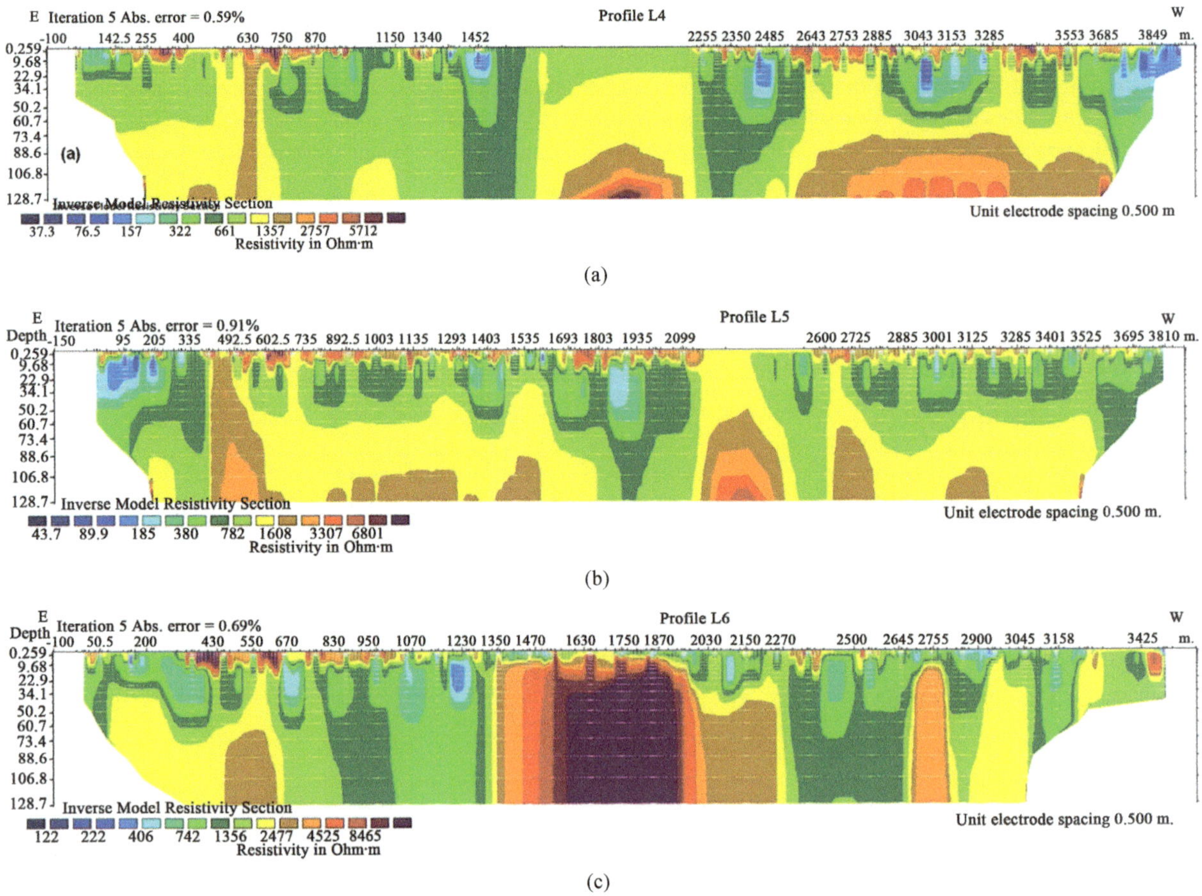

Figure 5. (a) Pseudo-section of apparent resistivity of Profile L4; (b) Pseudo-section of apparent resistivity of Profile L5; (c) Pseudo-section of apparent resistivity of Profile L6.

sistivity curves are significant to a weakness zone or conductive discontinuity. It may exhibit a fracture, a fault or a shear zone along the profile [16] [22] [23].

N-oriented, profiles L2 and L3 (**Figure 2** and **Figures 4(b)-(c)**) clearly highlight a strong resistivity contrast. The western part of the area under study ($0 < \mathbf{x}_l < 2100$ m) is generally low resistive (Rho < 2500 Ω·m). The $\boldsymbol{x}_l$ width increases progressively from profile L2 to profile L3. The electrical responses of these profiles (L2 and L3) look like resistive bedded layers (Rho > 3000 Ω·m) in the eastern side ($2500 < \mathbf{x}_l < 3800$ m and $\mathbf{x}_l > 3000$ m). Effects of these layers go downward to deep structures and they are evidences of fresh granitic structures which sometimes outcrop [22] [23]. The bedding and the resistance of layers disappear at the west; and a weakness, an accumulation or infiltrations' zone sets itself progressively. Thus, in some parts of the area, there are resistive shallow deposits (Rho > 3000 Ω·m) which may be laterite layer [24].

Along profiles L4 and L5 (**Figures 5(a)-(b)**) deep structures are low resistive. They are bounded top and down by more resistive structures. This structural geoelectrical morphology characterizes an accumulation zone of weathered structures or a fluids' infiltration zone [22] [23]. Meanwhile around 630 m and 450 m along profiles L4 and L5 respectively, an involvement of deep resistive materials on surface characterized by conic domes (**Figures 5(a)-(b)**) is observed.

On profile L6 (**Figure 5(c)**), granitic structures characterized by high resistivity values (Rho > 3000 Ω·m) cover the centre ($1250 < \mathbf{x}_l < 2250$ m). Weakness zones or discontinuities zones characterized by low resistivity values (Rho < 2500 Ω·m) cover the two ends of the profile ($0 < \mathbf{x}_l < 1250$ m) and ($1250 < \mathbf{x}_l < 3500$ m).

5.2. Resistivity Map

The **Figure 6** represents the space distributions of the resistivity computed by the surfer software for three depth

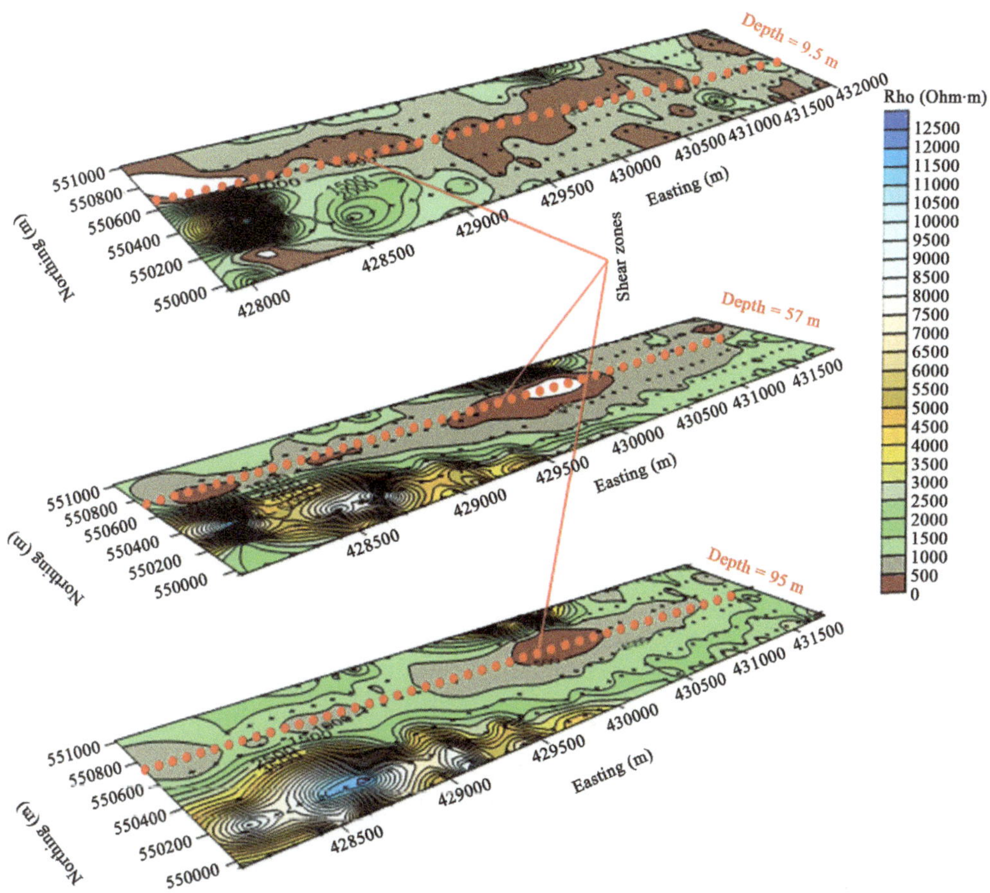

Figure 6. Resistivity maps of AB = 50 m, AB = 300 m and AB = 500 m.

levels corresponding to the three arrays used: AB = 50 m and MN = 5 m, AB = 300 m and MN = 30 m, AB = 500 m and MN = 50 m. These arrays represent depths 9.5 m, 57 m and 95 m respectively [16].

In the centre of the study area and along the E-W direction, resistivity maps highlight (**Figure 6**) a conductive discontinuity characterized by low resistivity values (Rho < 1500 Ω·m). The extension of this area decreases to deep structures (**Figure 6**). It is located at boundaries of resistive zones characterized by high resistivity values (Rho > 2500 Ω·m) and narrowing of iso-resistivity curves. This curves' narrowing prove an inner to outer variation of geological structures along the N-S trend. Central structures materialize decayed rocks, accumulation or infiltration structures [11] [22]. They show up the presence of weakness zones in the central part of the study area [11] [22]. Otherwise resistive structures bounding these zones characterize fresh granitic formations [25].

5.3. Geological Section

The interpretations of electrical sounding curves from the study area and geological surveys have permitted to realize the geological sections [26] of the subsurface crossed by profiles L2 and L4 (**Figure 7** and **Figure 8**). These geological sections quantitatively illustrate the geology of the area. East of profile L2 geological section, the geological profiling shows a four layered model structure (**Figure 7**). The superficial layer or topsoil lies on a lateritic cover with a variable thickness which remains less than 10 m (**Figure 7**). The third layer corresponds to a weathered or conductive layer. It lies on the basement. To the western part of profile L2 geological section, the geological profiling shows three layers (**Figure 7**). The superficial layer or topsoil lies on the slightly thick weathered granite. The third layer represents the rocky or granitic basement. The geological section of profile L4 (**Figure 8**) is similar to the eastern part of profile L2 geological section. The weathered layers or conductive layers evidenced by geological sections permits to highlight tectonic unevenness of the previous results [8] [9] [22] [25].

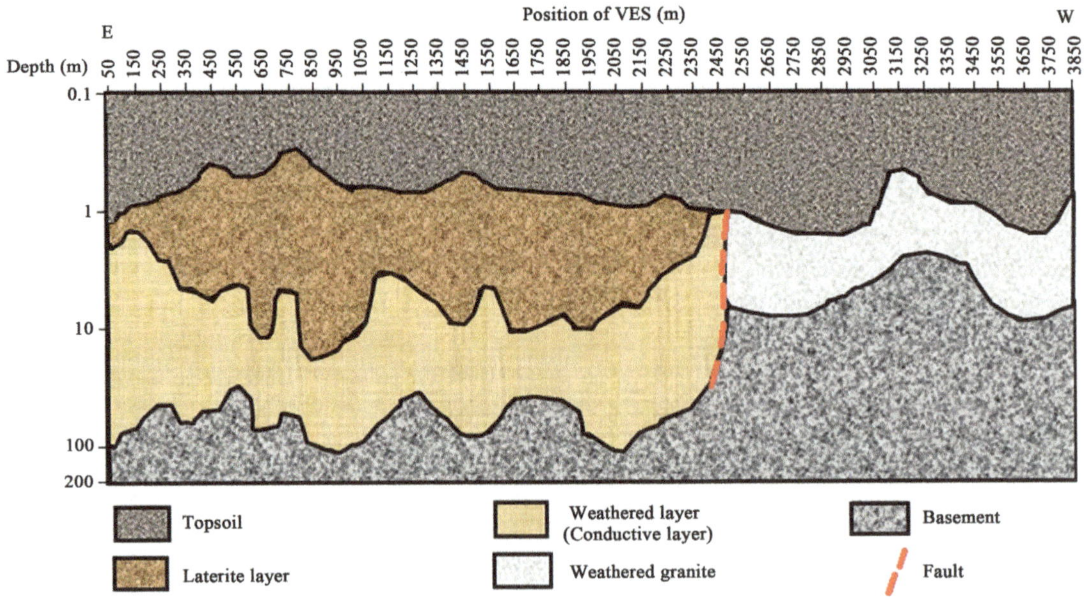

Figure 7. Geological section along of the profile L2.

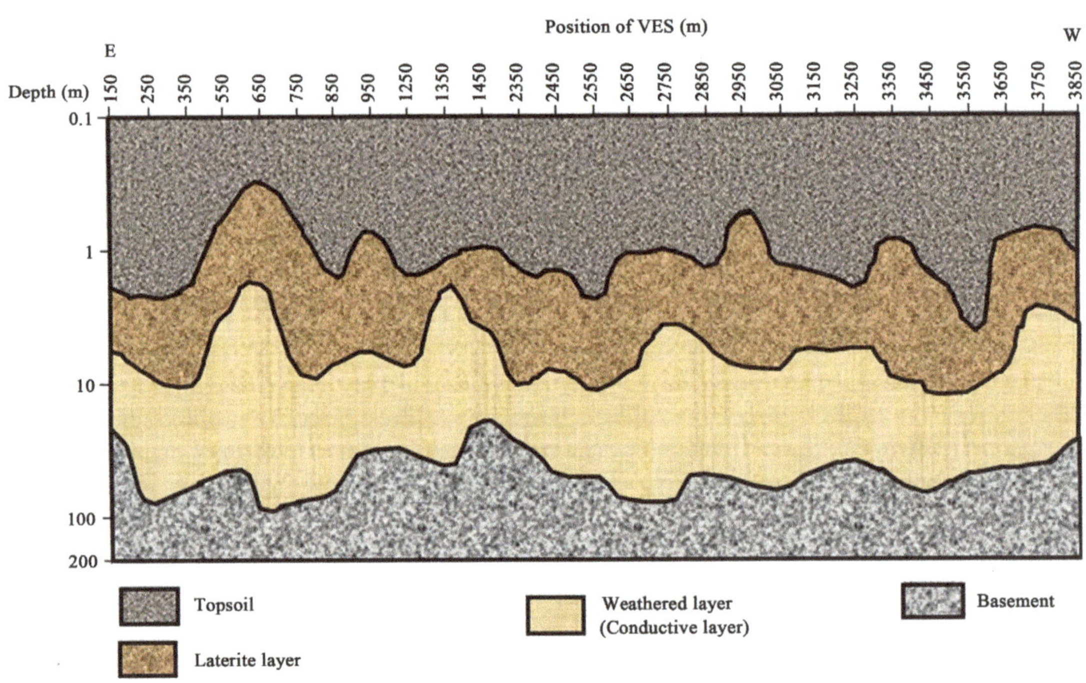

Figure 8. Geological section along of the profile L4.

6. Discussion

The analyses of resistivity variations through pseudo-sections provide quantitative and qualitative information upon the conductivity of the studied area's subsurface. During these analyses, we have delineated two different geological zones [27] [28]:

- The first zone is characterized by high resistive bedded structures. It fits the geological background of fresh granitic structures which sometimes outcrop [11].
- The second zone is characterized by low resistive structures. It fits the weakness zones or conductive zones [23]. These conductive zones or conductive discontinuities characterize clay minerals intrusions inside bar-

ren structures of the granitic basement which underwent shearing and weathering process of shallow structures [11].

The tectonic setting of the study area enables to match that tectonic unevenness with shear zones. There, fluids infiltration's ability is considerable, thus increasing the weathering ability of in situ metamorphic or magmatic structures [25].

In addition, analyses of the resistivity variations through resistivity map show that the space distributions of resistivity, for each subsurface level, are not uniform. The resistivity contrasts highlight weakness zones or fluids' infiltration zones within granitic structures [29] [30]. These weakness zones considered as tectonic unevenness also characterize shear zones with a high accumulation capability of in situ weathered or dissolved structures [22]. They are located at the middle of the study area and have an E-W strike. The mineralization characterized by conductive zones is located in shear zones [25] [28]. It reflects the presence of disseminated clay minerals [22] [27] [29].

In the mining and hydrogeological researches, weathered layers or conductive layers (**Figure 7** and **Figure 8**) characterize the target zones [31]. They are identified as mineralogical deposits and groundwater zones for the mining and hydrogeological researches respectively [22] [24] [26]. We suggest that conductive layers may represent target zones of clay minerals concentrations [28]-[30]. The gold washing activity exerted along stream banks in the area [32] [33] suggest these are gold bearing structures located in weathered quartz veins [34].

7. Conclusion

The electrical Schlumberger data acquisition (sounding and profiling) in the Tindikala-Boutou area along the Kadey River, using Direct Current (DC) method, has permitted to identify the tectonic unevenness and to characterize its mineralogical nature. Results which enabled this identification come from pseudo-sections of resistivity and the resistivity maps particularly. Interpretations made from different models are able to highlight weakness zones or conductive zones which characterize shear zones following an E-W strike. In these zones, the geological sections bring out weathered layers or conductive layers which constitute the target zones for mining and hydrogeological researches. According to the geological and tectonic background of the study area, clay minerals have been identified as the associated mineralization along shear zones within the Precambrian granitic basement. The presence of gold washers attests the existence of gold bearing clay mineralized structures disseminated in weathered quartz veins along the Kadey River.

Acknowledgements

The authors are grateful to the reviewers for their kind remarks making the manuscript clearer and more pertinent. The authors are also grateful to the Artisanal and Small Scale Unit of the Ministry of Mines, Industry & Technological Development for providing the Syscal Current Iris instrument to collect the data sets on site.

References

[1] Cornachia, M. and Dars, R. (1983) Un trait majeur du continent africain. Les Linéaments centrafricains du Cameroun au Golfe d'Aden. *Bulletin de la Societe Geologique de France*, **7**, 102-109.

[2] Rolin, P. (1995) La zone de décrochement panafricain des Oubanguides en République Centrafricaine. *Comptes Rendus de l'Académie des Sciences—Series I, Paris*, **320**, 63-69.

[3] Mvondo, H., Den-Brok, S.W.J. and Mvondo-Ondoa, J. (2003) Evidence for Symmetric Extension and Exhumation of the Yaoundé Nappe (Pan-African Fold Belt, Cameroon). *Journal of African Earth Sciences*, **35**, 215-231. http://dx.doi.org/10.1016/S0899-5362(03)00017-4

[4] Mvondo, H., Owona, S., Mvondo-Ondoa, J. and Essono, J. (2007) Tectonic Evolution of the Yaoundé Segment of the Neoproterozoic Central African Orogenic Belt in Southern Cameroon. *Canadian Journal of Earth Sciences*, **44**, 433-444. http://dx.doi.org/10.1139/e06-107

[5] Olinga, J.B., Mpesse, J.E., Minyem, D., Ngako, V., Ndougsa-Mbarga, T. and Ekodeck, G.E. (2010) The Awaé-Ayos Strike-Slip Shear Zones (Southern-Cameroon): Geometry, Kinematics and Significance in the Late Panafrican Tectonics. *Neues Jahrbuch für Geologie und Paläontologie—Abhandlungen*, **257**, 1-11. http://dx.doi.org/10.1127/0077-7749/2010/0042

[6] Regnoult, J.M. (1986) Synthèse géologique du Cameroun. DMG/MINMEE, 119 p.

[7] Gazel, J. and Giraudie, C. (1965) Notice explicative sur la region Abong-Mbang Ouest et Batourie de la carte géologique

de reconnaissance. Mémoire du BRGM, No 92, Direction des Mines et Géologie, Yaoundé, Cameroun, 29 p.

[8] Meying, A., Ndougsa Mbarga, T. and Manguelle-Dicoum, E. (2009) Evidence of Fractures from the Image of the Subsurface in the Akojolinga-Ayos Area (Cameroon) by Combining the Classical and the Bostick Approaches in the Interpretation of Audio-Magnetotelluric Data. *Journal of Geology and Mining Research*, **1**, 159-171.

[9] Mbom-Abane, S. (1997) Investigation géophysique en bordure du Craton du Congo (région d'Abong-Mbang/Akonolinga, Cameroun) et implications structurales. Thèse Doctorat d'Etat ès Sciences, Université de Yaoundé I, 187.

[10] Ndougsa-Mbarga, T., Meying, A., Bisso, D., Layu, D.Y., Sharma, K.K. and Manguelle-Dicoum, E. (2011) Audiomagnetotellurics (AMT) Soundings Based on the Bostick Approach and Evidence of Tectonic Features along the Northern Edge of the Congo Craton, in the Messamena/Abong-Mbang Area (Cameroon). *Journal of Indian Geophysical Union*, **15**, 145-159.

[11] Parasnis, D.S. (1997) Principles of Applied Geophysics. 5th Edition, Chapman and Hall, London, 104-176.

[12] Keary, P. and Brooks, M. (1991) An Introduction to Geophysical Exploration. 2nd Edition, Blackwell Scientific Publications, Oxford, 254 p.

[13] Ward, S.H. (1990) Resistivity and Induced Polarization Methods. In: Ward, S.H., Ed., *Geotechnical and Environmental Geophysics*, Society of Exploration Geophysicists, Tulsa, 147-190.

[14] Loke, M.H. (1994) The Inversion of Two Dimensional Resistivity Data. PhD Thesis, University of Birmingham, Birmingham, 122.

[15] Chapellier, D. (2000) Prospection électrique en surface. Cours de géophysique, Université de Lausanne, Institut Français de Pétrole, 98 p.

[16] Loke, M.H. (2000) Electrical Imaging Surveys for Environmental and Engineering Studies: A Practical Guide to 2-D and 3-D Surveys. 61.

[17] Burger, R.H., Sheehan, F.A. and Jones, C.H. (2006) Introduction to Applied Geophysics: Exploring the Shallow Subsurface, W.W. Norton & Company, Inc., New York, 265-347.

[18] Loke, M.H. and Barker, R.D. (1996) Rapid Least-Squares Inversion of Apparent Resistivity Pseudosections by a Quasi-Newton Method. *Geophysical Prospecting*, **44**, 131-152. http://dx.doi.org/10.1111/j.1365-2478.1996.tb00142.x

[19] Jenny, J. and Borreguero, M. (1999) Winsev 5, 1-D Inversion Software.

[20] Surfer. Version 7.00, Surface Mapping System Copyright © 1993-1999. Golden Software, Inc., Colorado.

[21] Res2Dinv ver. 3.59 for Windows XP/Vista/7 (2010) Rapid 2D Resistivity & IP Inversion Using the Least-Squares Method. Geoelectrical Imaging 2D & 3D. GEOTOMO Software, Malaysia.

[22] Meju, M.A. (2002) Geoelectromagnetic Exploration for Natural Resources: Models, Case Studies and Challenges. *Surveys in Geophysics*, **23**, 133-205. http://dx.doi.org/10.1023/A:1015052419222

[23] Ngando, A.M., Nouayou, R., Tabod, T.C. and Manguelle-Dicoum, E. (2011) Evidence for Precambrian Faulting in the Tibati Adamawa Region of Cameroon Using the Audiomagnetotelluric Method. *Geofisica Internacional*, **50**, 129-146

[24] Egbai, J.C. (2011) Vertical Electrical Sounding for the Investigation of Clay Deposit in Orerokpe, Delta State. *Journal of Emerging Trends in Engineering and Applied Sciences*, **2**, 260-265.

[25] Gouet, D.H., Ndougsa-Mbarga, T., Meying, A., Assembe, S.P. and Man-Mvele Pepogo, A.D. (2013) Gold Mineralization Channels Identification in the Tindikala-Boutou Area (Eastern-Cameroon) Using Geoelectrical (DC & IP) Methods: A Case Study. *International Journal of Geosciences*, **4**, 643-655. http://dx.doi.org/10.4236/ijg.2013.43059

[26] Coker, J.O. (2012) Vertical Electrical Sounding (VES) Methods to Delineate Potential Groundwater Aquifers in Akobo Area, Ibadan, South-western, Nigeria. *Journal of Geology and Mining Research*, **4**, 35-42.

[27] Salmirinne, H. and Turunen, P. (2007) Ground Geophysical Characteristics of Gold Targets in the Central Lapland Greenstone Belt. *Geological Survey of Finland, Special*, **44**, 209-223.

[28] Holliday, J.R. and Cooke, D.R. (2007) Advances in Geological Models and Exploration Methods for Copper ± Gold Porphyry Deposits. *Ore Deposits and Exploration Technology*, **53**, 791-809.

[29] Lowrie, W. (2007) Fundamentals of Geophysics. 2nd Edition, Cambridge University Press, Cambridge, 381. http://dx.doi.org/10.1017/CBO9780511807107

[30] Tijani, M.N., Osinowo, O.O. and Ogedengbe, O. (2009) Mapping of Sub-Surface Fracture Systems Using Integrated Electrical Resistivity Profiling and VLF-EM Methods: A Case Study of Suspected Gold Mineralization. *RMZ Materials and Geoenvironment*, **56**, 415-436.

[31] Campbell, D.L. and Fitterman, D.V. (2000) Geoelectrical Methods for Investigating Mine Dumps. *Proceedings of the 5th International Conference on Acid Rock Drainage (ICARD 2000), Denver, Colo*, 2, Society for Mining, Metallurgy, and Exploration, Inc., Littleton, 1513-1523.

[32] Kim, S., Yoon, M.J., Kim, S.G. and Kwon, D.J. (2009) Exploration Report Colomine Placer Gold Project Cameroon. Geotech Consultant Co., Ltd., Seoul, 87.

[33] Vairon, J., Edimo, A., Simeon, Y. and Vadala, P. (1986) Protocole d'accord pour la Recherche des minéralisations d'or dans la province aurifère de l'Est Cameroun. Rapport du BRGM. Mission Or Batouri, deuxième et troisième phase, 251 p.

[34] Biste, M.H., Bufler, R. and Friedrich, G. (1991) Geology and Exploration of Gold Placer Deposits the Precambrian Shield of Eastern Bolivia. In: Dpjdp-Dtcd, O., Ed., *International Symposium on Alluvial Gold Placers*, La Paz, 148-158.

19

Assessment of Heavy Metal Concentration in Hand Dug Well Water from Selected Land Uses in Wukari Town, Wukari, Taraba State, Nigeria

Kehinde T. Oyatayo[1], Godwin A. Songu[1], Greatest A. Amos[1], Christopher Ndabula[2]

[1]Department of Geography, Kwararafa University, Wukari, Nigeria
[2]Department of Geography and Regional Planning, Federal University, Dutsin-Ma, Nigeria
Email: kehindeoyatayo@gmail.com

Abstract

Hand dug well water remains the major source of domestic (drinking) water in the city of Wukari. This study was aimed at the assessment of heavy metal concentration in hand dug well water from four land uses of abattoir (A), waste dump site (B), residential (C) and commercial (D) in Wukari Town, Wukari, Taraba State. There are about 76 hand dug wells in the study area and for the purpose of this study. Yamane (1967) sample size technique was adopted. Based on this, 95% confidence level and an error limit of 5% were adopted. Thus, four (4) samples A, B, C and D of hand dug well water purposively drawn from the study area used sterilized bottles. The heavy metals tested for in the water samples include: zinc (Zn), copper (Cu), manganese (Mn), iron (Fe), potassium (K), phosphate, and chromium. The results of the analyses were compared with the NSDWQ*MPL and FEPA (2001) standards. The results shows that the concentration of magnesium and manganese in hand dug well water samples A, B, C and D had higher concentrations than the maximum permissible limits of NSDWQ*MPL and FEPA (2001) standards and the concentration of zinc, copper, iron, potassium, phosphate and chromium in hand dug well water sample A, B, C and D was at different compliance level with NSDWQ*MPL and FEPA (2001) standards. ANOVA (F-ratio) statistics at P_0.05 for variation in heavy metal concentration in hand dug well water from land uses A, B, C and D were insignificant. For now, it can be concluded that hand dug well water in Wukari Town is fit for domestic use, but that efforts should be made to reduce the magnesium and manganese concentration by having control on the anthropogenic factors that lead to such high concentration levels or else, soon this source of water may become unfit for domestic use. It was recommended that aggressive public awareness and enlightenment on possible impacts of hand dug well water pollution from human activities should be embarked upon by relevant agencies.

Keywords

Heavy Metals, Hand Dug Well Water, Water Pollution, Safety Standard

1. Introduction

Water is precious and necessary for a sustainable economic development of an area as it is the next major support to life after air. In the urban areas where pipe-borne water, bore-hole water and hand-dug wells are available, it is an indication that water is a vital component of human existence. Groundwater is of major importance and is intensively exploited for private, domestic and industrial uses. 90% of the population depends largely on hand-dug wells. Rapid growth in urban populations and industrial activities and commercial and agricultural developments result in increase in the search and uses of potable water [1]-[3].

A major source of groundwater abstraction for use is by means of a hand dug well. A well is a hole or shaft sunk into the earth for the purpose of obtaining water and other fluid from an underground supply. Such a hole can be constructed in different ways upon which wells are commonly classified as hand dug wells, borehole, among others. Hand dug wells are of vital importance as source of domestic water supply for the majority of Nigerians. Hand dug well water remain the major source of domestic (drinking) water in the city of Wukari. The quality of this hand dug well water across Nigeria is rarely analyzed nor treated before use. Until recently, most artificial wells in the country do not have pump. Hand dug wells are usually between 1 and 20 meter deep depending on the geological characteristics of the area, and lined with stone, bricks, and recently cemented rings [4] [5].

The spilling, leaking, improper disposal, or intentional application of chemicals at the land surface can result in runoff that contaminates nearby streams and lakes, or infiltration that contaminates underlying aquifers. The type and severity of water contamination often is directly related to human activity, which can be quantified in terms of the intensity and type of land use in the source areas of water to streams and aquifers.

A relatively simple way to study the effects of land use on groundwater quality is to compare the predominant land uses within a given area to the concentrations of selected contaminants in water drawn from shallow aquifers within that area. Analysis of the relation between land use and the magnitude of contamination in a specific area primarily is based on the following two assumptions [6]:

First, it is assumed that contaminated groundwater at a well originated as uncontaminated recharge (precipitation) that passed through a contaminated area before reaching the well. The area from which a well derives its water (and associated contaminants) is known as the well's groundwater "contributing area." Land use and land cover largely determine the type and amount of contaminants entering streams, lakes, and underground pathways, including aquifers. Some contaminants occur and move naturally, whereas others are produced by human activities and their movement often is accelerated as a result of rainfall that accentuates runoff and infiltration. Second, it is assumed that the contaminants detected in groundwater were present within the well's contributing area and were transported by groundwater flow to the well. The source(s) of contaminants within a contributing area, such as buried septic systems and leaking underground fuel tanks, can be difficult to identify and locate. In many instances, these sources can be inferred from the type and intensity of land use within the contributing area. Factors that can affect the movement of contaminants from source areas to wells are:

- The chemical nature of the contaminant;
- The physical properties of the soil and aquifer material;
- The amount and timing of recharge; and
- The direction and velocity of groundwater movement.

Heavy metals are elements having some atomic weight between 63.54 and 200.59, and a specific gravity greater than 4. Although trace amount of some heavy metals are required by living organisms, any excess amount of these metals can be detrimental to life [3]. The solubility of these metals in soils and groundwater is largely controlled by pH [1] [4] [7]; amount of metal and cation exchange capacity, organic carbon content and the oxidation state of mineral components as well as the redox potential of the system. Pollution of groundwater is an impairment of water quality by chemicals, heat or bacteria to a degree that does not necessarily create public health hazards, but does adversely affect such water for domestic, farm, municipal or industrial use. Trace elements are generally present in small concentration in natural water system. Their occurrence in groundwater and surface water can be due to natural sources such as dissolution of naturally occurring minerals containing trace elements in the soil zone or the aquifer material or to human activities such as mining, fuels, smelting of ores and improper disposal of industrial wastes [7]-[9] 2000 cited in: [2].

Heavy metals such as iron, manganese, cadmium, mercury, lead, arsenic, chromium, copper, Nickel, zinc cause a number of hazards to humans; and are indeed cofactors as activator biochemical reactions. Heavy metals are natural components of the earth's crust, as can neither be degraded nor destroyed. Through rivers and streams, the metals are transported as either dissolved species in water or as an integral part of suspended substances causing the most detrimental effects on aquatic life. Also, to a small extent, the metals enter the human bodies via food, drinking water and air. Though, some heavy metals (e.g. copper, selenium, zinc) are essential to maintain the metabolism of the human body, but their concentration above the desirable levels can be poisonous. Heavy metal poisoning could be from drinking-water contamination (e.g. lead pipes), high ambient air concentrations near emission sources, or intake via the food chain [2].

Heavy metals are dangerous because they tend to bioaccumulate. Bioaccumulation results when there is an increase in the concentration of a chemical in a biological organism over time, compared to the natural concentration of chemicals in the environment. Heavy metals may enter a water supply from industrial and household wastes, or acidic rain resulting in the breakdown of soils and releasing heavy metals into streams, lakes, rivers, and groundwater [8]. Investigation of heavy metals is very essential since slight changes in their concentration above the acceptable levels, whether due to natural or anthropogenic factors, can result in serious environmental and subsequent health problems. This study therefore seeks to investigate heavy metal concentration in hand dug wells in Wukari, with a view to ascertain whether this source of domestic water supply is safe for human consumption.

2. Study Area

2.1. Location of the Study Area

Taraba State is located in north eastern part of Nigeria as shown in **Figure 1**. Wukari is the headquarters of Wukari Local Government Area of Taraba State. It is located between latitude 7°51'N to 7°85'N and longitude 9°46'E to 9°78'E of the Greenwich meridian. Wukari Local Government area is situated in the southern part of Taraba State as shown in **Figure 1**. It is about two hundred kilometers away from Jalingo the state capital. The Local Government is bounded by Plateau State in the North, Benue State in the Southwest, Northeast by Karim Lamido, Bali, and Takum Local Government Area (LGA). It has an area of about 4308 km^2 (1663 sq mi). **Figure 2** is the map of Wukari Local Government Area showing the study area.

2.2. Physical Characteristics

Wukari LGA covers an area of approximately 6500 sq. Km. The whole area is a gently undulating plain, with a mean altitude of 200 m above sea level. The drainage systems drain northward and serve as tributaries network to the River Benue while eastward discharge of rivulets and other smaller tributaries from Wukari town drains towards the Donga River, which is a major tributary of River Benue.

2.3. Climate

The study area is characterized with a tropical continental climate. According to Koppen's climate classification scheme, the climate of the study area corresponds to the Aw type of climate which is characterized by marked distinctive wet and dry season. The mean annual rainfall value ranges from 1000 - 1500 mm. The unset of the raining season is usually around April while the offset period is October. This means the rainy season usually last for seven months and about five month of dryness from November to March. The mean maximum temperature is being experience around April at about 40°C while the mean minimum temperature occurs between the period of December and February at about 20°C. Relative humidity also exhibit temporal variability. It is higher during the night hours in the rainy season than during the day hours in the dry season respectively. By location, the climate of the area is being influenced by the prevalence of two opposing air masses, the tropical maritime air mass (MT, south westerly's trade) and the tropical continental air mass (CT, north easterlies trade). The tropical continental air mass (CT) is a dry air that blows across the Sahara desert to the West African region.

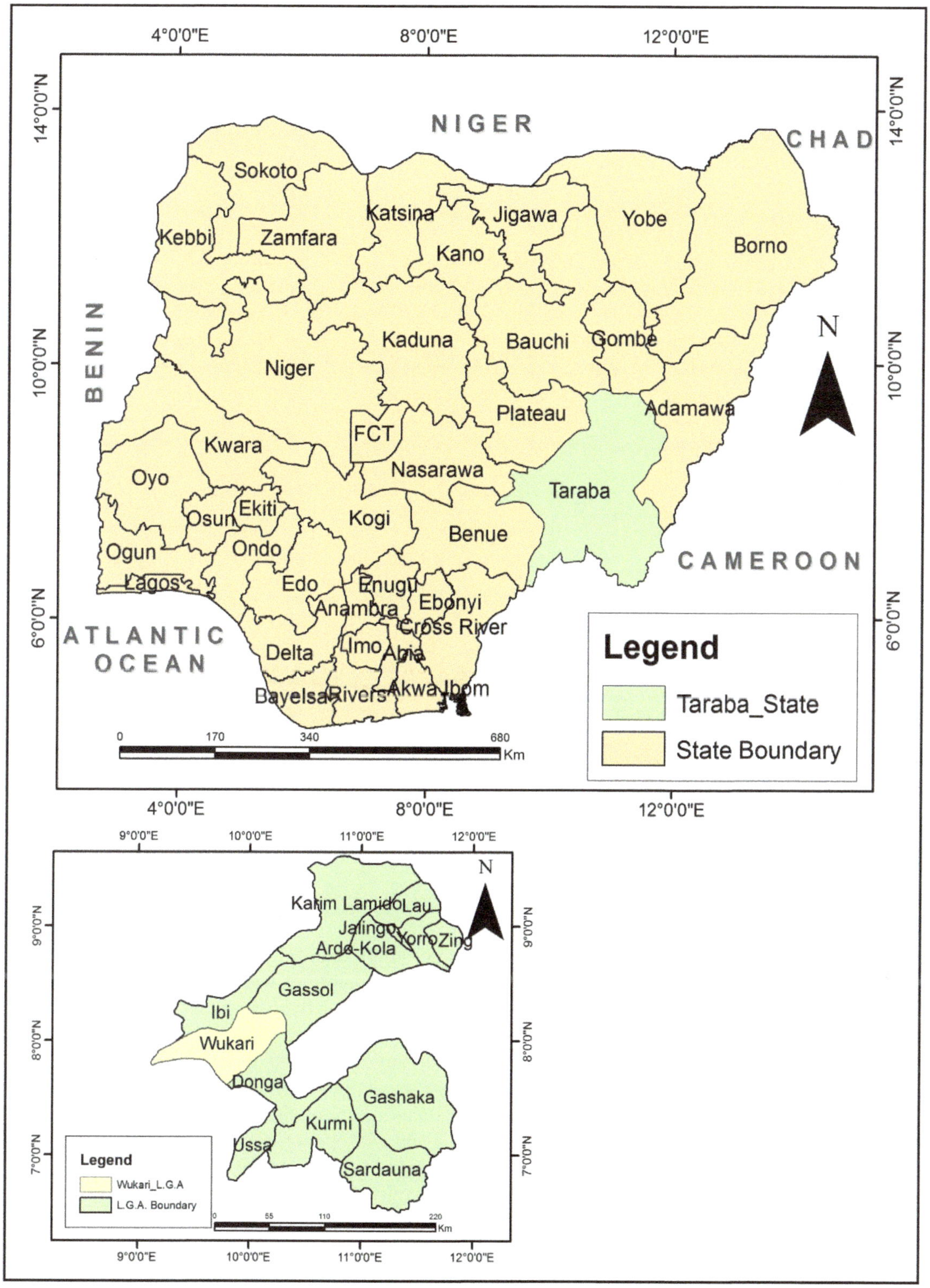

Figure 1. Map of Nigeria and Taraba State showing Wukari local Government Area. Source: Ministry of Land and Survey, Jalingo, Taraba State.

This air mass is commonly associated with cold, dry, and dusty condition. The tropical maritime air mass (MT) is characterized with warm, moist air from the 22 Atlantic Ocean south of the Nigeria. This air mass is responsible

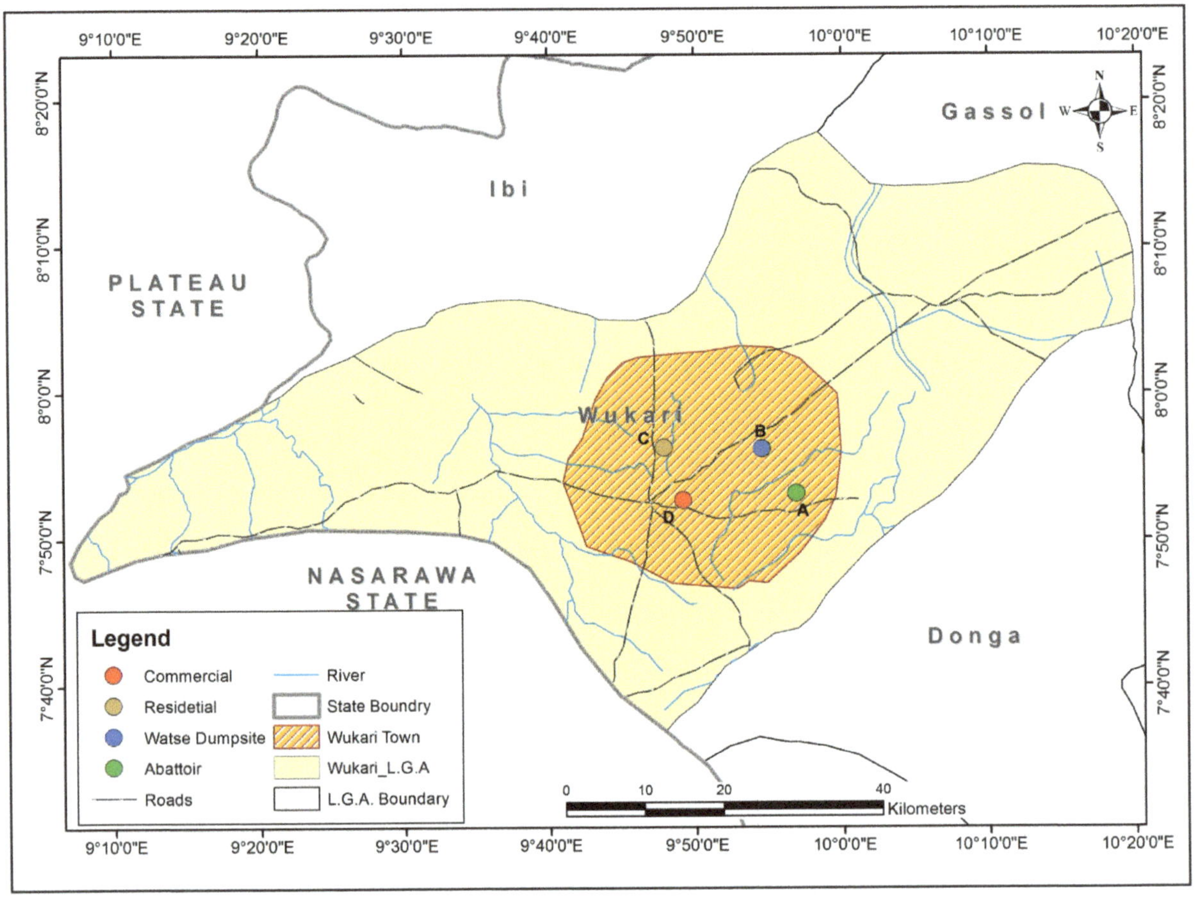

Figure 2. Map of the study area. Source: Ministry of Land and Survey, Jalingo, Taraba Sate.

for the rainy season and the former dry season. The confluence between these two air masses is known as the Inter tropical convergence zone (ITCZ).

2.4. Geology, Soils and Vegetation

Wukari is situated over cretaceous sandstone which is possibly Bima-sandstone. These rocks do not form adequate supplies of ground water for the township, so the public supply of water has to be supplemented through pumped water from the Donga River. The Wukari urban area is a relatively undissected, very gentle slope plain. The town is the beginning of the catchment areas for 6 major streams although no perennial streams intrude into the built-up area. The soils are deep, well drained and medium to coarse textured. The Wukari local government area falls within the southern guinea savannah zone. The vegetation manifest seasonal pattern, and it is mainly of tree savannah in which the dominant species is the dainiellia providing a limited amount of shade. The accompanying shrubs and grasses are of the hymenocardia and andropogon communities. There are also restricted areas of isoberlinia savannah woodland, which forms the forest reserves of the area. Other exotic species include eucalyptus, azadirachtaindica, and psidiumguajava, among others.

3. Materials and Methods

In order to achieve the aim and objectives of the study, the methodology that was employed in this study is subdivided into the following themes:

1) Reconnaissance survey.
2) Materials.
3) Methods and procedures of data collection.
4) Statistical analysis.

3.1. Reconnaissance Survey

A reconnaissance survey of the study area was carried out to give a real time site assessment and also to acquire information about the township which may not readily be available in the topographic map. The research work was carried out in two main phases:

Phase I: reconnaissance survey.

Phase II: Field work where ground water samples were collected.

3.2. Data Collection Procedure

A total of four hand dug wells were sampled for analyses within Wukari Township. Bottles used in the collection of water samples were sterilized in an autoclave at a temperature of 120°C for three hours before duplicate samples of 100 ml and 2 liters of each of the hand dug well water was collected for analyses, respectively. The heavy metals tested for in the water samples include: zinc (Zn), copper (Cu), manganese (Mn), iron (Fe), potassium (K), phosphate, and chromium. These samples were labeled in order to avoid mix-up and were taken to the laboratory for analyses. The laboratory analyses of the water samples were carried out in Bauchi State Rural water supply and sanitation (RUWASSA), Bauchi, Bauchi State, Nigeria.

3.3. Sampling Techniques

There are about 76 hand dug wells in the study area and for the purpose of this study; a representation sample with known confidence and risk level was selected, based on the work of [10]. Based on this, 95% confidence level and an error limit of 5% were adopted. Thus, four (4) hand dug wells were sampled for the purpose of water quality analysis (heavy metal concentration) in the study area. Purposive sampling was used in picking the sampled hand dug wells in the study area based on the following land uses: A (closeness to Abattoir), B (closeness to waste dumping site), C (residential area) and D (closeness to commercial).

3.4. Data Analysis

To examine heavy metal concentration in hand dug well water from selected land uses in Wukari, [11] methodology for analyzing chemical parameters were adopted as follows: Copper, zinc, manganese, Chromium, Iron, potassium and Manganese were determined using Atomic Absorption Spectrophotometer; the concentration was read using UV spectrophotometer (Model: 01-0960-00) at 510 nm. Finally, the laboratory results were compared with the Nigerian Standard for drinking water quality and FEPA minimum permissible limits.

3.5. Statistical Analysis

To examine whether there is variation in heavy metal concentration in hand dug well water from four land uses of A (abattoir), B (waste dump site), C (residential) and D (commercial) in Wukari town the following hypothesis was postulated:

H0: there is no significant variation in heavy metal concentration in hand dug well water from land uses of abattoir, waste dump site, residential and commercial in Wukari town.

The statistical test best suited to test the above hypothesis is the Analysis of Variance (ANOVA). This was done with the aid of the Statistical Package for Social Sciences (SPSS).

4. Results and Discussion

Magnesium: The amount of magnesium in the hand dug well water samples as obtained from the laboratory analysis is as follows: sample A, B, C, and D are 17.09, 20.76, 15.87 and 20.76 respectively as shown in **Table 1**. These values are higher than the NSDWQ*MPL and [12] effluent permission-able limit of 2.0 mg/l required for drinking water quality as shown in **Table 2**. High content of magnesium and calcium results to hardness of water. Magnesium has also been known to be essential for plants growth and development [2].

Copper: The amount of copper in the hand dug well water samples as obtained from the laboratory analysis is as follows: 0.09 mg/l, 0.07 mg/l, 0.12 mg/l and 0.02 mg/l for samples A, B, C and D respectively as shown in **Table 1**. There is consistency in the values obtained from laboratory analysis and these values fall below the permissible limit of 1.0 mg/l set by NSDWQ*MPL and [12] as shown in **Table 2**.

Table 1. Results of laboratory analysis of water samples A, B, C and D.

Parameters	A	B	C	D	NSDWQ*MPL
Calcium Ca (mg/1)	60	68	92	66	75
Magnesium Mg (mg/1	17.09	20.76	15.87	20.76	2
Floride (mg/1)	0.73	0.52	0.56	0.4	1
Copper Cu (mg/1)	0.09	0.07	0.12	0.02	1
Iron Fe (mg/1)	0.3	0.05	0.05	0	0.3
Potassium (mg/1)	11	24	20	9.5	200
Sulphate (mg/1)	7	7	30	4	100
Manganese Mn (mg/l)	0.6	0.5	0.2	0.3	0.05
Nitrate NO_3 (mg/l)	36.52	53.68	39.15	56.32	50
Nitrite NO (mg/l)	0.043	0.013	0.013	0.073	0.2
Zinc (mg/l)	0	0	0	0	5
Phosphate. (mg/l)	0.85	0.07	0.3	0.37	#NAME?
Chromium (mg/l)	0.01	0.01	0.01	0	0.05
Sodium. Na (mg/l)	41.25	48.13	50	32.41	200
Chloride Cl (mg/l)	104.97	122.46	139.96	82.47	250
HCO_3^- (mg^{-1})	30	40	45	35	100

Source: Field survey (2014).

Table 2. Comparison of the analyzed samples with the NSDWQ*MPL and FEPA acceptable limits.

Parameters (mg/l)	Mean Values	NSDWQ*MPL/FEPA	Deviation	Remark
Calcium Ca (mg/l)	71.50	75	−3.5	Below limit
Magnesium Mg^{2+} (mg/l)	18.6150	2.0	16.615	Acceptable limit
Fluoride(mg/l)	0.552500	1.0	−0.4475	Within limit
Copper Cu^{2+} (mg/l)	0.057000	1.0	−0.943	Within limit
Iron Fe (mg/l)	0.1000	0.3	−0.2	Below limit
Potassium (mg/l)	16.125	200	−183.875	Below limit
Sulphate (mg/l)	12.00	100	−88	Below limit
Manganese Mn (mg/l)	0.400	0.05	0.35	Acceptable limit
Nitrate NO_3 (mg/l)	46.4275	50	−3.5725	Below limit
Nitrite NO_2	0.035500	0.2	−0.1645	Below limit
Zinc Zn (mg/l)	0.00	5	−5	Below limit
Phosphate (mg/l)	0.397500	-	0.397500	Acceptable limit
Chromium (mg/l)	0.0075	0.05	−0.0425	Below limit
Sodium Na^+ (mg/l)	44.197500	200	−155.8025	Below limit
Chloride Cl^- (mg/l)	112.4650	250	−137.535	Below limit
HCO_3^- (mg/l)	37.50	100	−62.5	Below limit

Source: Field survey (2014). NSDWQ*MPL = Nigeria Standard for Water Quality and Minimum Permission Limit. FEPA = Federal Environment Protective Agency.

Though still within permissible limits, some people who drink water containing copper in excess of the action level in short term may experience gastrointestinal distress while those with long-term exposure may experience anemia, liver or kidney damage. People with Wilson's disease should consult their personal doctor if the amount of copper in their water exceeds the action level. Also, long term exposure to water that contains copper can cause irritation of nose, mouth and eyes which can lead to headache, dizziness, vomiting and diarrhea among other health hazards [13].

Iron content: The amount of iron in sample A hand dug well water is 0.30 mg/l , while samples B, C and D have values of 0.05 mg/l, 0.05 mg/l and 0.00 mg/l respectively as shown in **Table 1**. The beneficial effects of iron include; chlorophyll synthesis, oxidation-reduction in respiration, constituent of certain enzymes and proteins. Excess concentration of iron has a negative effect as it can cause gastrointestinal irritation and enhance the growth of iron bacteria that affects the water taste. The high concentration values of iron has the potential of staining laundry, metal pipes for reticulation and scaling in pipes. It can also result in gene mutation leading to haemochromatosis whose symptoms include fatigue, weight loss, joint pains and ultimate heart disease, liver problems and diabetes [13]. The iron concentration of the sample ranges between 0.00 - 0.30, when compared to NSDWQ*MPL and [12] limit, it is within the permissible limit of 0.3 mg/l as shown in **Table 2**.

Potassium: The concentration in the collected samples range between 9.5 - 24.0 as shown in **Table 1**. The values are quite below the NSDWQ*MPL and [10] recommended limit of 200mg/l as shown in **Table 2**.

Manganese: The manganese concentration of hand dug well water samples was 0.6, 0.5, 0.2 and 0.3 mg/l for sample A, B, C and D respectively as shown in **Table 1**. The values obtained far exceed SDWQ*MPL and [12] permissible limit of 0.05 mg/l as shown in **Table 2**.

Manganese impacts a bitter taste to water, stains cloths and metal parts and precipitate in foods when used for cooking and it also promotes the growth of algae in reservoirs [2].

Zinc: The amount of zinc in the hand dug well water samples as obtained from the laboratory analysis is as follows: 0.00 mg/l, 0.00 mg/l, 0.00 mg/l and 0.00 mg/l for samples A, B, C and D respectively as shown in **Table 1**. The laboratory values obtained all fall within the permissible limits of 5 mg/l as stipulated by [12] and NSDWQ*MPL as shown in **Table 2**.

Chromium: The amount of chromium in the hand dug well water samples as obtained from the laboratory analysis is as follows: 0.01 mg/l, 0.01 mg/l, 0.01 mg/l and 0.00 mg/l for samples A, B, C and D respectively as shown in **Table 1**. The laboratory values obtained all fall within the permissible limits of 0.05 mg/l as stipulated by [12] and NSDWQ*MPL as shown in **Table 2**. Since it is still below the permissible limit for drinking water quality, it shows that it has no implication on people who drink from the source. Water containing chromium in excess of the set limit may have an erythropoietin effect such as increased occurrence of goiter among humans and animals [1].

The findings from this study shows that the concentration of copper, iron, potassium, zinc and chromium in Wukari hand dug well water are all within the set permissible limit of [12] and NSDWQ*MPL as shown in **Table 2**. This confirms the findings of [2] which is supportive of the fact that heavy metals concentration in ground water in Obajana and environs are within the set limit of [12] and NSDWQ*MPL.

Also, the findings shows that the concentration of manganese and magnesium in Wukari hand dug well water exceeds the set permissible limit of [12] and NSDWQ*MPL, which corroborate the findings of [1] [3] [13]-[15] that asserts that ground water in some selected Nigerian settlements shows some elevated levels of toxic metals. This they observe is a source of worry considering the fact that typical settlementin Nigeria is devoid of basic infrastructure that makes life worth living.

5. Statistical Analysis

Statistical analyses using ANOVA single factor was conducted to test the null hypothesis that there is no significant variation in heavy metal concentration in hand dug well water from four different land uses of abattoir, waste dump site, residential and commercial in Wukari town. The summary of ANOVA single factor statistics is shown in **Table 3**. Using ANOVA single factor with a specified rejection level at α 0.05, the result indicates that the calculated value (F calculated) is 0.21, while the table value at $P < 0.05$ is 2.76 as shown in **Table 4**. This result is insignificant which implies that there is no significant variation in heavy metal concentration in hand dug well water from land uses A (abattoir), B (waste dump site), C (residential) and D (commercial) in Wukari town.

Table 3. Summary.

Groups	*Count*	*Sum*	*Average*	*Variance*
A	16	310.45	19.40	866.26
B	16	385.26	24.08	1213.22
C	16	433.23	27.08	1595.48
D	16	307.62	19.23	748.52

Table 3 shows the summary of ANOVA single factor analysis.

Table 4. ANOVA.

Source of Variation	*SS*	*df*	*MS*	*F*	*P-value*	*F crit*
Between Groups	699.7881	3	233.26	0.21	0.89	2.76
Within Groups	66352.16	60	1105.87			
Total	67051.95	63				

Using ANOVA single factor as shown in **Table 4** with a specified rejection level at α 0.05 the result is insignificant. The calculate F value is less than the table F value, thus, the Null hypothesis is to be accepted (There is no significant variation).

6. Conclusion

The result of the laboratory analysis shows that the concentration of iron, copper, chromium, zinc, potassium in hand dug well water in Wukari falls within the permissible limits of NSDWQ*MPL and [12]; while the concentration of manganese and magnesium far exceeds the permissible limits of NSDWQ*MPL and [12]. The results indicate that the drinking water quality is becoming deteriorated because most of the heavy metals tested for are found to be at minimum satisfactory level except for manganese and magnesium, which are found to be at unsatisfactory level because they are found to be at high levels in concentration. For now, it can be concluded that hand dug wells in Wukari Town are fit for domestic use, but that efforts should be made to reduce the magnesium and manganese concentration by having control on the anthropogenic factors (domestic waste water, agrochemicals, fumigation, local industries effluents) that lead to such high concentration levels or else, soon this source of water may become unfit for domestic use. The conclusion also is that there is no significant variation of heavy metal concentration in hand dug well water from land uses of abattoir, waste dump site, residential and commercial in Wukari town.

References

[1] Magaji, J.Y. (2009) The Impact of Waste Dump on Soil and Water Quality. A Case Study of Mpape Dumpsite, FCT, Abuja. An Unpublished Ph.D. Thesis Submitted to the Department of Geography and Environmental Management, University of Abuja, Abuja.

[2] Musa, O.K., Shuaibu, M.M. and Kudamnya, E.A. (2013) Heavy Metal Concentration in Groundwater around Obajana and Its Environs, Kogi State, North Central Nigeria. *American Journal of Contemporary Research*, **3**, 170-177.

[3] Ogunlaja, A. and Ogunlaja, O.O. (2007) Physico-Chemical Analysis of Water Sources in Ubeji Community and Their Histological Impact on Organs of Albino Mice. *Journal of Environmental Management*, **11**, 91-94.

[4] Huisman, L. (1986) Lecture Notes on Groundwater Recovery. Belgrade, Yugoslavia, Art B 8-3-8-4.

[5] Ajibade, L.T., Ogunkolu, A.B. and Ogundare, I.O. (2014) Physico-Chemical and Biological Characteristics of Hand Dug Well Water in Ilorin Metropolis, Nigeria. *Nigeria Geographical Journal*, **9**, 146-162.

[6] Eckhardt, D.A. and Stackelberg, P.E (1995) Relation of Ground-Water Quality to Landuse on Long Island, New York. *Ground Water*, **33**, 1019-1033. http://dx.doi.org/10.1111/j.1745-6584.1995.tb00047.x

[7] Kennish, M.J. (1992) Ecology of Estuaries Anthropogenic Effects. CRC Press, Inc., Boca Raton, 494 p.

[8] Ajibade, O.M., Omisanya, K.O. and Odunsi, G.O. (2011) Groundwater Portability and Flow Direction of Urban Aquifer, Ibadan Southwest Nigeria. *Journal of the Nigerian Association of Hydrologist*, **21**, 38-55.

[9] Martinez, C.E. and Motto, H.L. (2000) Suitability of Lead, Zinc and Copper Added to Mineral Soils. *Environmental Pollution*, **107**, 153-158. http://dx.doi.org/10.1016/S0269-7491(99)00111-6

[10] Yamane, T. (1967) Statistics: An Introductory Analysis. 2nd Edition, Harper and Row, New York.

[11] Makwe, E. and Chup, C.D. (2013) Spatial Variation in Ground Pollution by Effluents from Karu Abattoir Abuja, Nigeria. *Ethiopian Journal of Environmental studies and Management*, **6**, 489-497.

[12] FEPA (2001) Guidelines and Standards for Environmental Pollution Control in Nigeria. National Environmental Standards-Parts 2 and 3. Government Press, Lagos, 238.

[13] Nwachukwu, E., Ihedioha, J.N., Eze, I. and Agbazue, V.E. (2014) Health Risk Assessment in Relation to Heavy Metals in Water Resources in Rural Regions of South East Nigeria. *International Journal of Physical Science*, **9**, 109-116. http://dx.doi.org/10.5897/IJPS2014.4125

[14] Adekunle, M., Adetunji, M.T., Gbadebo, A.M. and Banjoko, B (2007) Assessment of Groundwater Quality in a Typical Rural Settlement in South West Nigeria. INT. *Journal of Environmental Research and Public Health*, **4**, 307-318.

[15] Yahaya, M.I., Mohammed, S. and Abdullahi, B.K. (2009) Seasonal Variation of Heavy Metals Concentration in Abattoir Dumping Site in Nigeria. *Journal of Applied Sciences and Environmental Management*, **13**, 9-13.

Assessment and Distribution of Heavy Metals Pollutants in Manzala Lake, Egypt

Mahmoud Abd El-Kawy Zahran[1], Yasser Ahmed El-Amier[1], Abdelhamid Ahmed Elnaggar[2], Hoda Abd El-Azim Mohamed[3], Muhammad Abd El-Hady El-Alfy[3]

[1]Botany Department, Faculty of Science, Mansoura University, Mansoura, Egypt
[2]Soils Department, Faculty of Agriculture, Mansoura University, Mansoura, Egypt
[3]Marine Pollution Department, National Institute of Oceanography and Fisheries, Suez, Egypt
Email: yasran@mans.edu.eg

Abstract

Contamination with heavy metals is one of the most serious problems in the aquatic environments. In Egypt, Manzala Lake is suffering from this problem. The objective of this work was to assess heavy metals pollutants and their spatial distribution in Manzala Lake using GIS technique. Georeferenced water and sediment samples were randomly collected from the lake. The detected heavy metals were: Fe, Pb, Cu, Cd, Cr, Zn and Co. The obtained results indicated that the highest concentrations of heavy metals were observed in the northeastern and the southern parts of the lake nearby drains. This could be attributed to industrial, agricultural and municipal wastes coming through the drains especially Bahr El-Baqar drain and the industrial wastes coming from Port Said drains. From the geo-accumulation index, it was noticed that the lake is more polluted with cadmium and lead in the hydrosoils samples. All metals in water are within the EPA standard limit except for cadmium. Geostatistics provides effective methods to quantify the contaminated waters and sediments which support decision-making about redevelopment scenarios or remediation techniques.

Keywords

Manzala Lake, Heavy Metals, Indices, Pollution, GIS

1. Introduction

The contamination by heavy metals in the aquatic environments has drawn particular attentions due to their toxicity, persistence and biological accumulation. Heavy metals have low solubility and primarily get absorbed and accumulated on bottom sediments. Bottom lake sediments are sensitive indicators for monitoring pollutants as they act as a sink and a carrier for contaminants in aquatic environment. The most anthropogenic sources of metals are industrial, petroleum contamination and sewage disposal [1]. Manzala Lake, the largest of Egypt's Mediterranean wetlands and the most productive for fisheries, is suffering from land reclamation, industrial and nutrient pollution and overgrowth by water hyacinth. In the last six decades Manzala Lake was subjected to various threats: agriculture drainage, municipal sewage and industrial waste water. These pollutants have turned the lake into polluted, unhealthy ecosystem affecting fish production and natural resources that are distributed within the lake. A total of fresh water (mostly from agricultural drainage) flow annually into Manzala Lake from nine major drains and canals. It receives and carries the greatest part of wastewater into the lake through a very densely populated area of the Eastern Delta passing through Qalubyia, Sharkia, Ismailia and Port Said Governorates, and contributing much to the deteriorating water quality of the lake. Due to the toxicity of heavy metals, accurate information about their concentrations in aquatic ecosystem is needed. Accordingly, the objective of this study is to assess and evaluate the heavy metals pollutants in Lake Manzala using heavy metals indices with integrating GIS techniques for distribution of heavy metals in water and hydrosoils and joining them with their sources.

2. Materials and Methods

2.1. Study Area

Manzala Lake is the largest lake in the northern region of Egypt and the most productive for fisheries. The lake, as illustrated in **Figure 1**, lies between 31°45', 32°15'E and 31°00', 31°35'N. It covers an area of about 52,611 hectares. It is bordered by Suez Canal from east, Nile-Damietta branch from west and Mediterranean Sea from north. It is a shallow lagoon where about 50% of the lake area has a depth ranged between 0.5 and 1.0 m. The lake receives polluted water from different drains namely; Bahr El-Baqar, Ramsis, El-Matria, Hadous, Faraskur, El-Serw and Lissa El-Gamalia.

2.1.1. Climate

The climate of the Mediterranean coastal region of Egypt is belonging to the dry arid climatic zone. The study area lies in Meig's warm coastal deserts in which summer is warmest season with mean temperature less than 30°C, and winter is the coldest season with mean temperature above 10°C. The mean minimum air temperature at Manzala Lake ranged between 8.4°C in January and 21.4°C in August with mean annual temperature of about 15.4°C, and the mean maximum air temperature varies from 18.3°C in January to 31°C in August with mean annual temperature of 24.9°C. The relative humidity ranges from 68% in May to 76% in August with a mean annual value of 72%. The total annual rainfall attains a value of 106.7 mm. while the evaporation rate varies from

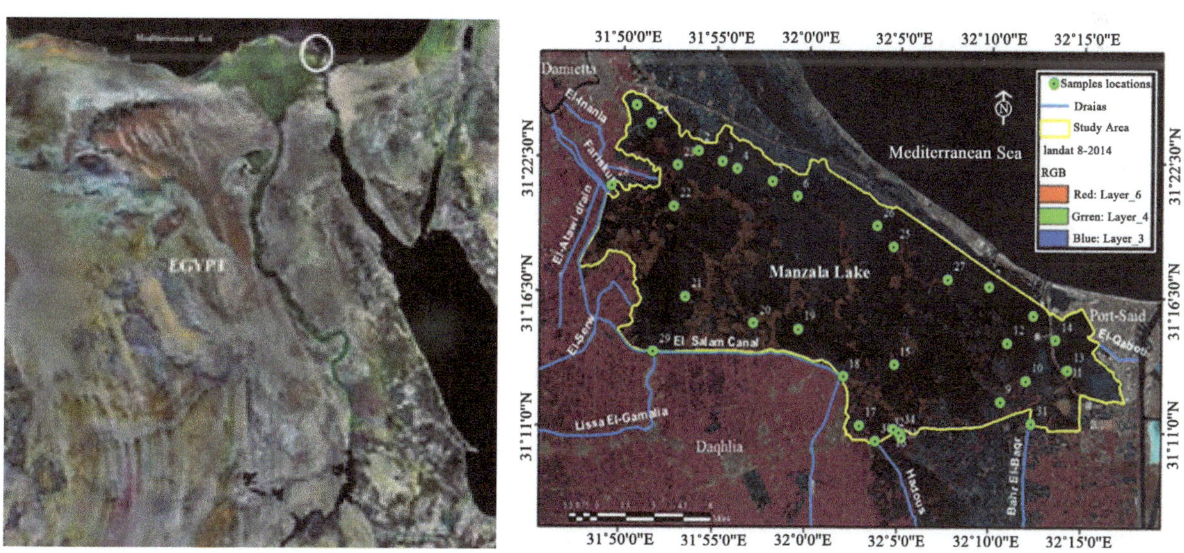

Figure 1. Location map of Manzala Lake and sampling locations.

2.8 mm/day in December and July to 5.4 mm/day in June. The prevailing wind is South westerly in January and February; North and north west in April, May, June, July, August and September; North easterly in October and November; and South westerly in December.

2.1.2. Water Resources

The lake is connected to the Mediterranean Sea via three outlets permitting exchange of water and biota between the lake and the sea, these outlets are El-Gamil, El-Boughdady and the new El-Gamil [2]. There is a small canal (El-Qaboty canal) that link between the lake and Suez Canal. There are major drains and canals flow into Lake Manzala as follows:

- Faraskur: Agricultural drain, serves 20,000 feddans, an area of about 44.48 km^2 (4% of total inflow).
- El-Sirw: Agriculture drain, serves 68,700 feddans (152.8 km^2), 13% of total inflow.
- Matariya: It serves 50,000 of land under agricultural reclamation (2%).
- Ramsis: Discharges a relatively small amount of water to Manzala Lake (24 km^2).
- Hadous: Is the largest drain in the eastern delta, serving some of agricultural land of about 1756.96 km^2 (49 %).
- Bahr El-Baqar: Serves an agricultural area of about 119.2 km^2, and receives about 300 million m^3/year of treated and untreated sewage from Cairo (25% of total inflow).

2.2. Samples

Thirty-four hydrosoil samples and thirty one water samples were randomly collected from different sites within the lake as illustrated in (**Figure 1**). Geographic positions of these samples were appointed using the GPS device (Garmin, etrex model). The collected samples were sent to the laboratory for heavy metal analysis. Seven essential heavy metals were studied in this work. These elements were Fe, Pb, Cu, Cd, Zn, Cr and Co.

2.3. Analytical Methods

2.3.1. The Analysis of the Dissolved Heavy Metals in Water Samples

Surface water samples were collected and then stored in acid-washed polyethylene bottles for analyses. Then, these samples were filtered using 0.45 um membrane filters. All the precautions occurred to minimize risks of sample contamination were followed during collection and treatment of samples. Solvent extraction was utilized using ammonium pyrrolidinedi-thiocarbamate (APDC) and methyl isobutyl ketone (MIBK). Where, water samples were pre-concentrated with APDC-MIBK extraction procedure according to the standard methods [3]. Heavy metals in the obtained solution were measured using the Flame Atomic Absorption Spectrophotometer (AAS: Perkin Elmer Analyst 100).

2.3.2. The Analysis of Heavy Metals in Hydrosoils Samples

Hydrosoil samples were collected using a Van-Veen grab coated with polyethylene. Sub-samples were taken from the central part of the grab to avoid contamination. The samples were kept in self-sealed acid pre-cleaned plastic bags, rinsed with metal-free water. The samples were deep-frozen until analysis. The samples were dried in the oven at 70°C and sieved using 0.75 mm plastic sieve and digested for about two hours in a mixture of 3:2:1 HNO_3, $HClO_4$ and HF acids, respectively as described by Oregioni and Astone [4].

2.4. Geostatistics

The Geostatistical analyst in ArcGIS (ver. 10.1) software package was used to develop the ordinary Kriging method and the semivariogram between each pairs of points versus their separation distances [5]. This semivariogram was used in predicting the studied heavy metals in both waters and sediments of Manzala Lake.

2.5. Heavy Metals Indices in Hydrosoil Samples

2.5.1. Enrichment Factor (EF)

Enrichment Factor is considered as an effective tool to evaluate the magnitude of contaminants in the environment. Iron (Fe) was chosen as the controlling element [6]. The EF values < 2 indicate that the metal is entirely from crustal materials or natural processes; whereas EF values > 2 suggest that the sources are more likely to be anthropogenic [7].

$$\text{Enrichment Factor} = (\text{M/Fe})\text{sample}/(\text{M/Fe})\text{background} \tag{1}$$

where, M is the concentration of metal. The background value is that of avera.ge shale. Six categories are recognized: ≤1 background concentration, 1 - 2 depletion to minimal enrichment, 2 - 5 moderate enrichment, 5 - 20 significant enrichment, 20 - 40 very high enrichment and >40 extremely high enrichment.

2.5.2. Contamination Factor (CF)

The CF is the ratio obtained by dividing the concentration of each metal in the sediment by the baseline or Background value [8].

$$\text{Contamination Factor}(\text{CF}) = \text{C metal/C background} \tag{2}$$

The following terminologies are used to describe the contamination factor: CF < 1 (low contamination factor); 1 ≤ CF < 3 (moderate contamination factors); 3 ≤ CF < 6 (considerable contamination factors) and CF ≥ 6 (very high contamination factor).

2.5.3. Pollution Load Index (PLI)

The PLI of a single site is the root of number (n) of multiplied together Contamination Factor (CF) values.

$$\text{PLI} = (\text{CF}_1 * \text{CF}_2 * \text{CF}_3 * \cdots * \text{CF}_n)^{1/n} \tag{3}$$

where, n is the number of metals (seven in the present study) and CF is the contamination factor. A PLI value of zero indicates perfection; a value of one indicates the presence of only baseline levels of pollutants, and values above one would indicate progressive deterioration of the site quality [9] [10].

2.5.4. Degree of Contamination (DC)

The degree of contamination (DC) defined as the sum of all contamination factors for a given site:

$$\text{Cd} = \sum_{i=1}^{n} \text{CF}i \tag{4}$$

where CF is the single contamination factor and n is the count of the elements present. DC values less than n would indicate low degree of contamination; $n \le \text{DC} < 2n$, moderate degree of contamination; $2n \le \text{DC} < 4n$, considerable degree of contamination; and DC > $4n$, very high degree of contamination. Where, n = 7 = the count of the studied heavy metals.

2.5.5. Geo-Accumulation Index (Igeo)

An index of geo-accumulation (Igeo) was originally defined by Müller [11] to determine and define the metal contamination in sediments by comparing current concentrations with pre-industrial levels.

$$\text{Igeo} = \text{Log2}\left(\frac{\text{Cn}}{1.5\text{Bn}}\right) \tag{5}$$

where, Cn is the measured concentration of heavy metals in sediments, Bn is the geochemical background value in average shale of element n and 1.5 is the background matrix correction due to anthropogenic influences. The geo-accumulation index (Igeo) was distinguished into seven classes by Buccolieri *et al.* [12]: Igeo ≤ 0, class 0, unpolluted; 0 < Igeo ≤ 1, class 1, from unpolluted to moderately polluted; 1 < Igeo ≤ 2, class 2, moderately polluted; 2 < Igeo ≤ 3, class 3, from moderately to strongly polluted; 3 < Igeo ≤ 4, class 4, strongly polluted; 4 < Igeo ≤ 5, class 5, from strongly to extremely polluted; and Igeo > 5, class 6, extremely polluted.

4. Results and Discussion

Metals generally enter the aquatic environment through erosion of the geological matrix, or due to anthropogenic activities caused by industrial effluents, domestic sewage, and mining wastes. Lead and Cadmium are toxic to living organisms even at quite low concentrations, whereas others, such as Zn and Cu, are biologically essential and natural constituents of aquatic ecosystems, and generally only become toxic at very high concentrations.

4.1. Heavy Metals in Water of Manzala Lake

The concentrations of dissolved heavy metals in water samples of Manzala Lake between different habitats are as shown in **Table 1**. **Figures 2(a)-(g)** illustrates the spatial distribution of dissolved heavy metals in the water samples collected from Manzala Lake. The mean concentration of the measured metals in water was found to be in the following sequence; Fe > Zn > Pb > Co > Cu > Cd > Cr.

Table 1. The concentrations of dissolved heavy metals in the water of Manzala Lake.

Site no.	X	Y	Fe	Zn	Pb	Cu	Co	Cr	Cd
1	31.402	31.845	13.65	5.773	3.808	1.229	1.269	0.33	0.932
2	31.372	31.901	34.73	7.423	6.463	2.728	1.597	0.63	0.887
3	31.365	31.923	17.57	8.612	4.712	1.868	1.896	0.416	0.891
4	31.360	31.936	24.58	8.519	4.871	1.892	3.077	0.492	0.902
5	31.352	31.968	17.71	7.292	4.954	1.617	1.673	0.454	0.941
6	31.342	31.990	20.16	6.741	2.694	1.683	1.645	0.951	0.725
7	31.282	32.166	14.06	9.976	5.482	1.468	1.621	0.721	0.785
8	31.262	32.206	20.89	9.506	3.794	2.113	1.689	0.359	0.802
9	31.204	32.177	11.405	5.375	5.5	1.15	0.975	1.27	0.225
10	31.218	32.200	41.52	3.265	3.93	0.655	0.775	1.1	0.23
11	31.225	32.237	32.115	5.58	9.15	0.98	1.395	1.98	5.5
12	31.244	32.183	20.56	3.935	5.225	1.38	1.765	1.66	4.67
13	31.225	32.237	14.17	5.98	3.75	1.265	1.775	22.1	3.93
14	31.246	32.226	17.4	3.15	5.19	1.24	1.925	4.015	9.15
15	31.228	32.080	12.315	1.18	5.95	0.64	0.85	0.46	2.09
16	31.184	32.080	19.87	1.635	1.925	1.14	1.005	0.3	3.75
17	31.186	32.049	17.05	1.305	1.805	0.72	0.995	1.84	5.19
18	31.220	32.034	12.395	1.365	2.21	0.605	0.805	0.095	0.245
19	31.251	31.993	7.38	1.495	1.595	0.515	0.705	0.34	0.27
20	31.255	31.952	4.785	1.14	3.25	1.03	0.685	0.45	0.29
21	31.273	31.890	9.055	1.17	2.505	0.695	0.71	0.24	0.28
22	31.334	31.879	11.4	1.34	4.56	1.05	2.324	0.23	0.275
23	31.362	31.882	12.98	1.32	5.63	1.65	0.86	0.25	0.275
24	31.390	31.858	13.34	1.12	6.46	1.76	1.552	0.35	0.285
25	31.308	32.078	15.62	5.605	8.234	1.776	1.765	0.65	0.924
26	31.322	32.063	17.95	6.34	3.992	1.392	2.103	0.86	0.802
27	31.286	32.129	21.48	4.56	7.374	3.793	2.369	1.23	1.212
28	31.348	31.823	183.3	11.77	6.12	4.879	6.557	0.973	0.987
29	31.235	31.862	167.4	11.91	6.73	4.775	5.937	0.633	0.325
30	31.176	32.064	42.73	7.069	3.58	2.996	6.557	0.252	0.805
31	31.189	32.205	47.74	9.465	4.77	2.947	7.15	0.592	0.748
	Mean		29.59	5.19	4.72	1.73	2.13	1.49	1.59
	SE (±)		7.24	0.61	0.33	0.20	0.33	0.70	0.37
	EPA, 2002		300	5000	50	50		100	2.37

EPA (2002): Environmental Protection Agency limit for water & values in (μg/l).

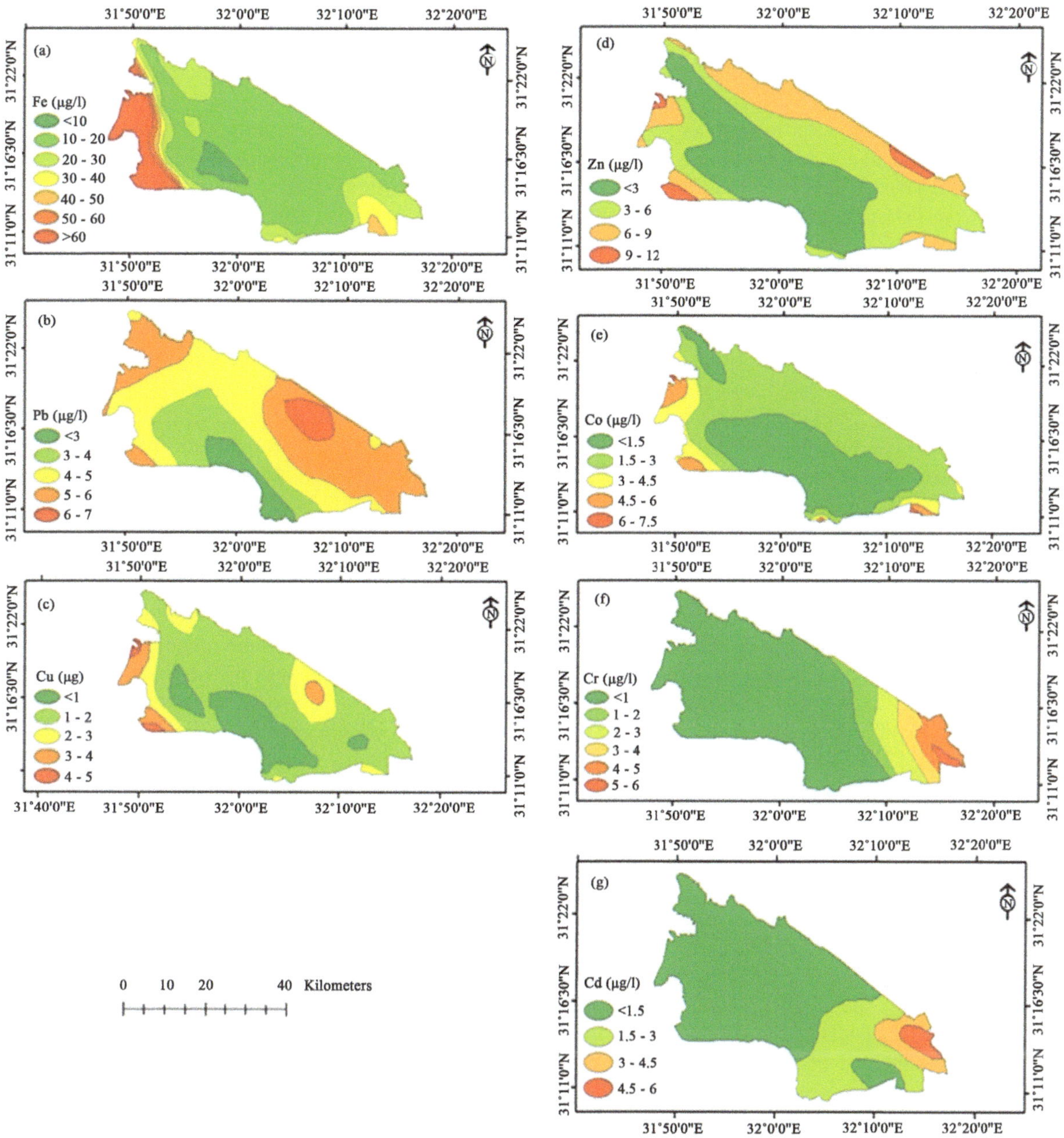

Figure 2. The spatial distribution of (a) iron; (b) lead; (c) copper; (d) zinc; (e) cobalt; (f) chromium and (g) cadmium in the water samples of Manzala Lake using GIS technique.

Iron varied from 4.78 µg/l nearby El-Gamalia city to 183.3 µg/l at Faraskur drain. This result is lower than those (1420 µg/l - 696 µg/l) recorded by Saeed and Shaker [1] and El-Alfy [13]. While the concentration of copper varied from 0.51 to 4.87 at Faraskur drain. The highest concentrations of iron and copper are distributed at the southwestern parts of lake especially nearby drains, where there are more agricultural wastes as revealed by Arain *et al.* [14]. The concentrations of dissolved Fe in the oxygen depleted waters were markedly higher than those found in the oxic northern waters of Manzala Lake "*i.e.* nearby El-Boughaz area" [9]. Higher values of Cu in Manzala Lake were observed in both water and hydrosoil by Ali and Abdel-Satar [15].

Zinc is one of the most abundant toxic heavy metals, the oral toxicity of humans to most zinc compounds is relatively low. High concentrations of zinc in the marine environment may exist from the discharge of industrial wastes especially electroplating and synthetic fiber production. It ranged between 1.12 and 11.91 µg/l. It's noticeable that the highest concentrations of zinc were observed as well at the drains and nearby industrial areas as a result to direct discharge of industrial wastes in these areas. However, the lowest concentration of zinc is distributed in the middle part of the lake far from drains. These values are lower than that recorded by Bahnsawy *et al.* [16].

Dissolved lead in water ranged between 1.59 to 8.23 μg/l. The lowest concentration of lead was recorded far away from drainage areas, but the highest concentration was observed at northeastern side nearby the industrial compound. These areas receive huge quantities of sewage and industrial wastes, as well as from spill of leaded petrol from fishing boats and dust which holds a huge amount of lead [10]. The present result of lead in water is within the limit of EPA [10]. The United States Environmental Protection Agency EPA [18] has classified Pb as being potentially hazardous and toxic to most forms of life.

Cobalt was in the range between 0.68 to 7.15 μg/l. The lower concentrations of cobalt are distributed at the middle and northern parts of the Lake. The highest concentration of cobalt was recorded nearby drains at the southern parts as a result of agricultural wastes especially Bahr El-Baqar drain. This result agreed with Nagpal, [19].

The sources of chromium entering the aquatic environment are from paint and chemical works, oil drilling and recovery rigs. Large quantities of chromium may be released from petrochemical industries and cement, fertilizer, power, and chlor-alkali plants. It varied from 0.09 to 22.10 μg/l. The highest concentration of Cr is found at Ashtoum El-Gamil area, near tourist villages and Company of Natural gas Petroget. Furthermore, untreated domestic sewage was discharged into the sea from activities at the coastal area [20]. The lowest concentration of Cr is obtained nearby Hadous drain. The highest concentration in Manzala is higher than the threshold limit (1.5 μg/l) of Environment Canada [21], but is within EPA limit [10].

Cadmium is one of the black listed elements, it can be considered as one of the most dangerous elements on marine life and humans. The dissolved cadmium in water of Manzala Lake is in the range between 0.22 to 9.15 μg/l. The lowest concentration was recorded at open water far from drains, but the highest concentration was recorded nearby El-Qaboty area south to port said city as a result of urban extension and agricultural disposal especially the phosphatic fertilizers, this is in agreement with Bahnasawy *et al.* [17]; El-Serehy *et al.* [20] and Hamed *et al.* [22]. The estimated Cd value in this study is more than the EPA limits [10]. It is one of the most toxic elements with widespread carcinogenic effects in humans and considered to be toxic if its concentration exceeds 0.01 mg/L both in drinking and irrigation water.

4.2. Heavy Metals in the Hydrosoils of Manzala Lake

Hydrosoil contamination poses one of the worst environmental problems in ecosystems, acting as sinks and sources of contaminants in aquatic systems. Hydrosoil analyses play an important role in assessing the pollution status of the environment. The concentrations of heavy metals in the hydrosoils of Manzala Lake are as given in **Table 2** and the spatial distribution is illustrated in **Figures 3(a)-(g)**. The concentrations of heavy metals in the hydrosoils of Manzala Lake take the following sequence; Fe > Cr > Zn > Pb > Cu > Co > Cd, which except Fe differs from these of the lake's water.

Iron in the hydrosoils of Manzala Lake was in the range between 250 and 666.37 μg/g. Where copper varied from 1.25 to 52.68 μg/g. The highest value of iron was recorded at Bahr El-Baqar drain. For copper the highest concentration was recorded at Hadous drain. These drains have agricultural drainage, industrial and domestic wastes. These results are in agreement with those obtained by Saeed and Shaker [1] and El-Alfy [13]. It was found that those drains are rich in organic carbon and some authors found a correlation between the concentration of heavy metals in sediment and the abundance of organic matter [13]. Whereas the lowest values of iron were recorded far away from drains, nearby El-Boughaz area and the northern parts of the Lake. For copper, the lowest value was recorded at El-Temsah area far away from drains as well. The values of copper are higher than that recorded by Hamed *et al.* [22]. The maximum values of iron and copper in hydrosoils exceed the limits set by EPA [10] where the guideline values are 15 and 25 μg/g for iron and copper, respectively.

The range of lead in the hydrosoils of Manzala Lake was between 3.44 and 65.53 μg/g. The maximum concentration of lead was recorded nearby the drains, as they receive huge quantities of sewage and industrial wastes, beside agricultural drainage water via Bahr El-Baqar drain and may attributed to the decaying of plankton and precipitation of organic matter associated with Pb and Cd to the sediment [17]. The lowest concentration of lead was noticed at El-Temsah area (at the northern parts of the lake far from drainage water). The maximum value of lead is higher than EPA limit [10], but within the limit of EU [23].

Table 2. The concentrations of heavy metals in the hydrosoils of Manzala Lake.

Site no.	X	Y	Fe	Zn	Cr	Cu	Pb	Co	Cd
1	31.402	31.845	301.60	26.85	27.51	8.80	11.27	8.51	1.40
2	31.372	31.901	250.00	4.76	4.52	1.25	3.44	2.97	0.66
3	31.365	31.923	306.10	6.75	66.24	8.21	14.11	13.91	1.46
4	31.360	31.936	271.70	5.75	8.80	2.21	5.53	3.32	0.74
5	31.352	31.968	305.40	12.79	42.72	10.72	12.18	12.34	1.37
6	31.342	31.990	298.50	22.22	27.86	6.59	9.62	7.50	1.16
7	31.282	32.166	290.50	2.98	54.46	10.05	19.01	9.70	2.04
8	31.262	32.206	306.50	78.06	39.73	7.08	14.36	14.48	1.56
9	31.204	32.177	310.30	3.38	38.80	30.14	17.27	20.24	1.31
10	31.218	32.200	294.10	75.65	12.41	9.72	65.53	6.79	1.19
11	31.225	32.237	306.80	4.07	62.98	2.50	3.80	19.01	1.23
12	31.244	32.183	307.20	28.10	28.76	9.46	63.12	14.36	2.65
13	31.225	32.237	303.20	6.01	5.78	3.46	12.52	12.52	1.91
14	31.246	32.226	306.40	3.85	67.49	9.46	4.70	15.16	2.71
15	31.228	32.080	308.80	3.18	21.19	28.61	15.36	17.20	1.94
16	31.184	32.080	306.90	3.02	16.27	24.40	23.70	15.92	2.17
17	31.186	32.049	307.50	3.02	16.88	26.33	36.51	18.04	1.78
18	31.220	32.034	304.90	2.51	9.41	21.63	22.29	16.37	2.38
19	31.251	31.993	308.10	2.59	9.31	25.52	19.97	17.91	2.56
20	31.255	31.952	301.50	3.35	2.64	16.85	21.21	14.05	2.30
21	31.273	31.890	309.20	3.51	24.88	33.25	14.82	19.88	2.33
22	31.334	31.879	305.40	3.08	22.23	17.62	21.10	15.13	1.96
23	31.362	31.882	305.30	2.43	22.62	17.14	12.50	14.88	1.49
24	31.390	31.858	305.20	2.84	21.87	21.59	20.66	5.51	2.06
25	31.308	32.078	308.80	69.36	31.50	12.64	24.52	16.61	2.82
26	31.322	32.063	309.10	77.13	29.73	13.85	25.57	15.65	2.82
27	31.286	32.129	309.30	157.70	37.85	14.14	23.36	16.02	2.73
28	31.348	31.823	630.44	50.84	37.48	10.81	22.06	6.56	0.56
29	31.235	31.862	661.21	85.77	86.09	42.47	26.45	5.94	0.66
30	31.176	32.064	654.77	72.61	101.50	52.68	20.44	6.56	0.88
31	31.189	32.205	666.37	71.21	101.40	50.36	20.70	7.15	1.07
32	31.180	32.087	299.50	19.00	14.51	18.80	33.50	7.59	1.16
33	31.178	32.087	301.30	19.50	16.00	14.44	47.54	11.76	2.62
34	31.182	32.085	299.40	54.81	15.23	12.14	18.65	7.49	1.26
	Mean		342.98	29.08	33.14	17.50	21.39	12.27	1.73
	SE (±)		19.84	6.34	4.51	2.20	2.44	0.86	0.12
	Average Shale		47200	95	90	45	20	19	0.30
	EPA, 2002		15	123	25	25	10	-	6
	EU, 2002		-	300	150	140	300	11.6	3

EPA: Environmental Protection Agency for sediment samples in (μg/g); EU: European Union Standard in (μg/g).

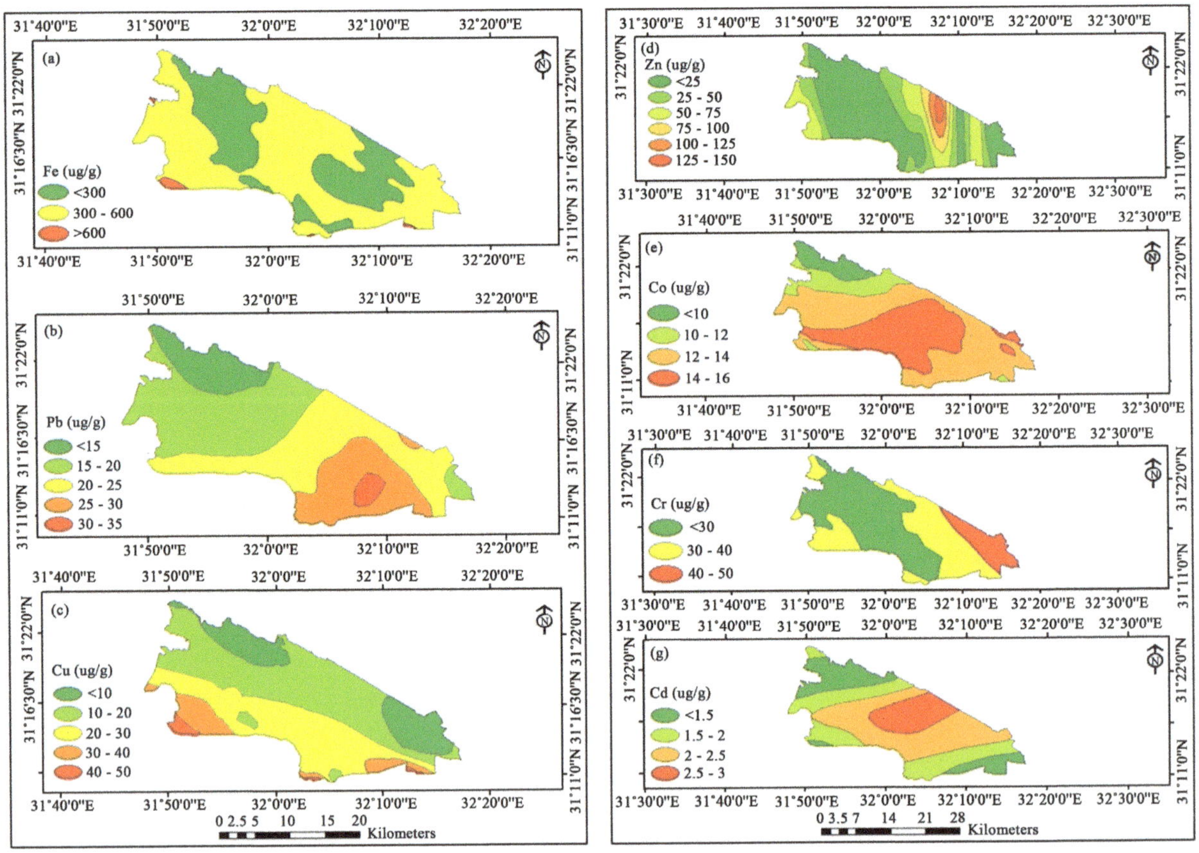

Figure 3. The spatial distribution of (a) iron; (b) lead; (c) copper; (d) zinc (e) cobalt; (f) chromium and (g) cadmium in the hydrosoils of Manzala Lake.

Zinc varied from 2.43 to 157.70 μg/g, its highest concentration may be as a result of industrial disposal spilled directly or indirectly to the lake. The maximum concentration of Zinc in hydrosoils was obtained nearby the industrial compound northern to Lake Manzala. This result is higher than those recorded by Hamed *et al.* [22] and El-Alfy [13]. The lowest value of Zn was estimated at the western part of the lake far from industrial wastes. It's obvious that Cd and Pb accumulation decreased in saline medium while the accumulation of Zn increased so the highest concentration of zinc in the hydrosoil especially in the northern parts where sea water intrusion occurs [24]. The maximum values of zinc are more than the EPA limit, but within the limit of EU [23].

The concentration of cobalt in the hydrosoils of Manzala Lake was in the range of (2.79 to 20.24 μg/g). Its maximum concentration was observed nearby Damietta city which may due to the irrigation of agricultural lands with untreated water and uses of agricultural fertilizers. Also high concentrations of Co were observed nearby Bahr El-Baqar drain. Its lowest value was recorded at the northwestern part of the lake far away from drains. These values were lower than (54.29 to 80.30 μg/g) recorded by El-Bady [25] at Bahr El-Baqar region. Cobalt in the hydrosoils of Manzala Lake recorded values more than the EU limit [23].

Chromium varied from 2.64 to 101.50 μg/g. The lowest concentration of Cr was recorded far away from drains. While the highest concentration was observed nearby drains (especially, Bahr El-Baqar and Hadous drains), the areas south to Port Said city near the industrial activities [19]. Also, high concentrations of Cr were found at Ashtoum El-Gamil area and near tourist villages and Company of Natural gas Petroget. Furthermore, untreated domestic sewage was discharged into this area from activities at the coastal area, the combustion of coal and oil. Chromium is within the limit of EU [23], but its maximum value is more than what be revealed by the EPA limit [10].

Cadmium in the hydrosoils varied from 0.56 to 2.82 μg/g. The lowest concentration was recorded at Faraskur drain. However, the higher concentrations were concentrated in the area nearby the industrial activities. The previous results are in agreement with those obtained by Bahnasawy *et al.* [17] and Hamed *et al.* [22]. They reported that the high concentrations in this area were due to the paint factories disposal without treatment directly to the lake. Cadmium recorded higher values than the sediment quality guidelines and agreed with Saeed and Shaker [1]. The values of cadmium in hydrosoils are within EPA [10] limit of a guide value (6 μg/g). Cadmium in uncontaminated soils must not exceed 0.7 ppm.

4.3. Indices of Heavy Metals in the Hydrosoils of Manzala Lake

4.3.1. The Enrichment Factor

The Enrichment Factors of heavy metals in the hydrosoils of Manzala Lake are as shown in **Table 3**; it's indicated that the EF of Cu ranged between 5.24 and 112.79. For Zn, the EF varied from 3.95 to 253.32 and EF of Cr varied from 4.59 to 115.53. The EF of Pb varied from 32.51 to 484.91 and for Co, it varied from 22 to 160 and EF of Cd varied from 140.25 to 1435.26. From the previous results, it's obvious that the EF of all metals is in the high significant categories and the sources of these metals in Lake Manzala are almost from anthropogenic activities. The sequence of EF for heavy metals in the hydrosoils of Manzala Lake is: Cd > Pb > Zn > Co > Cr > Cu. From that, cadmium is more abundant than other metals; whereas Cu showed the lowest appearance.

4.3.2. Contamination Factor

From **Table 4**, the CF values of Fe showed low contamination factor. It ranged between low at all sites for copper and moderate CF at Hadous and Bahr El-Baqar Drains as (CF > 1). For Cr, CF is lower than 1 (low CF) and >1 only in the northern part of the Lake. CF of lead varied from 0.17 to 3.28 (low to considerable CF). The values of CF for cobalt are <1 in all sites (low CF) and >1 only nearby Bahr El-Baqar drain (moderate CF). The CF of cadmium showed moderate, considerable and very high CF.

4.3.3. Pollution Load Index (PLI) and Degree of Contamination (DC)

The PLI and DC are indicated in **Table 4**, where the PLI in this study area of Manzala Lake ranged between 0.11 and 0.57. This showed that there is no appreciable pollution in Lake Manzala with those metals. From the results of contamination degree (**Table 4**), it ranged between low to moderate degree of pollution. The spatial distribution of DC is illustrated in **Figure 4**.

4.3.4. The Geo-Accumulation Index (Igeo)

The results of Igeo are as shown in **Table 5** and **Figure 5**. The negative values of Fe, Cr, Co, Cu and Zn and according to the classification of Muller [26], it indicated that the Lake isn't polluted with those metals. From the geo-accumulation index results, Igeo values of lead showed unpolluted to moderately polluted categories nearby the biological treatment project east to Port Said City, at the southern part of the Lake nearby Drains and at Ibn-Salam area (nearby Hadous drain). The Igeo results of cadmium also showed moderately polluted category in all locations.

4.4. The Correlation between Heavy Metals in Water and Hydrosoils

From **Table 6**, there are significant positive correlation between Fe in hydrosoil with Cu, Cr in hydrosoil (r = 0.7); Fe, Zn and Co in water (r = 0.7, 0.5 and 0.9) and correlates negatively with Co and Cd in hydrosoil as r = −0.4 for each. Copper in hydrosoil showed a positive significant correlation with Cr in hydrosoil and Co in water where (r = 0.5 for each). Zinc in hydrosoil correlated with Cr in hydrosoil, Fe, Co and Zn in water (r = 0.4) and with Cu in water (r = 0.6). Cr in hydrosoil correlated with Fe in water (r = 0.37), Cu in water (r = 0.49), Zn in water (r = 0.54) and Co in water (r = 0.7). Cobalt in hydrosoil showed only significant positive correlation with Cd in hydrosoil (r = 0.6) and negative with Fe, Cu, Zn and Co in water as (r = −0.44, −0.53, −0.54 and −0.53 respectively). Also cadmium in hydrosoil showed negative significant correlation with Fe, Cu, Zn and Co in water (r = −0.54, −0.47, −0.58 and −0.51 respectively). There are a significant positive correlation between iron in water and Cu, Zn and Co in water (r = 0.8, 0.6 and 0.7). Cu in water and Zn, Pb and Co in water (r = 0.7, 0.4 and 0.8). Between Zn and Co in water (r = 0.6) and Cr and Cd in water (r = 0.36). Other relations between elements showed low or no significant correlations. It's obvious that the heavy metals with positive correlation were considered to have similar sources as estimated by Dan *et al.* [27].

Table 3. The enrichment factor (EF) of heavy metals in the hydrosoils of Manzala Lake.

NO	Enrichment Factor (EF)						
	Fe	Cu	Zn	Cr	Pb	Co	Cd
1	1.00	30.60	44.23	47.84	88.19	70	727.72
2	1.00	5.24	9.46	9.49	32.51	30	412.84
3	1.00	28.13	10.96	113.49	108.79	113	748.37
4	1.00	8.53	10.51	16.98	48.02	30	427.35
5	1.00	36.82	20.81	73.36	94.12	100	705.78
6	1.00	23.16	36.98	48.95	76.05	62	609.83
7	1.00	36.29	5.10	98.32	154.43	83	1103.23
8	1.00	24.23	126.54	67.98	110.57	117	799.24
9	1.00	101.88	5.41	65.58	131.35	162	664.72
10	1.00	34.67	127.80	22.13	525.84	57	638.75
11	1.00	8.56	6.59	107.66	29.23	154	629.23
12	1.00	32.32	45.45	49.11	484.91	116	1357.10
13	1.00	11.99	9.86	9.99	97.49	103	991.53
14	1.00	32.40	6.25	115.53	36.19	123	1391.96
15	1.00	97.18	5.12	35.99	117.43	138	986.39
16	1.00	83.39	4.89	27.80	182.25	129	1112.97
17	1.00	89.81	4.88	28.79	280.21	146	909.21
18	1.00	74.41	4.09	16.19	172.53	133	1230.18
19	1.00	86.88	4.18	15.84	152.97	144	1304.73
20	1.00	58.62	5.52	4.59	166.02	116	1202.31
21	1.00	112.79	5.64	42.20	113.12	160	1183.56
22	1.00	60.52	5.01	38.17	163.05	123	1011.28
23	1.00	58.89	3.95	38.86	96.63	121	766.83
24	1.00	74.20	4.62	37.58	159.76	45	1062.98
25	1.00	42.93	111.60	53.50	187.39	134	1435.26
26	1.00	47.00	123.98	50.44	195.23	126	1433.87
27	1.00	47.95	253.32	64.18	178.24	129	1390.21
28	1.00	17.99	40.07	31.18	82.58	26	140.25
29	1.00	67.37	64.45	68.28	94.41	22	157.28
30	1.00	84.39	55.10	81.30	73.67	25	210.25
31	1.00	79.27	53.10	79.80	73.31	27	252.87
32	1.00	65.84	31.52	25.41	263.97	63	610.42
33	1.00	50.27	32.16	27.85	372.37	97	1365.51
34	1.00	42.53	90.95	26.68	147.01	62	659.50

Table 4. The contamination factor (CF), pollution load index (PLI) and degree of contamination (DC) for heavy metals in the hydrosoils of Manzala Lake.

NO	Contamination factor (CF)							PLI	DC
	Fe	Cu	Zn	Cr	Pb	Co	Cd		
1	0.006	0.20	0.29	0.30	0.56	0.45	4.65	0.28	6.45
2	0.005	0.03	0.05	0.05	0.17	0.16	2.19	0.08	2.65
3	0.006	0.18	0.70	0.08	0.71	0.73	4.85	0.29	7.25
4	0.006	0.05	0.09	0.06	0.28	0.17	2.46	0.11	3.12
5	0.006	0.24	0.45	0.14	0.61	0.65	4.57	0.29	6.66
6	0.006	0.15	0.29	0.25	0.48	0.39	3.86	0.24	5.43
7	0.006	0.22	0.57	0.03	0.95	0.51	6.79	0.26	9.09
8	0.006	0.16	0.42	0.87	0.72	0.76	5.19	0.38	8.12
9	0.007	0.67	0.41	0.04	0.86	1.07	4.37	0.31	7.42
10	0.006	0.22	0.13	0.84	3.28	0.36	3.98	0.35	8.81
11	0.007	0.06	0.66	0.05	0.19	1.00	4.09	0.19	6.05
12	0.007	0.21	0.30	0.31	3.16	0.76	8.83	0.43	13.58
13	0.006	0.08	0.06	0.07	0.63	0.66	6.37	0.18	7.87
14	0.006	0.21	0.71	0.04	0.23	0.80	9.04	0.26	11.04
15	0.007	0.64	0.22	0.04	0.77	0.91	6.45	0.28	9.03
16	0.007	0.54	0.17	0.03	1.19	0.84	7.24	0.28	10.01
17	0.007	0.59	0.18	0.03	1.83	0.95	5.92	0.30	9.50
18	0.006	0.48	0.10	0.03	1.11	0.86	7.95	0.25	10.54
19	0.007	0.57	0.10	0.03	1.00	0.94	8.52	0.26	11.16
20	0.006	0.37	0.03	0.04	1.06	0.74	7.68	0.20	9.93
21	0.007	0.74	0.26	0.04	0.74	1.05	7.75	0.31	10.59
22	0.006	0.39	0.23	0.03	1.06	0.80	6.54	0.27	9.06
23	0.006	0.38	0.24	0.03	0.63	0.78	4.96	0.23	7.02
24	0.006	0.48	0.23	0.03	1.03	0.29	6.87	0.24	8.94
25	0.007	0.28	0.33	0.77	1.23	0.87	9.39	0.47	12.88
26	0.007	0.31	0.31	0.86	1.28	0.82	9.39	0.47	12.98
27	0.007	0.31	0.40	1.75	1.17	0.84	9.11	0.54	13.59
28	0.013	0.24	0.39	0.56	1.10	0.35	1.87	0.34	4.53
29	0.014	0.94	0.91	0.95	1.32	0.31	2.20	0.52	6.66
30	0.014	1.17	1.07	0.81	1.02	0.35	2.92	0.55	7.34
31	0.014	1.12	1.07	0.79	1.04	0.38	3.57	0.57	7.97
32	0.006	0.42	0.15	0.21	1.68	0.40	3.87	0.30	6.74
33	0.006	0.32	0.17	0.22	2.38	0.62	8.72	0.37	12.42
34	0.006	0.27	0.16	0.61	0.93	0.39	4.18	0.31	6.56

Table 5. The geoaccmulation index (Igeo) of heavy metals in the hydrosoils of Manzala Lake.

NO	Geoaccmulation index (Igeo)						
	Fe	Cu	Zn	Cr	Pb	Co	Cd
1	−2.37	−0.88	−0.72	−0.69	−0.43	−0.53	0.49
2	−2.45	−1.73	−1.48	−1.48	−0.94	−0.98	0.16
3	−2.36	−0.91	−1.32	−0.31	−0.33	−0.31	0.51
4	−2.42	−1.48	−1.39	−1.19	−0.73	−0.93	0.21
5	−2.37	−0.80	−1.05	−0.50	−0.39	−0.36	0.48
6	−2.38	−1.01	−0.81	−0.69	−0.49	−0.58	0.41
7	−2.39	−0.83	−1.68	−0.39	−0.20	−0.47	0.66
8	−2.36	−0.98	−0.26	−0.53	−0.32	−0.29	0.54
9	−2.36	−0.35	−1.62	−0.54	−0.24	−0.15	0.46
10	−2.38	−0.84	−0.28	−1.04	0.34	−0.62	0.42
11	−2.36	−1.43	−1.54	−0.33	−0.90	−0.18	0.44
12	−2.36	−0.85	−0.71	−0.67	0.32	−0.30	0.77
13	−2.37	−1.29	−1.37	−1.37	−0.38	−0.36	0.63
14	−2.36	−0.85	−1.57	−0.30	−0.81	−0.27	0.78
15	−2.36	−0.37	−1.65	−0.80	−0.29	−0.22	0.63
16	−2.36	−0.44	−1.67	−0.92	−0.10	−0.25	0.68
17	−2.36	−0.41	−1.67	−0.90	0.09	−0.20	0.60
18	−2.37	−0.49	−1.75	−1.16	−0.13	−0.24	0.72
19	−2.36	−0.42	−1.74	−1.16	−0.18	−0.20	0.75
20	−2.37	−0.60	−1.63	−1.71	−0.15	−0.31	0.71
21	−2.36	−0.31	−1.61	−0.73	−0.31	−0.16	0.71
22	−2.37	−0.58	−1.67	−0.78	−0.15	−0.28	0.64
23	−2.37	−0.60	−1.77	−0.78	−0.38	−0.28	0.52
24	−2.37	−0.50	−1.70	−0.79	−0.16	−0.71	0.66
25	−2.36	−0.73	−0.31	−0.63	−0.09	−0.23	0.80
26	−2.36	−0.69	−0.27	−0.66	−0.07	−0.26	0.80
27	−2.36	−0.68	0.04	−0.55	−0.11	−0.25	0.78
28	−2.05	−0.80	−0.45	−0.56	−0.13	−0.64	0.10
29	−2.03	−0.20	−0.22	−0.20	−0.05	−0.68	0.17
30	−2.03	−0.11	−0.29	−0.12	−0.17	−0.64	0.29
31	−2.03	−0.13	−0.30	−0.12	−0.16	−0.60	0.38
32	−2.37	−0.56	−0.88	−0.97	0.05	−0.57	0.41
33	−2.37	−0.67	−0.86	−0.93	0.20	−0.38	0.76
34	−2.37	−0.75	−0.41	−0.95	−0.21	−0.58	0.45

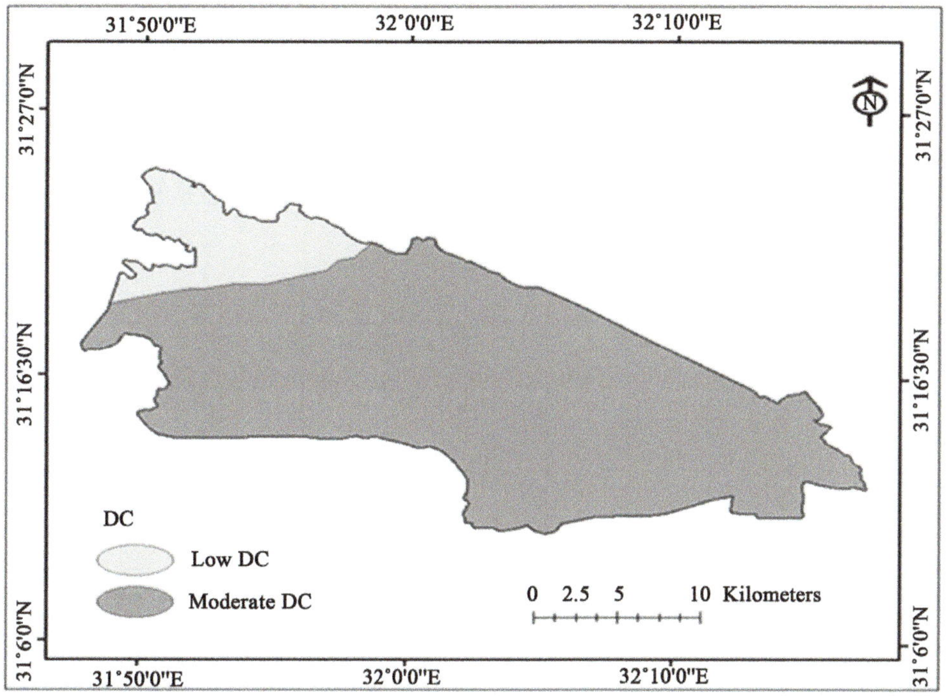

Figure 4. The spatial distribution of contamination degree (DC) of heavy metals in the hydrosoils of Manzala Lake.

Table 6. Pearson-moment correlation (r) between the concentrations of heavy metals in the hydrosoils and water of Manzala Lake.

	Variables	Hydrosoil (μg/g)							Water (μg/l)						
		Fe	Cu	Zn	Cr	Pb	Co	Cd	Fe	Cu	Zn	Cr	Pb	Co	Cd
Hydrosoil (μg/g)	Fe	1													
	Cu	0.669**	1												
	Zn	0.426*	0.180	1											
	Cr	0.714**	0.495**	0.362*	1										
	Pb	0.046	0.119	0.268	−0.154	1									
	Co	−0.352*	0.057	−0.185	−0.109	−0.014	1								
	Cd	−0.445**	−0.080	−0.022	−0.269	0.239	0.641**	1							
Water (μg/l)	Fe	0.755**	0.202	0.367*	0.372*	0.116	−0.449*	−0.549**	1						
	Cu	0.721**	0.193	0.614**	0.498**	−0.073	−0.539**	−0.479**	0.800**	1					
	Zn	0.510.9**	−0.079	0.395*	0.546**	−0.181	−0.541**	−0.589**	0.606**	0.709**	1				
	Cr	−0.084	−0.240	−0.108	−0.156	−0.089	0.035	0.081	−0.063	−0.085	0.036	1			
	Pb	0.108	−0.211	0.272	0.309	−0.151	−0.125	−0.152	0.285	0.432*	0.336	−0.048	1		
	Co	0.934**	0.480**	0.481**	0.703**	−0.010	−0.536**	−0.510**	0.723**	0.805**	0.660**	−0.044	0.216	1	
	Cd	−0.149	−0.222	−0.214	0.113	−0.018	0.285	0.254	−0.100	−0.206	−0.170	0.368*	0.077	−0.120	1

**Correlation is significant at the 0.01 level. *Correlation is significant at the 0.05 level.

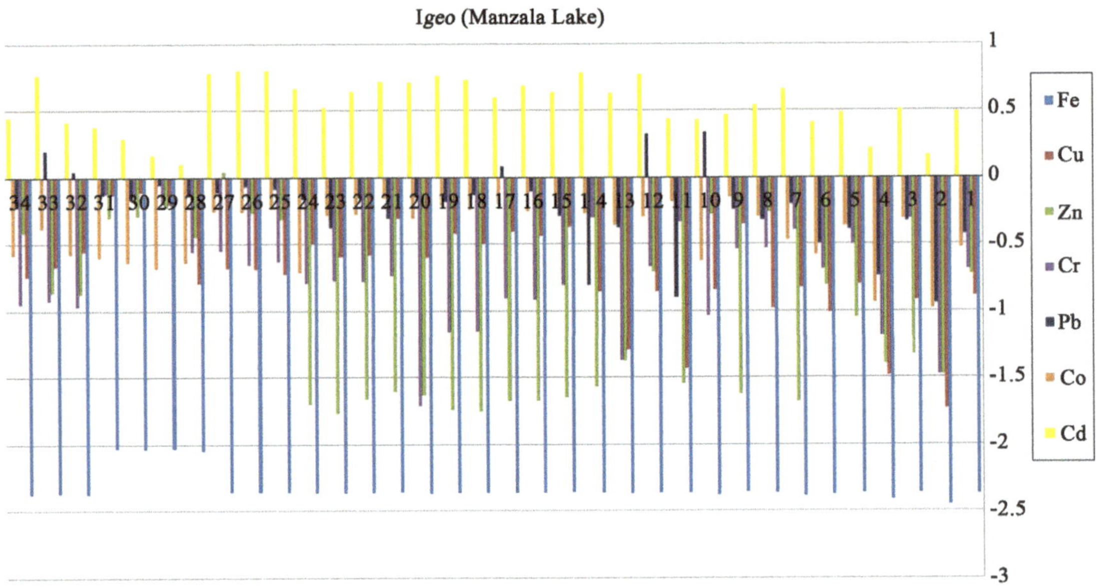

Figure 5. The geoaccumulation index for the heavy metals in the hydrosoils of Manzala Lake.

5. Conclusion

It could be concluded that the geospatial tools such as ordinary Kriging could be very helpful in evaluating and studying the spatial distribution of heavy metals in both water and sediment of Manzala Lake in Egypt. The obtained results clearly demonstrate that Manzala Lake is highly contaminated with Fe, Cd, Pb and Cr due to the continuous discharge of different pollutants into it. It can also be concluded that the southern drains namely, Bahr El-Baqar, Ramsis, El-Matria, Hadous, Faraskur, El-Serw and Lissa El-Gamalia play an important role in causing a severe pollution in Manzala Lake. This is especially in the southeastern parts of the lake, which receive great quantities of effluents from Bahr El-Baqar, Hadous and Ramsis drains. From the sequence of enrichment factor (EF), for heavy metals in the hydrosoils of Manzala Lake, cadmium is more abundant than other metals, whereas Cu shows the lowest appearance. Contamination factor (CF) showed that only nearby Bahr El-Baqar drain (moderate CF) and the CF of cadmium showed moderate, considerable and very high CF there. From the pollution load index (PLI) and degree of contamination (DC), it is showed that there is no appreciable pollution in Lake Manzala. Geo-accumulation index (Igeo) results of lead and cadmium showed moderately polluted category in all locations. Great efforts and cooperation between different authorities are needed to protect the lake from pollution and reduce the environmental risk. This can be achieved through the treatment of agricultural, industrial, and sewage discharge. Regular evaluation of pollutants in the lake is also very important.

References

[1] Saeed, S.M. and Shaker, I.M. (2008) Assessment of Heavy Metals Pollution in Water and Sediments and Their Effect on Oreochromis Niloticus in the Northern Delta Lakes, Egypt. *8th International Symposium on Tilapia in Aquaculture*, 475-490.

[2] Elewa, A.A., Saad, E.A., Shehata, M.B. and Ghallab, M.H. (2007) Studies on the Effect of Drain Effluents on the Water Quality of Lake Manzala. *Egyptian Journal of Aquatic Biology and Fisheries*, **11**, 65-78.

[3] APHA (1989) Standard Methods for the Examination of Water and Wastewater, Part 3, Determination of Metals. 17th, American Public Health Association, Washington DC, 164.

[4] Oregioni, B. and Aston, S.R. (1984) The Determination of Selected Trace Metals in Marine Sediments by Flameless/Flam-Atomic Absorption Spectrophotometry. *IAEA Monaco Laboratory Internal Report.*

[5] ESRI (2012) ArcGIS Geostatistical Analyst Tutorial (ArcGIS®10.1). ESRI, USA.

[6] Seshan, B.R.R., Natesan, U. and Deepthi, K. (2010) Geochemical and Statistical Approach for Evaluation of Heavy Metal Pollution in Core Sediments in Southeast Coast of India. *International Journal of Environmental Science and*

Technology, **7**, 291-306. http://dx.doi.org/10.1007/BF03326139

[7] Liaghati, T., Preda, M. and Cox, M. (2003) Heavy Metal Distribution and Controlling Factors within Coastal Plain Sediments, Bells Creek Catchments, Southeast Queensland, Australia. *Environment International*, **29**, 935-948. http://dx.doi.org/10.1016/S0160-4120(03)00060-6

[8] Tomilson, D.C., Wilson, D.J., Harris, C.R. and Jeffrey, D.W. (1980) Problem in Assessment of Heavy Metals in Estuaries and the Formation of Pollution Index. *Helgoländer Meeresuntersuchungen*, **33**, 566-575. http://dx.doi.org/10.1007/BF02414780

[9] Saad, M.A.H., Ezzat, A.A., El-Rayis, O.A. and Hafez, H. (1981) Occurrence and Distribution of Chemical Pollutants in Lake Mariut, Egypt. II. Heavy Metals. *Water, Air, and Soil Pollution*, **16**, 401-407. http://dx.doi.org/10.1007/BF01048131

[10] Environmental Protection Agency (2002) National Recommended Water Quality Criteria. EPA, USA, 822-R-02-047.

[11] Muller, G. (1969) Index of Geo-Accumulation in Sediments of the Rhine River. *GeoJournal*, **2**, 108-118.

[12] Buccolieri, A., Buccolieri, G. and Cardellicchio, N. (2006) Heavy Metals in Marine Sediments of Taranto Gulf (Ionian Sea, Southern Italy). *Marine Chemistry*, **99**, 227-235. http://dx.doi.org/10.1016/j.marchem.2005.09.009

[13] El-Alfy, M.A. (2011) An Integrated Approach for Monitoring the Impact of Industrial Activities on the Northeastern Part of Manzala Lagoon, Egypt. Master's Thesis, Environmental Sciences Department, Damietta Faculty of Science, Mansoura University, Mansoura.

[14] Arain, M.B., Kazi, T.G., Jamali, M.K., Afridi, H.I., Baig, J.A., Jalbani, N. and Shah, A. (2008) Evaluation of Physico-Chemical Parameters of Manchar Lake Water and Their Comparison with Other Global Published Values. *Pakistan Journal of Analytical & Environmental Chemistry*, **9**, 101-109.

[15] Ali, M. and Abdel-Satar, A. (2005) Studies of Some Heavy Metals in Water, Sediment, Fish and Fish Diets in Some Fish Farms in EL-Fayoum Province. *Egyptian Journal of Aquatic Research*, **31**, 261-273.

[16] Bahnasawy, M.H., Khidr, A.A. and Dheina, N.A. (2009) Assessment of Heavy Metals Concentrations in Water, Plankton and Fish of Lake Manzala, Egypt. *Egyptian Journal of Aquatic Biology and Fisheries*, **13**, 117-133.

[17] Bahnasawy, M., Khidr, A. and Dheina, N. (2011) Assessment of Heavy Metal Concentrations in Water, Plankton, and Fish of Lake Manzala, Egypt. *Turkish Journal of Zoology*, **35**, 271-280.

[18] EPA (1986) Quality Criteria for Water. EPA 440/5-86-001. Office of Water Regulations and Standards, Washington DC.

[19] Nagpal, N.K. (2004) Technical Report, Water Quality Guidelines for Cobalt. Water Quality—Standards—British Columbia. Ministry of Water, Land and Air Protection, 6.

[20] El-Serehy, H.A., Aboulela, H., Al-Misned, F., Kaiser, M., Al-Rasheid, K. and Ezz El-Din, H. (2012) Heavy Metals Contamination of a Mediterranean Coastal Ecosystem, Eastern Nile Delta, Egypt. *Turkish Journal of Fisheries and Aquatic Sciences*, **12**, 751-760.

[21] Environment Canada (1997) Canadian Water Quality Guidelines for Chromium. Supporting Document, Environment Canada, Science Policy and Environmental Quality Branch, Ottawa.

[22] Hamed, Y.A., Abdelmoneim, T.S., ElKiki, M.H., Hassan, M.A. and Berndtsson, R. (2013) Assessment of Heavy Metals Pollution and Microbial Contamination in Water, Sediments and Fish of Lake Manzala, Egypt. *Life Science Journal*, **10**, 86-99.

[23] European Union (2002) Heavy Metals in Wastes, European Commission on Environment. http://ec.europa.eu/environment/waste/studies/pdf/heavy_metalsreport.pdf

[24] Saeed, S.M. and Mohammed, M.A. (2012) Influence of Physico-Chemical Characteristics of Water on Metals Accumulation in Water and Tilapia Zillii Inhabiting Different Habitats in Egypt. *Journal of the Arabian Aquaculture Society*, **7**, 29-50.

[25] El-Bady, M.S.M. (2014) Spatial Distribution of Some Important Heavy Metals in the Soils South of Manzala Lake in Bahr El-Baqar Region, Egypt. *Nova Journal of Engineering and Applied Sciences*, **2**, 1-15.

[26] Müller, G. (1981) Die Schwermetallbelastung der sedimente des Neckars und seiner Nebenflusse: Eine Bestandsaufnahme. *Chemical Zeitung*, **105**, 157-164.

[27] Dan, S.F., Umoh, U.U. and Osabor, V.N. (2014) Seasonal Variation of Enrichment and Contamination of Heavy Metals in the Surface Water of Qua Iboe River Estuary and Adjoining Creeks South-South Nigeria. *Journal of Oceanography and Marine Science*, **5**, 45-54. http://dx.doi.org/10.5897/JOMS2013.0103

21

Water Analyses, Eco-Balance and Socio-Demographic Analyses as Prerequisites for Solutions of the Sewage Treatment Problems in Rural Areas

Christian Opp[1], Bastian Ziebolz[2]

[1]Department of Hydrology and Soil Science, Faculty of Geography, University of Marburg, Marburg, Germany
[2]KLT Consult, Hannover/Soest, Germany
Email: opp@staff.uni-marburg.de

Abstract

The implementation of the EU-water framework directive (EU-WFD) demands a financeable and adaptable sewage treatment, creating a need for action especially in rural areas, where the "good ecological status" is often not reached even though the sewage works fulfill the legal minimum requirements. Therefore the construction of new, more efficient wastewater treatment plants is advisable. Due to the precarious financial situation in most rural areas, many municipalities can however not master the necessary funds for an improved waste water infrastructure. A large part of the costs would have to be shouldered by the local population in the form of additional sewage levies, leading to massive socio-economic strain. Hence, different solutions, which are tailored to the socio-economic, demographic and environmental characteristics of rural areas and fulfill the legal requirements, need to be developed. One possible approach is presented in the form of the Dirlammen case study (municipality Lautertal, Hesse, Germany). It consists of 1) biological and physicochemical analyses of the waste waters, 2) a modified eco-balance for the construction of a new sewage works and 3) socio-economic and demographic analyses of the population. It serves as a tool for the assessment of the influencing ecological, socio-economic and demographic factors on the sewage problems in rural areas. These findings can easily be transferred to other rural areas with a similar configuration and thus deliver the basis for a generally applicable approach to these problems. The results of the comprehensive analysis have shown for the municipality Dirlammen/Lautertal that the continued operation of the existing pond sewage plant is, with regards to the ecological, socio-economic and demographic points of view, preferable to the construction of a new waste water treatment plant. With regard to the sustainability principles of the sewage treatment and the objectives of the EU-WFD, existing sewage works should be operated for as long as the sewage treatment is feasible within technical and legal boundaries. With the construction of new facilities on the other hand, the expected improvement of the water quality has to be weighed against the ecological impairments caused by the construction itself. Rural waste water plans have to be developed for the long-term, taking not only the ecology into account but also the demographic change. Possible approaches in this regard could be a shift towards a decentralized waste water treatment concept and–even more important–improvements of the river morphology. But also the reduction of the diffuse pollution from agricultural sources should be considered for the

development of strategies for the improvement of the ecological state. Short-term optimization measures in the existing sewage works might also improve the water quality until the long-term development plans can be implemented. In particular the challenges of the demographic change and linked with that the changing requirements for the rural waste water treatment must be addressed by any future plans. Only with the help of such a comprehensive approach the dangers of bad planning and unacceptable socio-economic burdens can be minimized.

Keywords

Water Analyses, Eco-Balance, Sewage Treatment Plant, Socio-Economic and Demographic Analyses

1. Introduction and Problem Setting

The EU-water framework directive (EU-WFD) demands a "good ecological state" for all ground and surface waters (rivers, lakes, transitional waters, and coastal waters) within the European community by the year 2015. The EU-WFD also has defined the "good ecological status" as values, determined by the chemical quality conditions (standards specify maximum concentrations for specific water pollutants), the biological quality conditions (by indicator species such as fish, benthic invertebrates, and aquatic flora) and the physicochemical quality conditions (temperature, oxygenation and nutrient concentrations) [1]. Besides these criteria the improvement of the hydro-morphological conditions of the water bodies will play an important role [2]. The water quality of the rivers is strongly influenced by diverse point and nonpoint impacts. Due to the fact that nonpoint impacts are extremely difficult to manage, the main activities to improve the water quality of the impacts are directed towards the reduction of the harmful pollutants from point sources.

Results of the above mentioned indicator monitoring often show problems in rural areas. Reasons for these problems are connected to the mostly small size of the rural rivers and creeks, which are less able for self-purification and dilution processes on the one hand and with the most common pond sewage treatment on the other hand. Pond sewage treatment plants are the most common technology for wastewater treatment in rural areas in Germany because they are based on the microbial self-purification processes and thus are especially cost-effective [3] [4]. Monitoring of the pond sewage treatment plants however show that they have purification deficits, and thus are often responsible for the a impairment of the water quality in little rivers and creeks downstream of the water-input coming from the pond sewage treatment. Due to this fact the construction of modern wastewater treatment plants is on the agenda. However, the rural population has already to pay above average per capita fees for the maintenance and operation of these wastewater treatment facilities and for the rehabilitation of the household's sewage pipe system. Additional payments of the rural population for the construction of a new sewage treatment plant (STP) are not reasonable, especially as the population density in rural areas is decreasing. That is why the investments for the construction of a new sewage treatment plant cannot be covered by the rural communities.

2. Study Area

The same problem situation was found in the community of Lautertal, in the Vogelsberg District, within the Federal State of Hesse (Germany). Seven villages with a total population of 2424 are part of this community. In one of the community parts, the location Dirlammen (419 inhabitants), the situation was studied in detail. The existing pond based STP of Dirlammen is operating since 1887 based on self-purification without denitrofication and nitrification, fulfilling all legal minimum requirements. The monitoring of pollutant input into the nearby creek shows an excess of the tolerable input, especially of the tolerable NH_3-concentration, which affects eutrophication.

3. Objectives and Research Approach

Based on the problem presented above, multiple research questions, objectives, and methods were selected for this study (**Figure 1**).

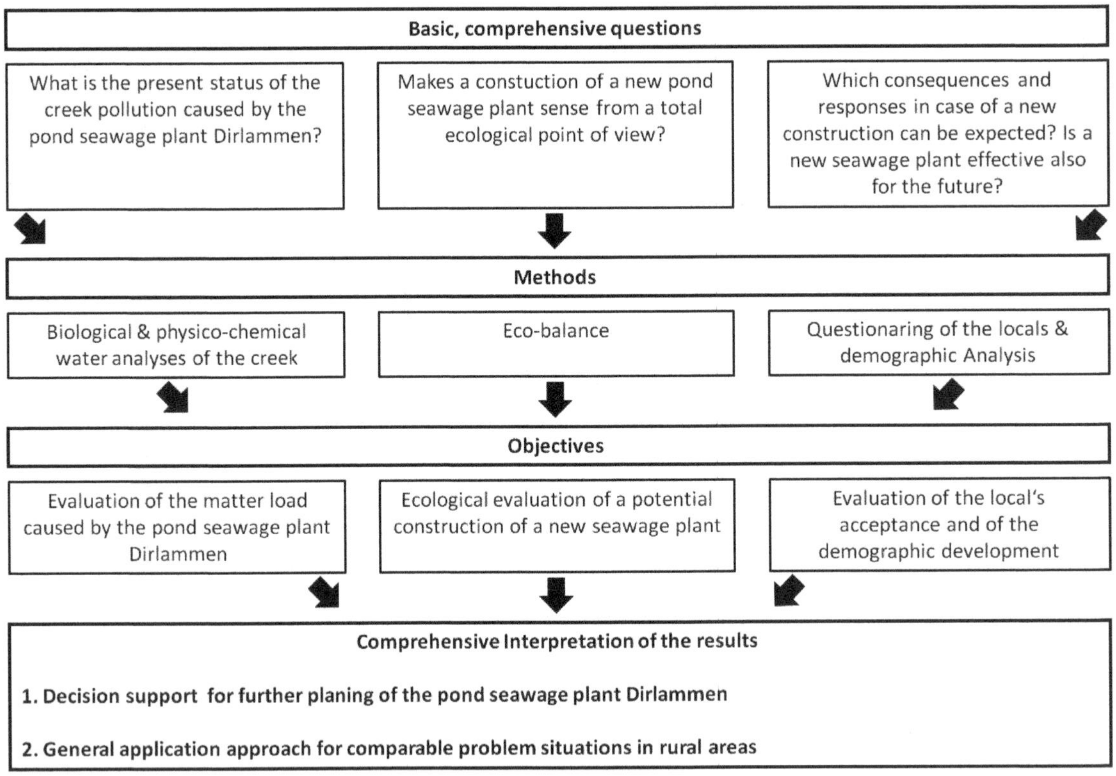

Figure 1. Concept and objectives of the Lautertal (Dirlammen) study.

4. Methods

Discharge, water temperature, pH-value, and electric conductivity measurements as well as saproby index, NH_3, NH_4, NO_2, NO_3 concentrations, oxygen saturation, oxygen content, total phosphor concentration determinations were carried out at four times during spring (March 22, April 26) and summer (June 21, July 12) 2010 at four sampling sites (SaSi) along the Brenderwasser creek near Dirlammen: (1) 200 m above of the STP input, (2) directly 50 m downstream of the STP input, (3) 900 m and (4) 2200 m downstream of the STP input. All analyses were carried out using WTW techniques [5] according to the federal water agreement [6].

To determine the environmental effects of the construction of a new sewage treatment plant in comparison to the existing pond sewage treatment, the eco-balance (or a life cycle assessment) method [7] was used. This life cycle assessment comprises the determination of the costs for the production of a new STP, including the costs for material, and the costs for the operation. These analysis and evaluation was carried out with the help of the computer software GEMIS (Global Emission Model Integrated System), Version 4.5. Different process modules for production and energy use have been analyzed according to [8]. Furthermore the eco-balance includes the comparison and evaluation of the additional load by selected environmental side effects, such as the greenhouse effect (with a very high ecological priority), the summer smog and the acidification (both with a high ecological priority), the energy demand and direct disturbances of the aquatic ecosystem (with a moderate ecological priority). All these parameters have been analyzed both for a new STP and the existing pond STP for operating times of 15 and 30 years. For the assessment of the age-related and employment-related population structure and their monetary potential to pay for higher wastewater treatment costs, socio-economic questioning [8] and interviews [9] [10] have been carried out.

5. Results and Discussion

5.1. Physicochemical and Biological Measurements

Results of the discharge measurements show that the value for March of 668.0 l/s represents a 27 times higher discharge than the long-term average of 25.0 l/s, while the April (9 l/s), June (5 l/s) and July (1 l/s) measurements represent low water situations. The different discharges are a good prerequisite for evaluation of the dilution effects. The water temperature increased from March (7.3°C), April (15.2°C), June (12.4°C) to July (19.9°C).

The saprobic index values show a good ecological quality at SaSi's 1, 3 and 4, while SaSi 2 has only a moderate saprobic index value. Responsible for this is the mass occurrence of Asselus aquaticus, an indicator species, who nourishes on decomposed substances, which occur in the surrounding of the STP outflow. Higher NO_3 concentrations during March are a result of de-icing salt input or of the input by melting snow into the creek. NH_3 and NH_4 have low or very low values during March, April and June at all SaSi's, but high (15 times higher values in April) and 31 times higher during the July measuring campaign at SaSi 2, The samplings in April and June were realized under comparable water temperature and low discharge conditions.

The electric conductivity, an indicator of the sweeping ion concentration has an average value of about 350 µS/cm at the SaSi (1), and shows a strong increasing value at SaSi (2) near by the STP outflow, after which it is decreasing continuously in the direction of flow until SaSi (4), where it achieves the average value again. All pH values at all SaSi's and during all measuring dates ranged between 7.2 and 8.0, which means that they are unproblematic. The lowest pH values were determined in March and in June nearby the STP outflow. The highest phosphorus values were measured during all sampling dates at SaSi (2). While the phosphorus concentrations during March, April and June were fluctuating here between 0.3 and 0.6 mg/l, the corresponding July value during the extreme low water situation is with 3.2 mg/l quite higher. This is a consequence of the low dilution capacity during the very low discharge. Also the values of oxygen content and oxygen saturation have their lowest extremes at SaSi (2), near the STP outflow. In March at all SaSi's registered the highest (12.0 mg/l and 100%) and in July the lowest O_2 values.

5.2. Eco Balance

Due to the fact that the costs of a newly constructed STP are very high, they are not comparable with the further ongoing operation of an old pond STP. However, it is interesting to study the environmental side-effects of an old STP and comparing them with a new one. All a.m. ecologic side-effects without the exception of the direct disturbances of the aquatic ecosystem have a distinct additional load through the construction and operation of a new STP. The additional load by construction and operation of a STP is regarding the greenhouse effect (CO_2-equivalents) 299% (after a 15 years operation) and 184% (after a 30 years operation) higher than the further operation of the old STP. The according values regarding the summer smog effect are 852% (after a 15 years operation) and 461% (after a 30 years operation) higher than the further operation of the old STP. The according values regarding the acidification (SO_2-equivalents) are 567% (after a 15 years operation) and 318% (after a 30 years operation) higher than the further operation of the old STP. Also the values for the energy demand are 212% (after a 15 years operation) and 141% (after a 30 years operation) higher than the further operation of the old STP. But the longer the operation time of the new STP is, the smaller are these additional loads. Only the direct disturbances of the aquatic ecosystem are 18% higher through the further operation of the old pond STP compared to a new one.

5.3. Socio-Economic and Demographic Developments

54% of all inhabitants (total 419) of the village Dirlammen are retired. 17% are working in the service sector, 13% in the public survey sector, 10% in the production sector, 3% in the construction sector, and 3% in agriculture. This is a typical employment structure for rural areas in Germany. 30 domestic homes or 1/3 of the inhabitants took part in the survey. 93% of all interviewed persons expressed their protest against additional costs caused by a new STP. 3% would leave the village as they could not afford the additional financial burden. 4% did not answer this question. According to the latest demographic development, a continuous population decrease can be expected for the future. This implicates not only higher wastewater costs for the smaller number of inhabitants. This also implicates an under-utilization of a new STP and thus questions the need for a new wastewater treatment facility.

6. Conclusion

Summing up the above mentioned results, a continuation of the pond STP operation is recommendable. It seems

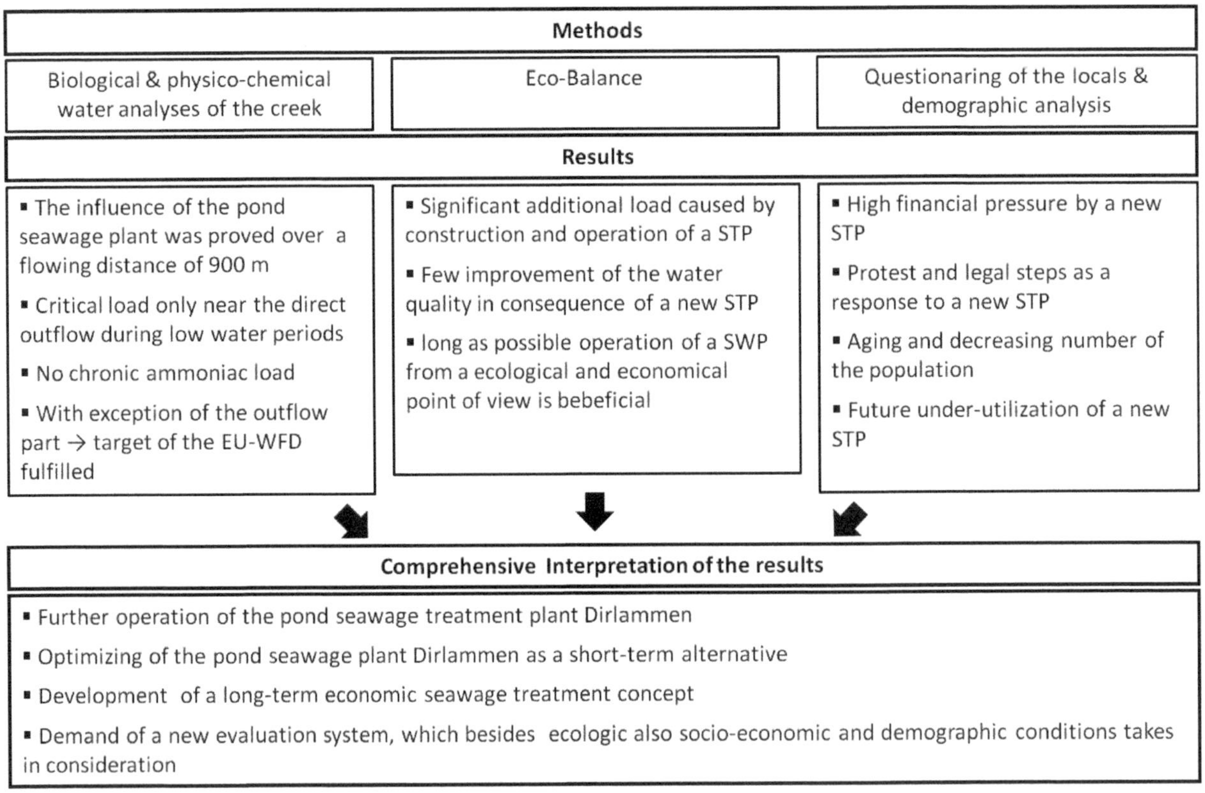

Figure 2. Concept and objectives of the Lautertal (Dirlammen) study.

to be more effective to improve the purification technology of the existing pond STP. The elaboration of a long-term affordable wastewater treatment concept is a present task. For decision finding of the right wastewater treatment for rural areas, socio-economic and demographic analyses have to be taken into consideration besides water analyses and eco-balance application (**Figure 2**).

Acknowledgements

The authors want to thank the administration and the inhabitants of the Dirlammen village within the Lautertal community for their cooperation, and Frank Reißig for his advices regarding the eco-balance application.

References

[1] European Waterframework Directive (EU-WFD) (2000) Directive 2000/60/EC of the European Parliament and the Council of 23 October 2000 Establishing a Framework for Community Action in the Field of Water Policy. Official Jounal L327, 22/12/2000 P 0001-0073.

[2] Groll, M. and Opp, Ch. (2007) Gewässerbettmorphologie und Habitate in einem renaturierten Abschnitt der Lahn. Exemplarische Anwendung des TRiSHa-Verfahrens. *Naturschutz und Landschaftsplanung*, **39**, 369-376.

[3] Landesamt für Umweltschutz Sachsen-Anhalt (LfU) (2006) Abwasserteichanlagen zur kommunalen Abwasserreinigung (Hinweise zu Planung, Bau, Betrieb und Optimierung). Fachbereich 2 Abfallwirtschaft, Bodenschutz, Anlagentechnik Wasserwirtschaft, Halle.

[4] Borchard, D. and Menadi, F. (2001) Empfehlungen für die Errichtung und den Betrieb von belüfteten Teichkläranlagen. Institut für Gewässerforschung und Gewässerschutz (IAG) & Universität Gesamthochschule Kassel.

[5] http://www.wtw.de/downloads/manuals/

[6] LAWA-AO (2007) Landesarbeitsgemeinschaft Wasser-Ausschuss Oberirdische Gewässer und Küstengewässer Rahmenkonzeption Monitoring. Teil B: Bewertungsgrundlagen und Methodenbeschreibung. Arbeitspapier 2. Hintergrund- und Orientierungswerte für physikalisch-chemische Komponenten.

[7] Klöpfer, W. and Grahl, B. (2009) Ökobilanz (LCA)—Ein Leitfaden für Ausbildung und Beruf. Weinheim. http://dx.doi.org/10.1002/9783527627158

[8] Institut für angewandte Ökologie (2010) Globales Emissions-Modell integrierter Systeme. Version 4.5.

[9] Schnell, R. (1999) Methoden der empirischen Sozialforschung. München.

[10] Gebhardt, H., Glaser, R., Radtke, U. and Reuber, P. (2007) Geographie. Physische Geographieund Humangeographie. München.

22

Impact of Climate Change on Water Resources Management in the Lower Chao Phraya Basin, Thailand

Sanit Wongsa

Department of Civil Technology Education, King Mongkut's University of Technology Thonburi (KMUTT), Bangkok, Thailand

Email: sanit.won@kmutt.ac.th

Abstract

Climate change and associated rising in sea water level have affected the salinity in many rivers around the world. It has an effect on the embouchure adjacent with the sea, which is called the salinity intrusion problem. This study investigated the effects of climate change on sea water level that affects the hydraulic conditions, salinity, water supply and agricultural areas in the lower Chao Phraya River by MIKE 11 model has been used. The study covered the area from Chao Phraya Dam (barrage), Chai Nat Province to the river estuary, Samut Prakan Province. The model was divided into two parts, hydrodynamic module and advection-dispersion model. Calibration of each part was done by adjusting its important coefficients. It was observed that the Manning's coefficient (n) and coefficient dispersion of mass were in the range of 0.025 - 0.40 and 800 - 1600 m^2/s, respectively. The results of comparison between models and observation data revealed order of forecasting error (R^2) with the range of 0.76-0.99 for water level and 0.73 - 0.86 for salinity. The RCP 8.5 scenario from IPCC report was simulated. It was found that sea water level rising in was 1.16 m in the year of 2100, and salinity at SamlaePump Station was risen to 0.37 - 0.75 g/l. The value of 0.25 g/l exceeding standard and the pointed tip of salinity was at Koh Rain District, Ayutthaya Province (137 km from Chao Phraya Dam: CPD). For agricultural sectors, the value of 0.20 g/l exceeding standard and the pointed tip of salinity was at Ban Mai District, Ayutthaya Province (123 km from CPD). Results obtained from this study will give the guideline in raw water resources management for water supply and agricultural in Chao Phraya River Basin.

Keywords

Climate Change, Water Resources Management, Lower Chao Phraya River, Salinity Intrusion

1. Introduction

Climate change has been observed having local and global effects. Most of the effects were negative. The notable phenomena effected by extreme weather events, are heavy rains, heat waves, and draught. Climate change causes serious risks to the well-being of nature and people all over the world. Within the estuaries, sea water can be the important controls of water level, salinity and coastal erosion. Although the precise effect of climate change on estuaries dynamics and its processes in the alluvial river system is still not clear, there seems to be no doubt that it influences sea water level and salinity intrusion. Recently, numerical modelling has been shown to answer some of these problems. A number of works have used numerical models attempting to simulate river catchment hydrological processes of rainfall-runoff, sediment transport, salinity intrusion and coastal erosion processes as well as to study the impact of climate change. In addition, the numerical model has been applied to predict hill slopes and river channels [1]-[5]. Thus, it is clear that the numerical models appear to have considerable potential as tools for investigating hydrodynamic, sediment transport and water quality over long period simulation. [6] have predicted that sea water level affected rice production in Mekong Delta, Vietnam, during the flood season. However, there are a few application studies in Thailand. [7]-[10] have exploited MIKE11 model to predict sea level affecting salinity intrusion and agricultural production in Tha Chin, Mae Klong and Chao Phraya rivers. Metropolitan Waterworks Authority (MWA) reported that salinity intrusion and raw water supply in Chao Phraya River was affected by climate change.

This paper addresses these issues by using a pro-posed MIKE11 numerical modelling to simulate the effects on sea water level change and salinity intrusion. Performance of the numerical model was applied to simulate flow events in 2100, which are water level change and salinity intrusion on raw water supply in the Lower Chao Phraya River, Thailand.

2. Material and Methods

2.1. Study Area

Chao Phraya River is the most important and largest river flowing from Chai Nat Province to estuary at the Gulf of Thailand in SamutPrakan Province, located at central part of Thailand. The climate of the Chao Phraya River has a tropical wet and dry or savanna climate, which generates wet and dry seasons of more or less equal length. The monsoon season is usually from May until late September and/or early October. In the wet season, averagely 1 - 2 tropical depressions occur over much of the area from August to October of the year. The average annual discharge is 718 m^3/s and rainfall varied between 1122 to 1511 mm, depending on monsoon direction and elevation. However, some parts of the catchment continue to suffer from drought problems due to the uneven distribution of rainfall. Some areas experience both flooding and drought conditions in a single year, due to temporal and spatial uncertainties in the monthly rainfall and/or the poor management of the conveyance infrastructure. The common practice in Thailand is to manage the risks after considering which areas are likely to be vulnerable to either flood or drought. Failure to manage risk by addressing one aspect at a time can lead to adverse results. Therefore, climate change and an association with managing flood and drought risks are new challenge in Thailand and becoming increasingly important.

2.2. MIKE11 Model

The MIKE11 model has been used. This numerical model simulates water flow and salinity as a consequence of low flow conditions. The shallow water equation was used for simulation of 1-D unsteady water flow and transportation of mass was used for salinity. To modelling the river network of the Lower Chao Phraya, a digital elevation map (DEM) was applied. The model input data were cross-section, flow discharge, water level, side flow and salinity. The MIKE11 program, 6-points Abbott's finite difference scheme was used to solve governing equations, consisting of separate modules each representing a different procedure in calculation process. A first module calculated hydrodynamics of river flow (HD module), and the next module of transportation of mass (salinity intrusion; AD module). The model setup of plan view and longitudinal profile of the Lower Chao Phraya are shown in **Figure 1**.

3. Results and Discussion

The Lower Chao Phraya River was used for calibration and verification of the proposed model. The calibration and validation has focused on the applicability of water flow and salinity intrusion by using flow conditions in the year of 2008 and 2010. Performance of the foregoing numerical model was applied to simulate scenario of IPCC SRES (RCP8.5), predicted from global average sea level rising from 1990 to 2100 by using GCMs. RCP8.5 scenario is a scenario of comparatively high greenhouse gas emissions [11]. For model simulation, flow

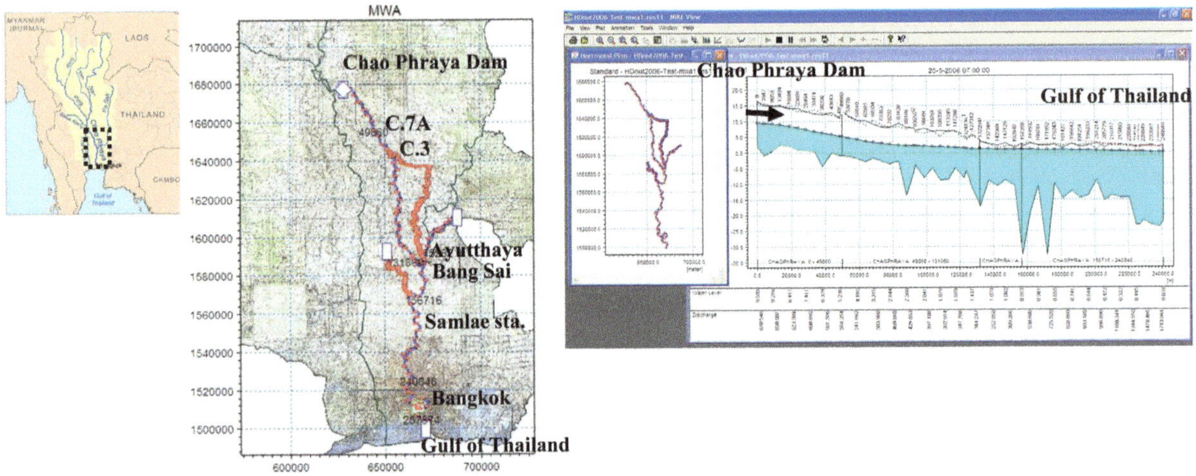

Figure 1. Study area of the Lower Chao Phraya River and model setup.

discharge at station C.13 (Chao Phraya Dam: CPD) and water level at Chao Phraya River Estuary (Fort Chula) were adopted for upstream and downstream boundaries, respectively. Before the water flow and salinity calculation was carried out, the model was run to provide the steady state of necessary flow variables.

3.1. Model Calibration and Verification

The comparison of time series of measured and simulated water surface level and salinity at two major gauge stations and Manning's roughness coefficient (n) are shown in **Figure 2** and **Figure 3**. Good agreement between the simulated and measured hydrographs for the low flow events was achieved by considering side flow and pumps in the areas. The coefficient of determinant (R^2) and Index of agreement (IA) have been used as the main criteria to judge whether the data fitted between measurement and simulation. The comparison model was followed by adjusting important coefficient of two parts. The results of manning (n), global dispersion factor, global exponent and K_{mix} were in the range of 0.022, 500 - 1800 m^2/s, 0.1 - 0.8 and 500 - 1800 hr^{-1}, respectively. The results of comparison between models and observation data revealed the order of forecasting error, R^2 and IA in the range of 0.88 - 0.91, 0.94 - 0.97 for water level and 0.86 - 0.96, 0.75 - 0.94 for salinity. These indicate well fitted between measured data and this proposed model. Good agreement between the simulated and measured water surface level was achieved by considering side flow and pumps in the areas for model verification. It was observed that the values of R^2 and IA for two major gauge stations in verification period were between 0.93 - 0.94, 0.97 - 0.98 for water level and 0.71 - 0.91, 0.72 - 0.87 for salinity, indicating well fitted between measured data and this proposed model (**Figure 4** and **Figure 5**). Good performance of simulated results was observed in both water flow and salinity intrusion characteristics, therefore, indicating that model simulation is reasonable.

3.2. Model Application

For model application, a scenario RCP8.5 from IPCC report was simulated. The comparison of time series of measured and simulated flow discharge at the major gauge stations are shown in **Figure 6** and **Figure 7**. It was found that sea water level at the Chao Phraya estuary had rising and betaking tendency to intrusion of sea water level. Salinity was shown to be in the same tendency. For IPCC SRES in the year of 2100, sea water level rising in RCP8.5 scenario was 1.16 m, and salinity values and intrusion distant were in the range of 0.38 - 0.45 g/l and 142.0 - 157.0 km (from Chao Phraya Dam), and salinity at Samlae Station were 0.29 - 0.37 g/l. The value of 0.25 g/l exceeding standard and the pointed tip of salinity was at Bang Sai District, Ayutthaya Province. The worst case scenario, a constant value of salinity of 30 g/l was adopted at the Gulf of Thailand. It was found that salinity at Samlae Station was 0.99 - 1.70 g/l, the value of 0.25 g/l was at Pratuchai Sub-district, Ayutthaya Province, 133.0 km from Chao Phraya Dam. We could also observe from the simulation that the climate change would have more conspicuous effects in rising up sea water. However, Metropolitan Waterworks Authority reported that climate change affected salinity intrusion and raw water supply in Lower Chao Phraya River. This observation demonstrated that Samlae Pumping Station will not be available when salinity is over 0.25 and/or

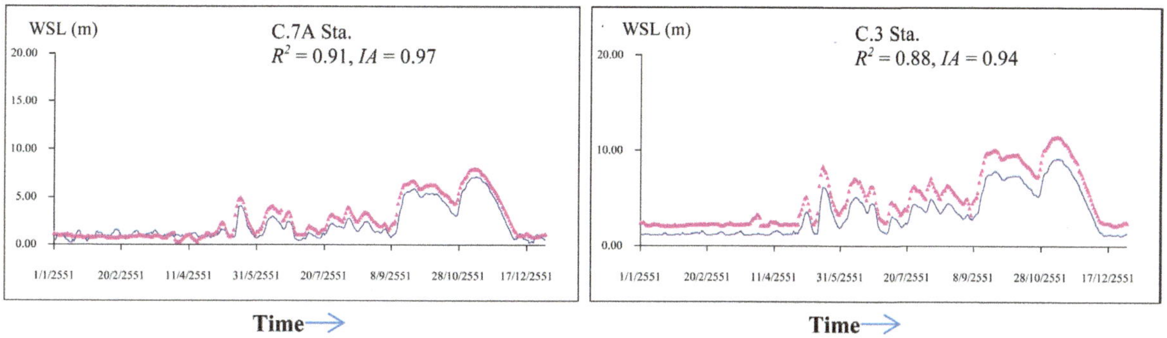

Figure 2. Comparison of time series of stage hydrograph between simulated results (smoothed-line) and measured data (dotted) at the major gauge stations (model calibration).

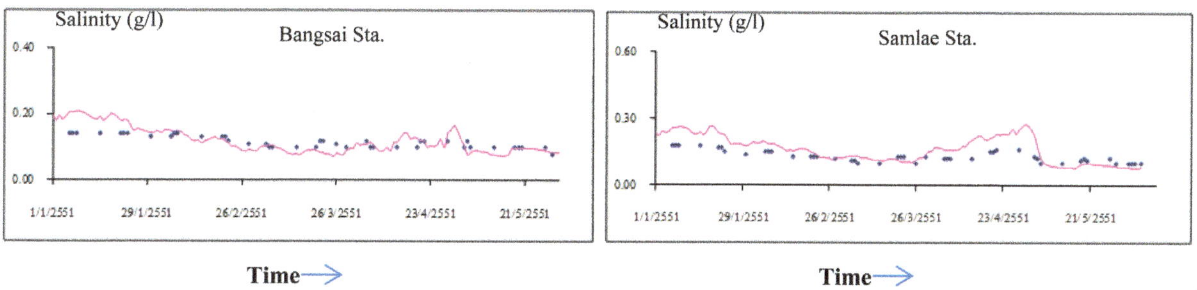

Figure 3. Comparison of time series of salinity between simulated results (smoothed-line) and measured data (dotted) at the major gauge stations (model calibration).

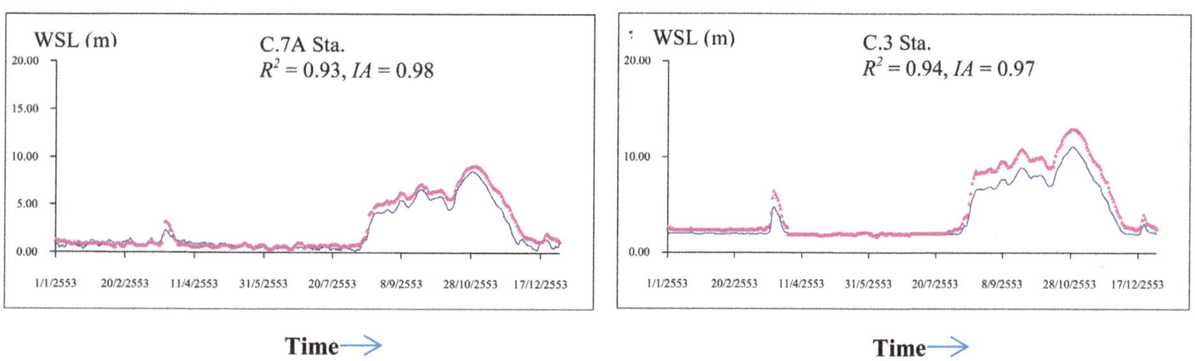

Figure 4. Comparison of time series of stage hydrograph between simulated results (smoothed-line) and measured data (dotted) at the major gauge stations (model verification).

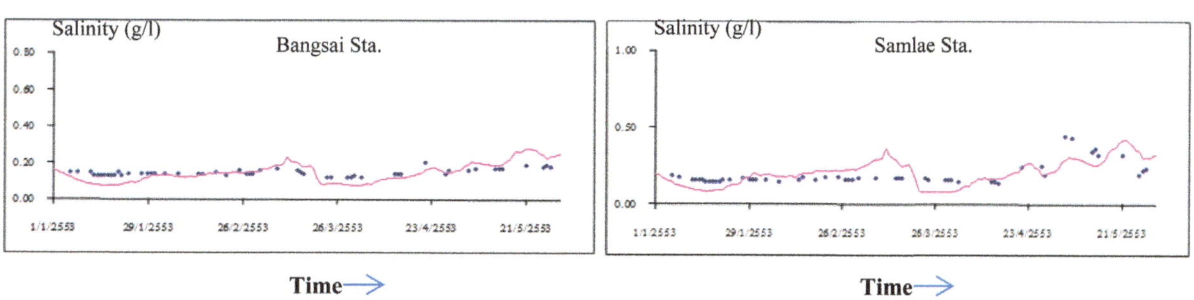

Figure 5. Comparison of time series of salinity between simulated results (smoothed-line) and measured data (dotted) at the major gauge stations (model verification).

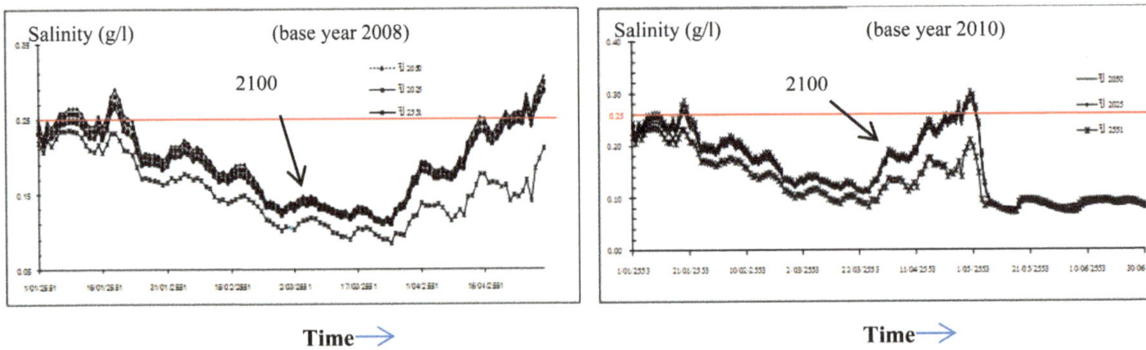

Figure 6. Salinity in the year of 2100 (IPCC), with 1.16 m sea water level rising in RCP8.5 scenario (model application).

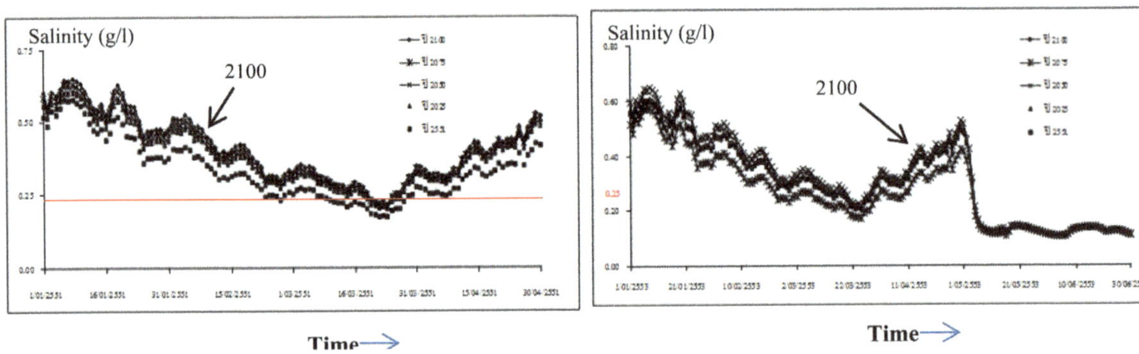

Figure 7. Salinity in the year of 2100 (IPCC), with 1.16 m sea water level rising in RCP8.5 scenario and salinity 30 g/l at downstream (model application).

0.50 g/l, and water supply shortage. The mitigation plan has suggested the diversion flow from Mae Klong River and construction of new pump station for short-term and long-term, respectively. The proposed locations of new raw water pump station along the Chao Phraya River from Ayutthaya to Singburi Province have been investigated. It was found that the most suitable sites when evaluation by multi-criteria analysis (MCA) is at Klong-Pongpheng, Ayutthaya Province (site 4).

Water level change will affected the water supply in urban, and agricultural areas. Salinity level less than 0.2 g/l was not impacted; it can use irrigation for any planting. Salinity level from 0.2 had little impact, while salinity level from 0.5, and salinity level more than 1.5 g/l were highly impacted (**Figure 8**). For agricultural sectors, the value of 0.20 g/l exceeding standard and the pointed tip of salinity was at Ban Mai District, Ayutthaya Province (123 km from Chao Phraya Dam).

4. Conclusion

In this study, the MIKE11 model was exploited to simulate the effects of climate and sea level changes on the raw water supply of MWA and agricultural areas in the lower part of Chao Phraya River in the year 2100. The study covered the area from Chao Phraya Dam, Chai Nat Province to the Gulf of Thailand at Samut Prakan Province. IPCC SRES predicted that in the year 2100, sea water level rising in RCP8.5 scenario was 1.16 m. Based on this information, the simulation showed that salinity at Samlae will be exceeded 0.25 g/l. Therefore, it is suggested to add pump station at Samlae to site (4) at KlongPongpheng, borderline of Ayutthaya and Ang Thong Province. This will help to balance level of salinity in river not over 0.25 g/l. For agricultural sectors, the value of 0.20 g/l exceeding standard and the pointed tip of salinity was at Ban Mai District, Ayutthaya Province (123 km from Chao Phraya Dam). Results obtained from this study will give guideline in raw water resources management for water supply and agricultural in Chao Phraya River Basin.

Acknowledgements

The authors would like to thank NRU-National Research University and King Mongkut's University of Tech-

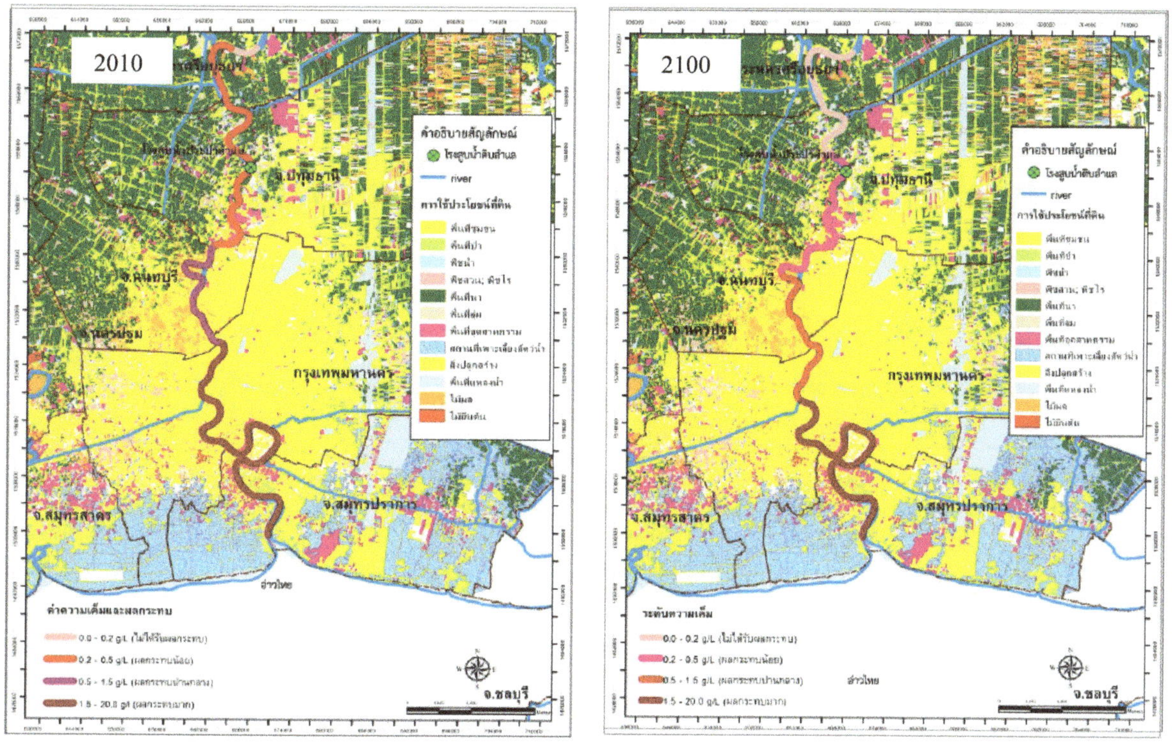

Figure 8. Salinity profile in the year of 2010 and 2100 (IPCC SRES).

nology Thonburi (KMUTT). We are also indebted to Metropolitan Waterworks Authority (MWA), Royal Irrigation Department (RID) and Pollution Control Department (PCD) for providing fields data.

References

[1] Kirby, M.J. (1987) Modelling Some Influences of Soil Erosion, Landslides and Valley Gradient on Drainage Density and Hollow Development. *Catena Supplement*, **10**, 1-11.

[2] Wicks, J.M. and Bathurst, J.C. (1996) SHESED: A Physical Based, Distributed Erosion and Sediment Yield Component for the SHE Hydrological Modelling System. *Journal of Hydrology*, **175**, 213-238. http://dx.doi.org/10.1016/S0022-1694(96)80012-6

[3] DHI Water and Environment (2000) MIKE11: A Modeling Sys-Tem for Rivers and Channels—User Guide, Denmark.

[4] Wongsa, S. and Shimizu, Y. (2004) Modelling Artificial Channel and Landuse Changes and Their Impact on Floods and Sediment Yield to the Ishikari Basin. *Hydrological Processes*, **18**, 1837-1852. http://dx.doi.org/10.1002/hyp.1450

[5] Wongsa, S. (2004) Distributed Model for Catchment Scale Simulation of Runoff and Sediment Transport. Ph.D. Thesis, Hokkaido University, Japan.

[6] Wassmann, R., Hein, N.X., Hoanh, C.T. and Tuong, T.P. (2004) Sea Level Affecting Vietnamese Mekong Delta: Water Elevation in the Flood Season and Implications for Rice Production. *Climate Change*, **66**, 89-107. http://dx.doi.org/10.1023/B:CLIM.0000043144.69736.b7

[7] Wongsa, S. and Ekkawatapanit, C. (2009) Effect of Sea Water Level Change on the Management in the Lower Thachin River. *Report for ESS Program*, King Mongkut's University of Technology, Thonburi, Thailand.

[8] Wongsa, S., Ekkawatapanit, C. and Treerittiwitaya, K. (2010) Effect of Sea Water Level Change on the Management in the Lower Thachin River. *Proceedings of the International Symposium on Coastal Zones and Climate Change*: *Assessing the Impacts and Developing Adaptation Strategies*, Australia.

[9] Wongsa, S., Kamolsin, P. and Inkliang, K. (2010) Effect of Sea Water Level Change on the Estuaries Management in the Lower Thachin and Mae Klong Rivers. *Proceedings of GMSTEC* 2010: *International Conference for a Sustainable Greater Mekong Subregion*, Bangkok, Thailand.

[10] Sueathung, S. and Wongsa, S. (2015) Effect of Sea Water Level Change on the Water Resources Management and Agriculture in Chao Phraya River. *Proceedings of NCCE*2015, Chonburi, Thailand.

[11] http://www.ipcc.ch/

23

Analysis on Pollution Factors of Urban River

Zhiyi Lei

College of Harbour, Coastal and Offshore Engineering, Hohai University, Nanjing, China
Email: leizhiyi@hhu.edu.cn

Abstract

Main pollutants of the urban scenic river in Nanjing City are studied in this paper. In the study area a total of 39 monitoring points are set in natural water, around pumping stations and near the tail water of sewage treatment plant. Through the monitoring of pollution sources of receiving conventional index, the pollution sources distribution and river pollution factors are detailed analyzed, as nutrient salts, heavy metals, and environmental endocrine disruptors. And sources of the pollution factors are analyzed by principal component analysis to get the main pollution factors in this channel.

Keywords

Source Apportionment, Distribution, Water Environment, Principal Component Analysis

1. Introduction

At present, city river catchment comes mainly from tail water of wastewater treatment plant, straight row of sewage, and the upstream catchment [1] [2]. Besides conventional pollutants, the catchment water also contains toxic and harmful pollutants, including detergent, pathogenic microorganism, chlorination by-product, endocrine disruptors, high concentration heavy metal and pesticide etc. [3] [4].

A large number of epidemiological data showed that, environmental endocrine disruptors (environmental endocrine disruptors/endocrine disrupting chemicals, ECDs) can cause reproductive disorders, birth defects, developmental abnormalities, metabolic disorders and certain cancers, and is a major risk factor for the water environment in the world [5] [6]. The endocrine disrupting chemicals in sewage are often detected with polycyclic aromatic hydrocarbons (PAHs), phthalic acid esters (two PAEs), phenols, organic chlorine pesticides (OCPs), polychlorinated biphenyls (polychilodnated-bipheneyls, referred to as PCBs) and two phenyl ketones [7]. Among them, the adjacent benzene two formic acid esters, PAHs, PCBs are detected in the city river [8] [9].

2. Distribution of the Pollution Factors in Urban River

2.1. Measuring Points

According to the analysis of the characteristics of water environment and investigation of pollution sources in the Qinhuai River, in the study area a total of 39 monitoring points are set for water quality monitoring, among them there are 24 points in natural water, 11 points around pumping station, and 4 points in the tail water of sewage treatment plant (**Figure 1**).

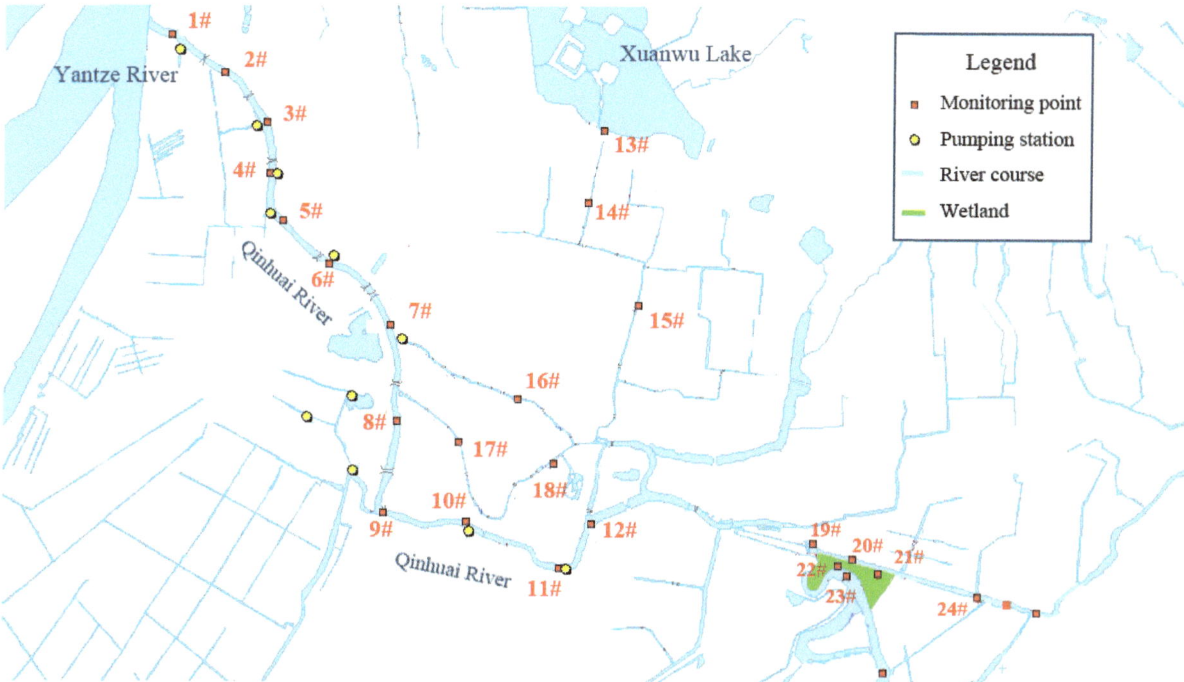

Figure 1. Distribution of monitoring points.

Detection of indicators includes the general physical and chemical indicators, eutrophication and nutrient index, health index, toxicology indicators (including heavy metals, endocrine disrupting chemicals etc.).

2.2. Characteristics and Sources of Heavy Metals

Distribution of heavy metals in natural waters is shown in **Figure 2**. It can be seen, as is ranking in natural waters: the Yunliang River > the WaiQinhuai River > the NeiQinhuai River; Sequencing of *Zn* content is the Yunliang River > the Qinhuai River > the Qinhuai River, and is the same as the sort of *Pb* content, *Cd* content and Mn content. The content of *Cr* and *Cu* distribute evenly; sorting the *Fe* content is the WaiQinhuai River > the NeiQinhuai River > the Yunliang river.

The distribution of heavy metal content in pollution source is shown in **Figure 3**. The heavy metal content in Yunliang River is $Zn > Pb > As > Cd > Mn > Cr > Cu = Fe$, in which the content of *Zn* is the highest of 1878 mg/L; the content in upstream of Qinhuai River is $Zn > Pb > As > Fe > Mn > Cd = Cr > Cu$, and $Pb > As > Fe > Zn > Mn > Cd > Cr = Cu$ in Qingjiangqiao; and $Pb > As > Fe > Zn > Mn > Cd > Cr = Cu$ in Xiaodoumen; and $Pb > As > Fe > Zn > Mn = Cd = Cr > Cu$ in Shitoucheng; and $Pb > As > Fe > Mn > Zn > Cr > Cd = Cu$ in Tiechuangleng; $Pb > As > Zn > Fe > Mn = Cr > Cd > Cu$ in South Yudai River; $Pb > As > Fe = Cr > Mn = Zn = Cd > Cu$ in Qingliangmen; $Zn > Pb > As > Cd > Mn > Cr > Cu > Fe$ in Chengdong Sewage plant.

2.3. Distribution of Endocrine Disruptors

Phthalate esters (PAEs). The distribution of phthalic esters in natural water is shown in **Figure 4**. Thus it can be seen, that the contents of ∑PAEs in natural water is the WaiQinhuai River > the NeiQinhuai River > the Yunliang River > the city lakes. The average contribution rate of 15 PAEs concentration is DiBP > DEEP > DMP > DBP > DEP > DOP > DMEP > BBP > DBEP > DCHP > DPP > DHP > DEHP > BMPP > DINP. The concentration range of ∑PAEs is 108.04 ng/L - 1121.82 ng/L with the highest point in the Fengtai bridge (7#), followed by 5# and the Shangfu bridge, the concentrations of 1070.42 ng/L and 1023.94 ng/L.

The distribution of phthalate esters content in pollution source is shown in **Figure 5**. The content sequence is Qinjiang Bridge > Shitoucheng > Tiechuangleng > Xiaodoumen > Qingliangmen > Yunliang River > South YudaiRiver > Jingxinzhou > Chengdong sewage plant > Xianlin sewage plant > upstream of QinhuaiRiver > Chengbei sewage plant. ∑*PAEs* concentration ranges from 92.68 ng/L to 2698.50 ng/L with the main factor of DMP, DEP, DiBP, DBP and BBP.

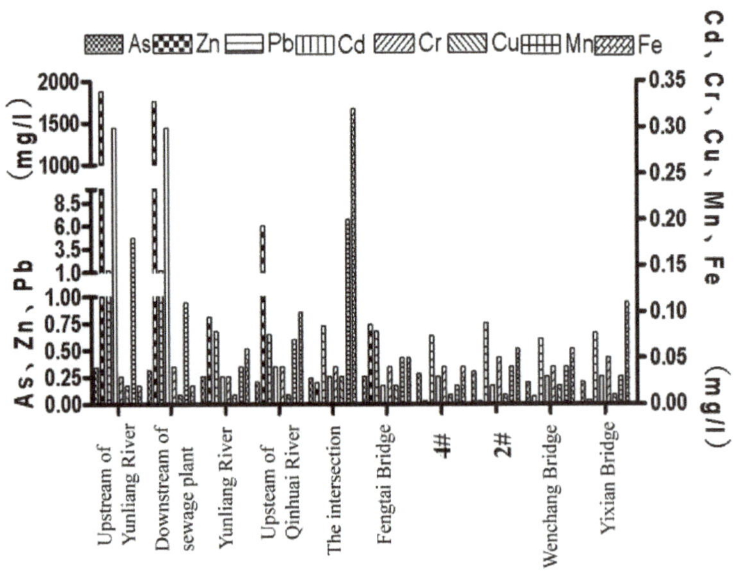

Figure 2. Distribution of heavy metal in natural water.

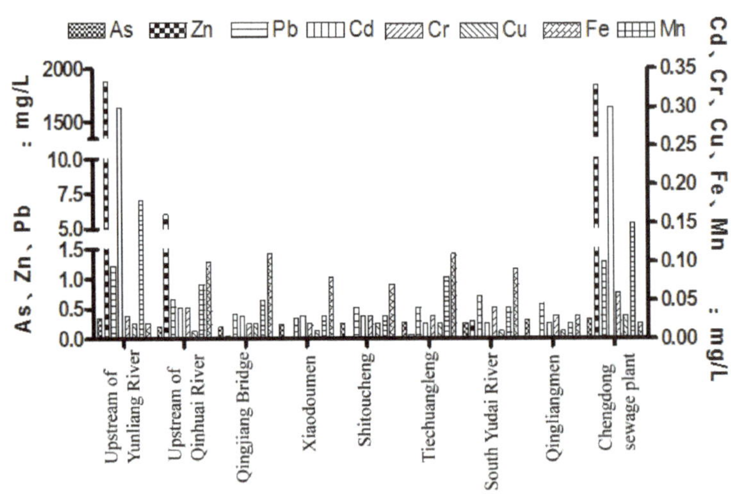

Figure 3. Distribution of heavy metal in pollutant source.

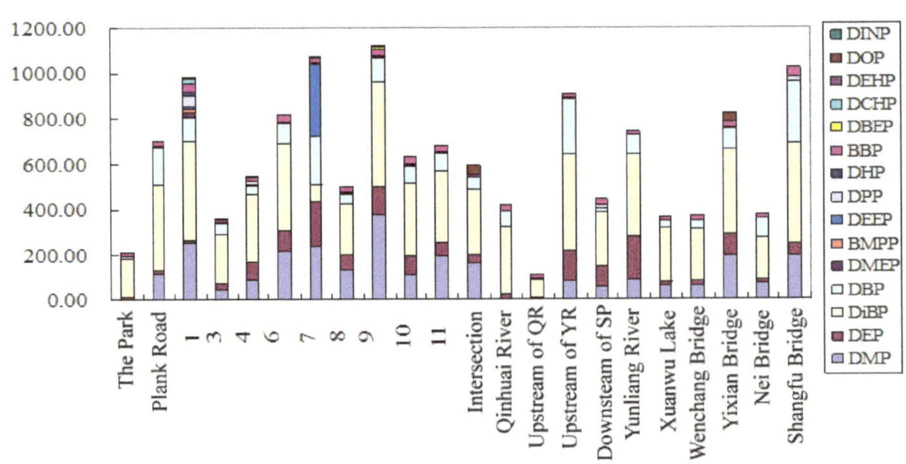

Figure 4. Distribution of PAEs in natural water (ng/L).

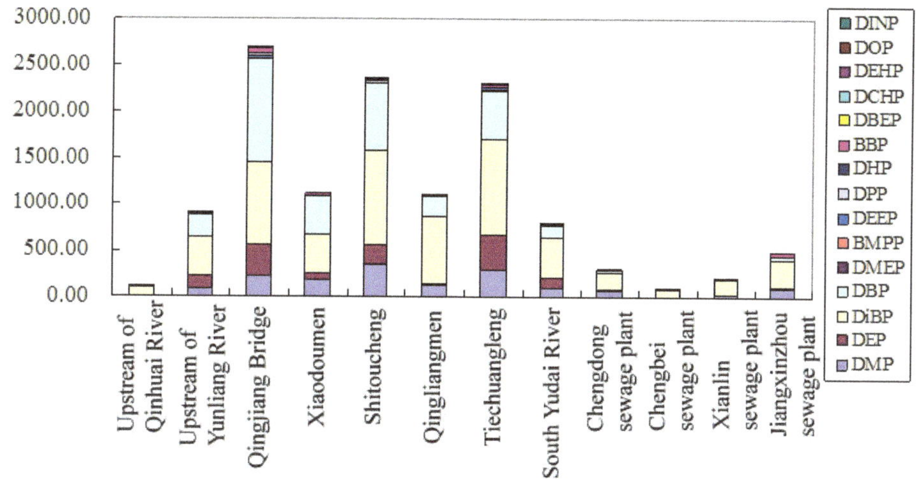

Figure 5. Distribution of PAEs in pollutant source (ng/L).

Polychlorinated biphenyls (PCBs). The distribution of PCBs in natural water is shown in **Figure 6**. It's clearly that the contents of ∑*PCBs* in natural waters is the NeiQinhuai River > the Yunliang River > the Wai-Qinhuai River > the city lakes. The average contribution rate of 7 PCBs concentration is PCB52 > PCB180 > PCB101 > PCB118 > PCB153 > PCB138 > PCB28. ∑PCBs concentration ranges from 2.66 ng/L to 474.30 ng/L with the highest point in the Shangfu Bridge, followed by the downstream of sewage treatment plant and point 1#, the concentrations of 290.01 ng/L and 252.04 ng/L.

The distribution of PCBs in pollutant source is shown in **Figure 7**. It's clearly that the contents of ∑PCBs is Qingjiang bridge > Tiechuangleng > upstream of Yunliang River > South Yudai River > Xiaodoumen > Shitoucheng > Jiangxinzhou sewage plant > Qingliangmen > Chengdong sewage plant > Upstream of Qinhuai River > Chengbei sewage plant > Xianlin sewage plant. ∑PCBs concentration ranges from 1.04 ng/L to 614.42 ng/L with the main factor of PCB52, PCB118 and PCB180.

Polycyclic aromatic hydrocarbon (PAH). The distribution of PCBs in natural water is shown in **Figure 8**. It's clearly that the contents of ∑*PAHs* in natural waters is the WaiQinhuai River > the NeiQinhuai River > the Yunliang River > the city lakes. The average contribution rate of 16 PAHs concentration is Naphthalene > Acenaphthene (Ace) > Phenanthrene > Fluorene > Indeno [1,2,3-cd] pyrene (InP) > Benzo [b] fluoranthene (BbF) > Acenaphthene > Benzo [a] anthracene (BaA) > Two benzo [a, H] anthracene > Fluoranthene > Pyrene > Chrysene (Chr) > Benzo[g,h,i]perylene (BPR) > Benzo [k] fluoranthene (BkF) > Benzo[a]pyrene (BaP) > Anthracene (ANT). ∑*PAHs* concentration ranges from 16.93 ng/L to 455.95 ng/L with the highest point in the Fengtai bridge, followed by point 1# and 5#, the concentrations of 226.24 ng/L and 220.70 ng/L.

The distribution of PCBs in pollutant source is shown in **Figure 9**. It's clearly that the contents of ∑*PAHs* is Tiechuangleng > Qingjiangqiao > Shitoucheng > Upsteam of Yunliang River > Qingliangmen > Xiaodoumen > Chengdong sewage plant > South Yudai River > Jiangxinzhou > Xianlin sewage plant > Upsteam of Qinhuai River > Chengbei sewage plant. ∑*PAHs* concentration ranges from 11.90 ng/L to 1064.68 ng/L with fluoranthene,the main factor of naphthalene, Ace, fluorine, phenanthrene, BaA, BbF and InP.

From the comparison of heavy metal and environmental endocrine disruptors in different types of natural water and the pollution source, we can get that there are different concentration of *Zn*, *Pb*, *As*, *Fe*, *Mn*, *Cd*, *Cr* and *Cu* in natural water and pollution source, in which the main characteristic factors in upstream and tail-water of sewage plant; The main phthalic acid esters in studied natural waters are DMP, DEP, DiBP, DBP, BMPP, DPP, BBP, DCHP, DEHP. And the concentrations of DMP, DiBP and DBP are higher. PCB52, PCB180 are mainly in natural water, and the concentrations are higher in upstream and storm runoff.

The PAHs in natural water conclude of naphthalene, Acenaphthene (Ace), phenanthrene, fluorine, Indeno (1,2,3-cd) pyrene (InP), Benzo (b) fluoranthene (BbF), acenaphthene, Benzo[a] anthracene (BaA),Two benzo (a, H) anthracene, fluoranthene, pyrene, chrysene (Chr), benzo[g,h,i]perylene (BPR), benzo(k) fluoranthene (BkF), benzo[a]pyrene (BaP), anthracene(ANT). The concentration of naphthalene, fluorene, BaA, BaP and InP are higher in upstream water, and those of naphthalene, acenaphthene, fluorene, BaP and InP are higher in storm runoff.

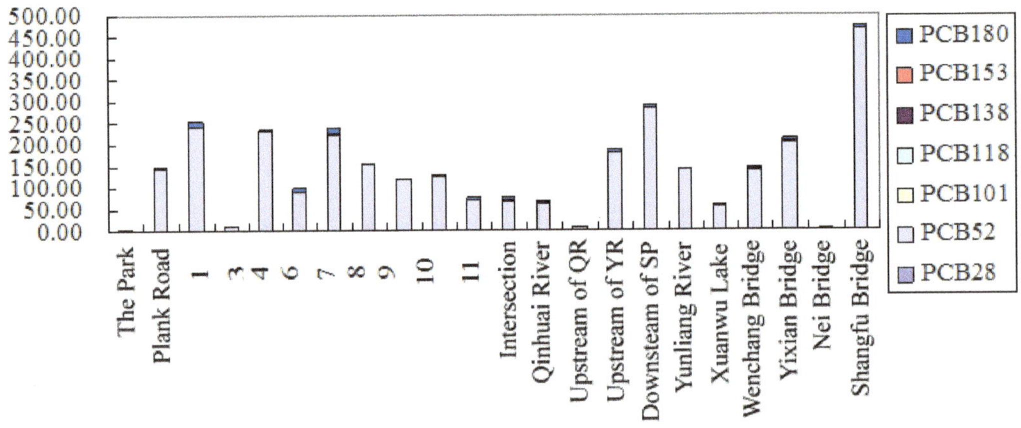

Figure 6. Distribution of PCBs in natural water (ng/L).

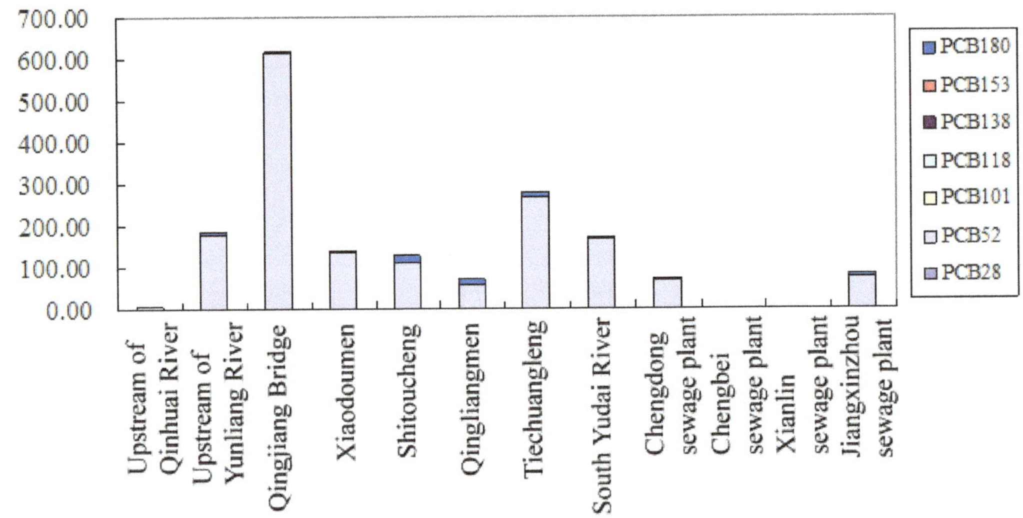

Figure 7. Distribution of PCBs in pollutant source (ng/L).

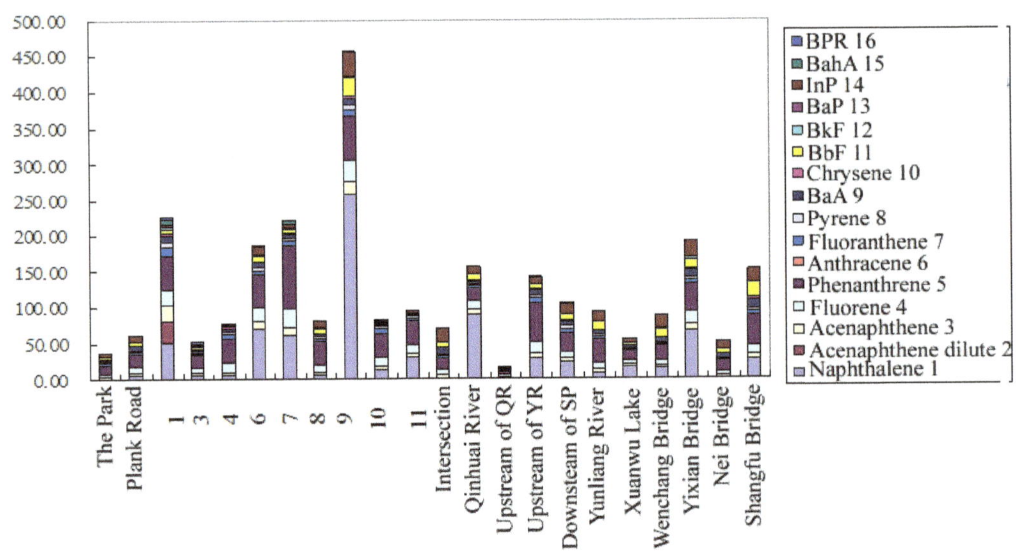

Figure 8. Distribution of PAHs in natural water (ng/L).

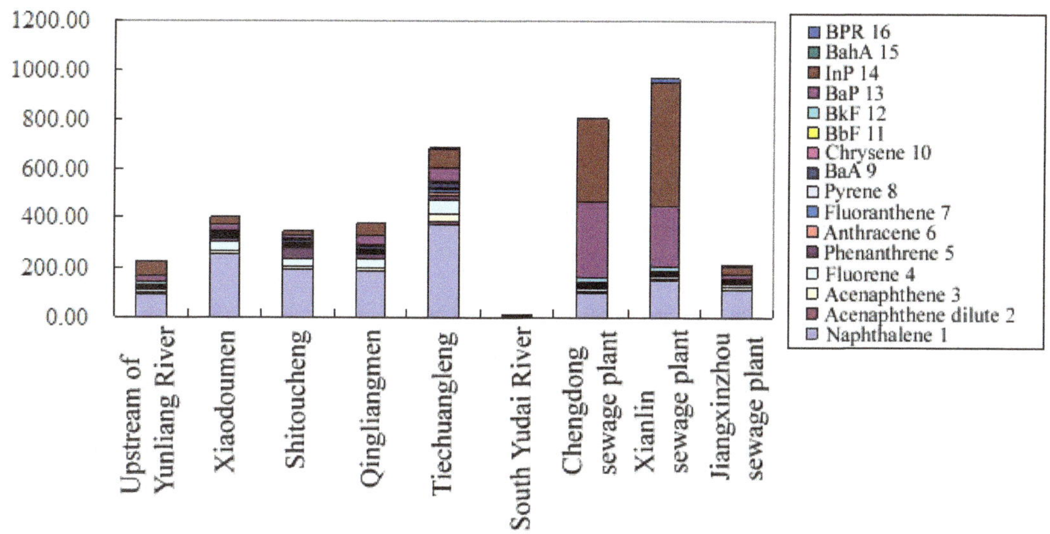

Figure 9. Distribution of PAHs in pollutant source (ng/L).

3. Source Apportionment of Pollution Factors

3.1. Principal Component Analysis of Water Quality

Using SPSS13.0, select C, N, S and P, some hygiene factors and main risk factor to be statistical analyzed. Based on the rule of cumulative variance of principal component number should be more than 70%, the maximal variance rotation method (Varimax), select the factors whose load is greater than 0.6. Thus the results are the KMO value is 0.760, Bartlett's sphericity is 563.89, and concomitant probability is 0, so that the use of principal component analysis method is feasible.

The eigenvalue and contribution rate of principal component analysis are shown in **Table 1**, in which the four factors can reflect 71.07% of total information.

Table 1 shows that the contribution rate of the first principal component feature, F1, is 22.74%, associated with the index of BMPP, DPP, BBP, DiBP, acenaphthene, pyrene, fluoranthene, phenanthrene, flexion. It mainly reflects the pollution status of macromolecular phthalate esters and small molecules of PAHS.

The contribution rate of the second principal component feature, F2, is 21.91%, associated with NH_4^+-N, TN, TOC, DBP, DEP. It mainly reflects the population status of the nutrient salts, organic pollution and macromolecular of PAES.

The contribution rate of the third principal component feature, F3, is 14.69%, associated with naphthalene, BaF, BbF, InP. It mainly reflects the population status of the macromolecular of PAHS.

The contribution rate of the forth principal component feature, F4, is 11.73%, associated with nitrite, nitrate, sulfate and fecal coliform bacteria. It mainly reflects the population status of inorganic toxicology indicators and health indicators.

3.2. Principal Component Analysis of Pollutant Source

The method and rules are the same to the analysis of water quality. Thus the results are the KMO value is 0.773, Bartlett's sphericity is 500.01, concomitant probability is 0, so that the use of principal component analysis method is feasible.

The eigenvalue and contribution rate of principal component analysis are shown in **Table 2**, in which the three factors can reflect 72.58% of total information.

Table 2 shows the main components in pollutant source are PAHs, polychlorinated biphenyls (PCB52, PCB180), PAEs (BMPP, DPP, DEP, DBP, DMP, DiBP), TOC, TN, SO_4^{2-}, fecal coliform bacteria. Principal component analysis results shows that the main pollution group and water quality have a good corresponding relationship.

Table 1. Characteristic value and contribution rate.

	Characteristic Value	Contribution Rate (%)	Cumulative Contribution Rate (%)
F1	6.821	22.74	22.74
F2	6.574	21.91	44.65
F3	4.407	14.69	59.34
F4	3.520	11.73	71.07

Table 2. Characteristic value and contribution rate in pollutant source.

	Characteristic Value	Contribution Rate (%)	Cumulative Contribution Rate (%)
F1	10.297	34.33	34.33
F2	7.491	24.97	59.29
F3	3.986	13.29	72.58

4. Conclusions

Heavy metal, PAEs, PCBs and PAHs are all detected. Factor with relevance ratio higher than 0.5 are Zn, Pb, As, Cd, Cr, Cu, DMP, DEP, DiBP, DBP, BMPP, DPP, BBP, DCHP, DEHP, PCB52, PCB101, PCB118, PCB153, PCB180, naphthalene, Ace, phenanthrene, fluorine, ANT, pyrene, fluoranthene, BaA, Chr, BbF, BaP, InP, Benzo(ghi)perylene .

Sources of pollution in the study area includes forebay sewage of pumping station, storm runoff, tail water of sewage plant and the water upstream. Among them, the indicators are very serious in forebay sewage, the concentrations of nitrite, nitrate and sulfate are higher in the tail water of sewage plant, and the concentration of nitrate nitrogen is higher in the upstream water.

The main pollution group and water quality have a good corresponding relationship. The main pollution factors with potential sources are PAHs, PAEs, TN, NH_4^+-N, TOC, SO_4^{2-} and fecal coliforms.

Acknowledgements

This research is financially supported by the National Natural Science Foundation of China, Grant No. 51409088.

References

[1] Smith, S.V., Renwick, W.H., *et al.* (2012) Distribution and Significance of Small, Artificial Water Bodies across the United States Landscape. *The Science of the Total Environment*, 21-30.

[2] WHO (2010) Guidelines for Safe Recreational Water Environments, Volume 2: Swimming Pools and Similar Environments.

[3] Hermann, F., Thomas, K., Thomas, O., *et al.* (2002) Occurrence of Phthalate and Bisphenol A and F in the Environment. *Water Research*, 1429-1438.

[4] Zhao, X.K., Yang, G.P. and Wang, Y.J. (2004) Adsorption of Diethyl Phthalate on Marine Sediments. *Water Air and Soil Pollution*, **157**, 179-192. http://dx.doi.org/10.1023/B:WATE.0000038880.57430.c3

[5] Mitra, S., Dellapenna, T.M., Dickhut, R.M., *et al.* (2009) Polycyclic Aromatic Hydrocarbons Distribution within Lower Hudson River Estuary Sediments: Physical Mixing vs Sediment Geochemistry. *Estuarine, Coastal and Shelf Science*, 311-326.

[6] Rosa, M.V., Fernandez, P., Martinez, C., *et al.* (2010) Polycyclic Aromatic Hydrocarbons in Remote Mountain Lake Waters. *Water Research*, 3916-3926.

[7] Mitra, S. and Bianchi, T.S. (2003) A Preliminary Assessment of Polycyclic Aromatic Hydrocarbon Distributions in the Lower Mississippi River and Gulf of Mexico. *Marine Chemistry*, 273-288. http://dx.doi.org/10.1016/S0304-4203(03)00074-4

[8] Chen, Y.Y., Zhu, L.Z. and Zhou, R.B. (2006) Characterization and Distribution of Polycyclic Aromatic Hydrocarbon in Surface Water and Sediment from Qiantang River, China. *Journal of Hazardous Materials*, 311-326.

[9] Bremle, G. and Larsson, P. (1998) PCBs in River Ema Ecosystem. *Human and Environment Magazine*, **27**, 384-391.

24

Seasonal Variation in Physicochemical Parameters of Lekki Lagoon and the Conservation of Its Ecosystem

Isaiah O. Opadokun*, **A. E. Falaye, E. K. Ajani**

Department of Aquaculture and Fisheries Management, University of Ibadan, Ibadan, Nigeria
Email: *opadokundipo@gmail.com

Abstract

Some physicochemical parameters like temperature, transparency, dissolved solid, suspended materials, turbidity, conductivity, pH, total alkalinity, dissolved oxygen, nitrate, phosphate and salinity of water samples collected at Lekki Lagoon from November 2010 to September 2012 were evaluated. Fifteen locations were selected for water sampling according to the activities that took place in the lagoon. Physical water quality parameters such as temperature, transparency, total dissolved solids and total suspended solid showed strong seasonal and spatial fluctuation. Parameters like conductivity, pH and total alkalinity had relatively constant values among stations though they varied seasonally. Dissolved oxygen, nitrate and phosphate values were within permissible range for fish. Higher salinity values were recorded in wet seasons than dry seasons suggesting that salinity variation at Lekki Lagoon was due to salt incursion from the adjourning Lagos Lagoon and Mahin Creek during rainy season rather than evaporation factor. Salinity range of 1.70 - 6.00 psu recorded in this study also shows that Lekki Lagoon is tending towards brackish and this may affect the ichthyofauna composition of the lagoon.

Keywords

Water Quality Parameters, Brackish, Spatial Fluctuation, Ichthyofauna

1. Introduction

Larger percentage of fish being consumed in the country comes from brackish-marine environment [1]. Conditions of the lagoon environment differ greatly from that of the freshwater environment due to continuous varia-

*Corresponding author.

tion in the physiochemical and hydrological characteristics of the brackish environment.

Lekki Lagoon, located in Lagos and Ogun states of Nigeria covers an area of nearly 247 km^2 and lies between longitude 4˚00'E and 4˚15'E and latitude 6˚22'N and 6˚37'N [2] [3]. Lagos Lagoon links Lekki Lagoon at the upper part while Mahin creeks link the lagoon at the lower part. River Oni flows into Lekki Lagoon in the north eastern side while Rivers Oshun and Saga flow in from the north western parts [4] [5]. The lagoon is surrounded by many beaches.

Lekki Lagoon is the most important source of affordable animal protein in the form of fish in Epe and Ibeju-Lekki Local Government Areas of Lagos State [6]. This lagoon is the most important source of freshwater fish in Lagos State. The fishery is diverse, highly dispersed and fragmented with about 35 landing sites and more than 3000 fisherfolks [6]. The main occupation of the people in the surrounding communities of the lagoon is fishing [7].

The mean depth of Lekki Lagoon is 3.1 m, and the maximum depth is 6.4m at low tide [3] [4] [8]. The highest salinity recorded for Lekki Lagoon in 1981 was 0.3 psu [9] which had made some authors to describe the lagoon as fresh [4] [10]. However, recent works [11] on Lekki Lagoon between year 2006 and 2008 recorded fifty-six fish species that had fresh water origin, ten species of marine water origin and eleven species that were of euryhaline origin and concluded that this lagoon formed a link between the brackish and freshwater systems. There is a possibility that changes in the physicochemical parameters of the water among others things are responsible for the changes in the fish species composition of the lagoon. Since there is paucity of information on the current physicochemical composition of the lagoon and its possible influence on the conservation of the ichthyofauna composition of the lagoon, this study becomes imperative.

2. Materials and Methods

The study area, Lekki Lagoon (**Figure 1**) was monitored from November 2010 to September 2012. Bimonthly field trips were embarked on to collect water samples between the morning hour 9.00 a.m. and 10.00 a.m. in order to determine the physicochemical parameters at the marked sampling points in **Figure 1**. The following physico-chemical parameters were investigated: temperature, transparency, total dissolved solid, total suspended

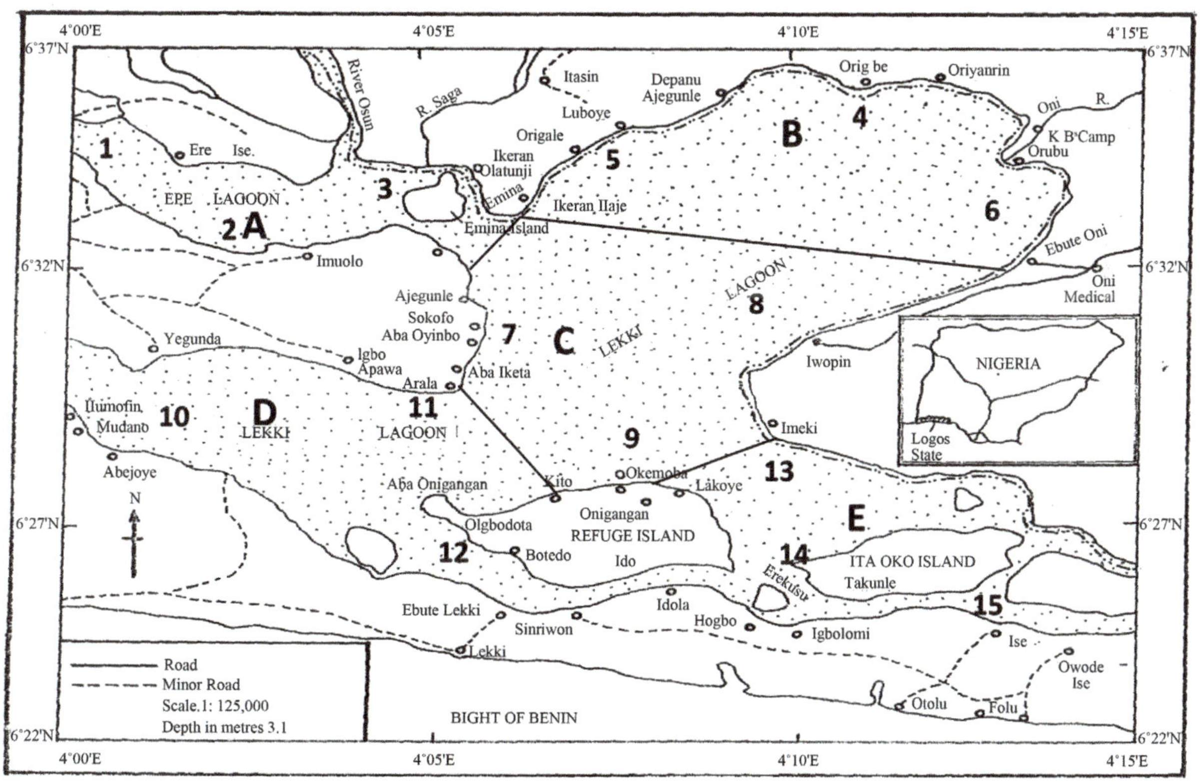

Figure 1. Map of Lekki Lagoon showing Sampling Stations.

solid, turbidity, conductivity, pH, total alkalinity, dissolved oxygen, nitrate, phosphate and salinity. The analysis was done using a combination of *in situ* and laboratory method [12].

The sampling methods were time and space stratified. Time stratification covered two dry seasons and two wet seasons. Space stratification was based on the method of Ajani [13] in which the entire lagoon was divided into five zones-mouth, upper, middle, lower and head. Three sampling stations were chosen in each of the five zones giving a total of fifteen sampling stations within the lagoon (**Figure 1**). The sampling stations were located using bearings such as major landing sites, shoreline features, fishermen poles and wrecks on the lagoon. The exact sampling locations were fixed using Global Positioning System (GPS) as stated in **Table 1**.

Basic statistical analyses such as means, range, standard deviation, t-test and probability levels were used to test the physicochemical parameter data of Lekki Lagoon. One way and two ways analysis of variance were used to find level of significance on seasonal variation while water quality parameters were subjected to Chi-square test of homogeneity.

3. Results and Discussion

Temperature was relatively high during the study period with a mean of 28.73˚C ± 1.83˚C. This may be due to the shallow nature of Lekki Lagoon (3.0 - 6.4 m) according to Kusemiju [8], which permits the heat of the sun to penetrate deep into the water as well as the regular tidal motion of the lagoon. A similar trend was observed on Lagos Lagoon by Ajani [13]. Dry seasons' temperatures were slightly higher than the wet seasons although the fluctuation was just 1.03˚C and 0.33˚C in the first and second sampling years respectively. Actually this agrees with earlier works [13]-[15] who asserted a slight variation of about 0.5˚C only in the dry and wet seasons' temperature of the lagoon of Southwestern Nigeria.

Temperature range recorded for the sampling period was 26.5˚C - 32.1˚C. This is similar to observation of Lawal-Are [16] who recorded a temperature range of 26.5˚C - 31.2˚C for Lekki Lagoon, The relatively small temperature range observed in this study also tallied with the observations of some authors [6] [9] [13]. They all noticed that the temperature is a stable environmental factor in the shallow brackish environments of West Africa.

Turner and Millward [17] stated the importance of water transparency in the Lagoons as a main determinant factor of the condition and productivity of an aquatic system and the docility of water for human consumption and recreation. Mean water transparency recorded for the sampling period was 0.65 ± 0.08 m with a range of 0.40 - 0.80 m. This range was higher than the observation of Lawal-are [16] who recorded a transparency range

Table 1. GPS location of sampling stations.

s/n	Zone	Sampling station	Gps coordinates
1	A	Ere ise	06˚34.753'N/004˚01.525'E
2	A	Imogo	06˚32.690'N/004˚04.232'E
3	A	Emina	06˚32.765'N/004˚04.972'E
4	B	Origbe	06˚35.690'N/004˚10.845'E
5	B	Iluboye	06˚34.292'N/004˚07.772'E
6	B	Ebute oni	06˚32.004'N/004˚13.313'E
7	C	Aba oyinbo	06˚30.487'N/004˚05.894'E
8	C	Iwopin	06˚30.614'N/004˚10.259'E
9	C	Onigangan	06˚27.472'N/004˚08.214'E
10	D	Ilumofin	06˚29.478'N/004˚01.009'E
11	D	Arala	06˚29.750'N/004˚05.523'E
12	D	Batedo	06˚26.443'N/004˚06.212'E
13	E	Lakoye	06˚27.621'N/004˚08.743'E
14	E	Takunle	06˚25.757'N/004˚11.234'E
15	E	Ise	06˚25.011'N/004˚12.865'E

of 0.10 to 0.47 m on the same Lekki Lagoon, but in line with works of Kusemiju [8] and Emmanuel [18] who recorded a range of 0.52 - 1.04 m and 0.80 - 2.13 m respectively. Different readings by all these authors are directly related to the rainfall pattern during the sampling period. Notwithstanding, all these previous works agreed with this study that higher transparency was recorded in dry season and lowest transparency in the rainy season. This could be attributed to higher turbidity during the wet season caused by the discharge of rain water into the Lekki Lagoon which brings debris, suspended particles and disturbance at the bottom [8].

Total dissolved solid (TDS) and Total suspended solid (TSS) were both higher in the wet season than the dry season. High values of TDS and TSS must have contributed greatly to the low transparency experienced during wet season. Ajani [13] who also recorded low transparency in Lagos Lagoon during wet season attributed this to high quantity of suspended solid materials washed into the lagoon by the run-off from the surrounding rivers. Although this was contrary to the findings of Emmanuel [18] who recorded a higher TDS for dry season than wet season in Lagos Lagoon and attributed this to the proximity of Lagos Lagoon to the sea. Both TDS and TSS had negative correlation with transparency $r = -0.26$ and -0.28 for both TDS and TSS respectively.

In this study, turbidity was lowest in the dry season (6 NTU) and had highest value in the rainy season (22 NTU). Thus, turbidity and transparency had an inverse relationship. This was in line with the work of Babalola [19] on Kuramo Lagoon in Lagos state who stated that the higher the turbidity, the lower the transparency, and vice versa.

Water conductivity is a measure of the concentration of dissolved solids in water and it gives an indirect dissolved salt content of fluid. The mean water conductivity for the sampling period was 133.41 ± 15.32 μScm^{-1} with a range of 94 - 160 μScm^{-1}. The lowest water conductivity (94.00 μScm^{-1}) was recorded in May 2011 at the onset of rainy season in Station 3; while the highest conductivity reading (160 μScm^{-1}) was recorded in September, 2012 at Station 7. Lagoon was relatively rough at this period. A weak positive correlation exist between conductivity and total dissolved solid content of the water ($r = 0.22$) during the period under review. Seasonal variations in the lagoon water conductivity was significant ($P < 0.05$), whereas spatial variations was not.

Washington State Department of Ecology [20] reported a conductivity of 56,000 μScm^{-1} for sea water and stated that conductivity is an indirect index of water hardness. 6 - 25 μScm^{-1} was reported for conductivity in Cross River System [21], 44 - 96 μScm^{-1} for River Oshun [22] and 46 - 96 μScm^{-1} in the upper Niger River [23], respectively. Emmanuel [18] reported a conductivity mean value of 27500 μScm^{-1} for wet season in Lagos Lagoon and 51,400 μScm^{-1} for dry season because of the lagoon proximity to the sea. He however recorded 466.0 μScm^{-1} for Lekki Lagoon in dry season, a value which was higher than the one obtained in this study. Babalola and Agbebi [19] reported that the conductivity of Kuramo Lagoon water fluctuate from 582 μScm^{-1} in wet season to 820 μScm^{-1} in dry season.

pH is a measure of the acidity or basicity of an aqueous solution. An aqueous solution with pH value less than 7 is classified as acidic while the one with a pH value greater than 7 is said to be basic or alkaline. The pH varied between 5.00 and 8.00 with a mean of 6.45 ± 0.89 during the sampling period. This range falls within permissible range for fish culture. The lowest water pH was 5.00 and it was recorded in January 2012 at the peak of dry season in Stations 2 and 4; while the highest pH reading (8.00) was recorded in May and July, 2011 at the peak of rainy season. Seasonal pH variation was significant ($P < 0.05$) whereas there was no significant difference ($P > 0.05$) in the mean water pH among different Stations. pH is a very important physicochemical parameter that influences fish performance in water. Several fish will stop growing at a lethal pH and eventually die off in extreme cases.

The measurement of the capacity of water to neutralize acids is referred to as alkalinity [24]. The value of total alkalinity of a water body is very important in determining the ability of such water body to neutralize acidic pollution from waste water, industrial effluents or rainfall. Total alkalinity value for the sampling period ranged between 16.0 - 45.0 mg/L with a mean of 20.90 ± 5.70 mg/L. Wet seasons recorded higher alkalinity values than dry seasons; this may be due to run offs from surrounding rocks and soils during rainfall as well as certain plant activities [24]. Mean total alkalinities for 2010-2011 and 2011-2012 wet seasons respectively were 17.90 ± 1.13 mg/L and 26.38 ± 9.32 mg/L. Dry seasons total alkalinities values were 19.73 ± 0.81 mg/L and 19.58 ± 0.77 mg/L for 2010-2011 and 2011-2012 sampling periods respectively. Seasonal variation was significant ($P < 0.05$) while spatial variation was not ($P > 0.05$).

The dissolved oxygen (DO) range for the study period was 4.7 - 8.0 mg/L. This value still falls within acceptable range for optimal performance of fish. DO ranges of 4.9 - 11.2 mg/L, 4.0 - 11.2 mg/L and 5.2 - 11.8 mg/L were reported for Badagry, Lagos and Lekki Lagoons respectively [16]; whereas a range of 4.03 - 8.93

mg/L was reported for Lagos Lagoon [13]. Thus it appear as if DO is usually high in lagoon probably due to constant tidal waves that ensures mixing of lagoon water, thereby constantly exposing the bottom water to atmospheric oxygen thus ensuring constant diffusion of atmospheric oxygen into the lagoon water.

Fluctuation in the DO was much in the second sampling year (2011-2012) than the first sampling year (2010-2011). Likewise, DO values for the wet seasons were higher than the dry seasons (**Table 2**), an observation similar to what was reported on Lagos Lagoon [13]. These relationships between the DO and rainfall pattern could be as a result of highly oxygenated rainwater that mix with the lagoon water during wet season thus increasing the DO content of the lagoon water. Babalola and Agbebi [19] state that during the dry season, the dissolved oxygen escaped to the atmosphere from the surface of Kuramo Lagoon water due to high temperature because oxygen is sensitive to high temperature. During this period, according to them, aquatic plants compete for dissolved oxygen in the water for respiration although this can be gotten back as a product of photosynthesis during the day time. During the wet season however, the dissolved oxygen in Kuramo Lagoon increased as a result of dissolved atmospheric oxygen from rain water and high wind current.

Phosphate and nitrate did not show distinct seasonal variation but did show spatial variation. Highest value of Nitrate 0.39 ± 0.22 mg/L and Phosphate 0.36 ± 0.18 mg/L were recorded in Zone E while Zone A had the lowest Nitrate 0.18 ± 0.15 mg/L and Phosphate 0.15 ± 0.09 mg/L. Nitrate 0.12 ± 0.02 - 0.48 ± 0.15 mg/L and phosphate 0.14 ± 0.09 mg/L - 0.29 ± 0.14 mg/L were recorded in dry and wet seasons respectively. Bulk of the nitrogen might be from agricultural fields or deposition of raw sewage into the lagoon while phosphate addition could be from detergent rich sewage effluents. Phosphate and nitrate showed almost positive relationship in all locations and the relationship was significant.

Salinity range for the study period was between 1.70 and 6.00 psu. The lowest salinity (1.7 psu) was recorded in November 2011 at Station 5 while the highest salinity (6.00 psu) was obtained at Station 13 in July 2011. The highest salinity recorded by Kusemiju [9] for Lekki Lagoon was 0.3 psu and Williams [25] as cited by Trewavas [10] and Fawole [4] described the lagoon as fresh. However, this recent study confirms the works of Lawal-Are [16] and Emmanuel [18] who reported a range of 0.00 to 3.10 psu and 0.007 to 4.70 psu respectively for Lekki Lagoon. They identified two major sources of saline water incursion into Lekki Lagoon which were Lagos Lagoon and Mahin Creek. Waljeski [26] in a similar work on Rodeo Lagoon in California reported salt water intrusion into Rodeo Lagoon by subsurface flow through the barrier beach from the ocean and leaching of ions through lagoon bottom sediments.

The seasonal salinity variation during the wet and dry seasons was quite high (**Table 2**). Analysis of the seasonal salinity variation showed that the mean salinity for dry seasons were 2.53 ± 0.47 psu and 2.63 ± 0.63 psu in 2010-2011 and 2011-2012 sampling periods respectively while wet seasons mean values were 3.83 ± 0.93 psu

Table 2. Means of the major physicochemical parameters recorded in the two sampling years (2010-2011 and 2011-2012).

Parameters	2010-2011	2011	2011-2012	2012
	Dry	Wet	Dry	Wet
Temperature (˚C)	28.33 ± 0.91	27.30 ± 1.74	29.80 ± 1.32	29.47 ± 2.01
Transparency (m)	0.71 ± 0.04	0.60 ± 0.90	0.70 ± 0.03	0.61 ± 0.08
Salinity (psu)	2.53 ± 0.47	3.83 ± 0.93	2.63 ± 0.63	4.43 ± 0.52
Dissolved oxygen (mg/L)	6.33 ± 0.40	6.63 ± 0.45	5.98 ± 0.65	4.43 ± 0.52
pH	5.68 ± 0.38	7.08 ± 0.82	5.96 ± 0.55	7.07 ± 0.65
Nitrate (mg/L)	0.18 ± 0.04	0.39 ± 0.15	0.20 ± 0.09	0.52 ± 0.28
Phosphate (mg/L)	0.16 ± 0.03	0.37 ± 0.16	0.22 ± 0.10	0.29 ± 0.13
Total alkalinity (mg/L)	19.73 ± 0.81	17.90 ± 1.13	19.58 ± 0.77	26.38 ± 9.32
Conductivity (μScm^{-1})	131.67 ± 7.05	126.33 ± 20.50	125.23 ± 5.99	150.40 ± 5.63
Total dissolved solid (mg/L)	48.82 ± 5.63	52.07 ± 8.13	46.67 ± 3.07	45.66 ± 6.70
Total suspended solid (mg/L)	44.83 ± 4.21	50.17 ± 8.07	45.07 ± 3.09	43.57 ± 7.37
Turbidity (m)	11.83 ± 1.89	16.00 ± 3.26	11.67 ± 2.37	13.50 ± 3.70

and 4.43 ± 0.52 psu for years 2010-2011 and 2011-2012 respectively. These suggest that the inflow of water from Lagos Lagoon and Mahin Creek in the rainy season was the main source of increase in salinity level at Lekki Lagoon. Indeed, Lawal-Are [16] concluded that it would appear that Lekki Lagoon is beginning to experience a low brackish water regime instead of being completely fresh water.

Salinity as an important ecological factor has been known to have a great effect on the biotic life in aquatic ecosystem because each fish species has a salinity range where they can operate. Blaber [27] reported that egg development of some fish species occurred at certain salinity ranges and any departure from such ranges will have serious negative effect on the growth and development of the fish. Thus, freshwater fish species like *Gymnarchus niloticus*, *Mormyrus rume* and *Parachana obscura* which are predominant in Lekki Lagoon [11] could have their existence in the lagoon threatened if the saltwater incursion into the lagoon should continue. ANOVA showed that although the seasonal variation was significant ($p < 0.05$), spatial variation (among different stations) was insignificant ($p > 0.05$). This means that there were not much variation in the salinity values at different sampling locations during the sampling period although much variation existed within the sampling months (period).

4. Conclusions

This study confirms stability in the physical and chemical factors of Lekki Lagoon. The spatial distributions were relatively stable while there was a distinct seasonal variation in most cases. The highest salinity value of 6.00 psu recorded in this study is more than the highest values reported by earlier works which are 0.3 psu, 3.10 psu and 4.70 psu respectively [9] [16] [18]. This could mean that Lekki Lagoon is now experiencing a low brackish water regime instead of being completely fresh water. This brackish level tends to be increasing every year apparently due to continuous saline water incursion from the two ends of the lagoon-Lagos Lagoon and Mahin Creek. This would eventually change the ichthyofauna composition of the lagoon.

There should be regular monitoring and study of the physical and chemical state of the Lekki Lagoon environment in order to provide a basis for an informed decision on sustainable management of the lagoon.

References

[1] FAO (1995) FAO Yearbook: Fishery Statistics—Catches and Landings 1993. Vol. 76. Food and Agriculture Organization of the United Nations, Rome, Italy, 687 p.

[2] Opadokun, I.O and Ajani, E.K. (2015) Some Aspects of the Reproductive Biology of *Gymnarchus niloticus* Cuvier, 1829 (Knifefish) in Lekki Lagoon, Nigeria. *International Journal of Fisheries and Aquatic Studies*, **2**, 166-170.

[3] Akinsanya B., Otubanjo, O.A. and Ibidapo, C.A. (2007) Helminth Bioload of *Chrysichthys nigrodigitatus* (Lacepede 1802) from Lekki Lagoon Lagos, Nigeria. *Turkish Journal of Fisheries and Aquatic Science*, **7**, 83-87.

[4] Fawole, O.O. (2002) Morphometry and Diet of *Momyrus rume* in the Lekki Lagoon Lagos, Nigeria. *International Journal of Tropical Biology and Conservation*, **50**, 689-694.

[5] Kuton, M.P. (2006) The Biology of Cichlid Fishes from Badagry, Lagos and Lekki Lagoons, Southwest Nigeria. Ph.D. Dissertation, University of Lagos, Lagos.

[6] Emmanuel, B.E (2009) The Artisanal Fishing Gears, Crafts Technology and Their Efficiency in the Lekki Lagoon, Nigeria. Ph.D. Dissertation, University of Lagos, Lagos.

[7] Lagos State Facts and Figures (1996) No 20, Akinlade L. Ed. 30 p.

[8] Kusemiju, K. (1973) A Study of the Catfishes of Lekki Lagoon with Particular Reference to the Species *Chrysichthys walkeri* (Bagridae). Ph.D. Dissertation, University of Lagos, Lagos.

[9] Kusemiju, K. (1981) The Hydrobiology and Fishes of the Lekki Lagoon, Nigeria. *Nigerian Journal of Natural Sciences*, **3**, 135-145.

[10] Trewavas, E. (1983) Tilapine Fishes of the Genera *Sarotherodon*, *Oreochromis* and *Danakilia*. London British Museum (Natural History), 583 p.

[11] Emmanuel, B.E. and Osibona, A.O. (2013) Ichthyofauna Characteristics of a Tropical Low Brackish Open Lagoon in South-Western Nigeria. *International Journal of Fisheries and Aquaculture*, **5**, 122-135.

[12] APHA (1998) Standard Methods for the Examination of Water and Wastewater. 20th Edition, American Public Health Association, Washington DC.

[13] Ajani, E.K. (2000) Effects of Biotic and Abiotic Components of the Habitat on Fish Productivity in Lagos, Nigeria. PhD Dissertation, University of Ibadan, Ibadan.

[14] Olaniyan, C.I.O. (1969) The Seasonal Variations in the Hydrography and Total Plankton of the Lagoons of Southwest Nigeria. *Nigerian Journal of Science*, **3**, 101-119.

[15] Ajao, E.A. (1990) The Influence of Domestic and Industrial Effluents on Populations of Sessile and Benthic Organisms in Lagos Lagoon. PhD Dissertation, University of Ibadan, Ibadan.

[16] Lawal-Are, A.O. (2001) Aspects of the Biology of the Lagoon Crab, *Callinectes amnicola* (Derocheburne) in Badagry, Lagos and Lekki Lagoons, Nigeria. *Journal of Scientific Research and Development*, **5**, 215-220.

[17] Turner, A. and Millward, G.E. (2002) Suspended Particles: Their Role in Estuarine Biogeochemical Cycles. *Estuarine, Coastal and Shelf Science*, **55**, 857-883. http://dx.doi.org/10.1006/ecss.2002.1033

[18] Emmanuel, B.E. and Chukwu, L.O. (2010) Spatial Distribution of Saline Water and Possible Sources of Intrusion into a Tropical Freshwater Lagoon and the Transitional Effects on the Lacustrine Ichthyofaunal Diversity. *African Journal of Environmental Science and Technology*, **4**, 480-491.

[19] Babalola, O.A. and Agbebi, F.O. (2013) Physico-Chemical Characteristics and Water Quality Assessment from Kuramo Lagoon, Lagos, Nigeria. *Society for Science and Nature*, **3**, 98-102.

[20] Washington State Department of Ecology (1991) Lakes: Temperature in Lakes. In: Blaber, S.J.M., Ed., *A Citizen's Guide to Understanding and Monitoring Lakes and Streams*, Fish and Fisheries of Tropical Estuaries, Chapman Hall, London, 367. http://www.ecy.wa.gov/programs/wq/plants/management/joysmanual/temperature.html

[21] Akinrotimi, O.A. and Edun, O.M. (2013) Impact of Climate Change on Brackish Water Aquaculture Development in the Coastal Areas of the Niger Delta. *Proceedings of the* 28*th Annual Conference of the Fisheries Society of Nigeria*, Abuja, 25-28 November 2013, 140-145.

[22] Egborge, B.M. (1979) The Effect of Impoundment on the Water Chemistry of Lake Asejire, Nigeria. *Freshwater Biology*, **9**, 403-412. http://dx.doi.org/10.1111/j.1365-2427.1979.tb01525.x

[23] Welcomme, R.L. (1985) River Fisheries. FAO Technical Paper No. 262, Rome, 320 p.

[24] United States Environmental Protection Agency (USEPA) (2012) Total Alkalinity, Monitoring and Assessment. http://water.epa.gov/type/rsl/monitoring/vms510.cfm

[25] Williams, N.V. (1962) The Seasonal Distribution of the Teleost Fauna in Lagos Harbor, Creek and Lagoon in Relation to Salt Tolerance. M.Sc. Thesis, University College of North Wales, Bangor.

[26] Waljeski, C.A. and Williams, J.L. (2004) Spatial Distribution and Possible Sources of Saline Waters in Rodeo Lagoon, Golden Gate National Recreation Area, Mavin County, California. Water Resources Center, Archives, 23.

[27] Blaber, S.J.M. (1997) Fish and Fisheries of Tropical Estuaries. Chapman Hall, London, 367.

25

Assessment of Arsenic Contamination in Deep Groundwater Resources of the Kathmandu Valley, Nepal

Suman Man Shrestha[1], Kedar Rijal[1], Megh Raj Pokhrel[2]

[1]Central Department of Environmental Science, Tribhuvan University, Kirtipur, Kathmandu, Nepal
[2]Central Department of Chemistry, Tribhuvan University, Kirtipur, Kathmandu, Nepal
Email: sshrestha@cdes.edu.np

Abstract

This study was carried out to assess arsenic in deep groundwater resources in the Kathmandu Valley, Nepal and to predict arsenic mobilization process in relation to iron, manganese, pH and ORP. Forty-one deep groundwater samples were collected during pre monsoon and post monsoon in 2013. The depths of the wells were ranged from 84 to 304 m. In pre monsoon and post monsoon, arsenic concentration in 17% and 26% of examined groundwater wells, respectively exceeded permissible World Health Organization (WHO) guideline value of 0.010 mg/L for drinking water. The concentrations of arsenic were in the range between < 0.003 to 0.137 mg/L. The study demonstrated elevated concentrations of iron and manganese in the groundwater. Arsenic is highly correlated with iron and manganese. The strong negative correlation between arsenic and ORP indicates that arsenic mobilization occurs under reducing condition. These distinct relationships indicate that arsenic release is considered to be affected by the reductive dissolution of Fe/Mn oxides in the groundwater. Arsenic has very weak negative correlation with pH suggesting less effect of pH on arsenic mobilization. Arsenic is not significantly correlated with the season which infers similar distribution of arsenic in both seasons. Arsenic varies spatially in groundwater of the valley showing high concentrations in central groundwater district.

Keywords

Arsenic, Deep Groundwater, Reductive Dissolution, Kathmandu Valley

1. Introduction

Heavy metal contamination of groundwater is one of the major problems in the world. The heavy metals that occur as natural contaminants of groundwater and are potentially bio-hazardous include manganese (Mn), lead (Pb), cadmium (Cd), mercury (Hg) and arsenic (As) [1] [2]. Arsenic is recognized as a toxic element and has been classified as a human carcinogen affecting skin and lungs [3]. Arsenic has strong toxicity at even low concentrations and can accumulate in body tissues over long periods of time and is nonessential for human health [4] [5]. So, elevated levels of arsenic constitute problems for water supplies around the world [6]. In recent past years, the occurrence of high concentrations of arsenic has been detected in groundwater from a number of regions across the world. The problem has increased greatly in recent years in several regions of Southeast Asia. In this region, countries affected with arsenic in groundwater include Bangladesh, several states of India (West Bengal, Bihar, Uttar Pradesh, Jharkhand, Assam, Chhattisgarh and Manipur), Nepal, Myanmar, Pakistan, Vietnam, Lao People's Democratic Republic, Cambodia and China [7] [8].

The demand for water is increasing due to rapid growth of urban population and industrial activities in the Kathmandu Valley. There is a high decadal population growth rate found in Kathmandu district *i.e.*, 61.23% [9]. As a result there is an immense pressure on groundwater resources in the valley. Groundwater was first exploited for water supply in 1970 in the valley. Mechanized extraction of groundwater resources began in 1984. Groundwater is an important water resource in the Kathmandu Valley. It contributes about 50% of the total water supply in the valley [10]. In dry season, 60% to 70% of the water supply is met by ground water [11].

There have been several studies of assessment of arsenic in groundwater of the Kathmandu Valley. Groundwater survey of the valley reported the presence of arsenic and the concentrations were below WHO (1993) guideline values [12]. Similar study revealed that the ground water resource in the valley is contaminated with arsenic in deep aquifers [13]. Likewise, other studies have reported elevated levels of arsenic in groundwater of the valley [14]-[19]. Arsenic in groundwater wells in the Kathmandu Valley is one of the major environmental issues due to its negative health impact and more than 50% of water supply in the valley is derived from groundwater resources. So, this paper presents an overall study on the occurrence, mobilization and distribution of arsenic in deep groundwater resources in the valley. The study attempted to visualize spatial distribution pattern of arsenic in northern groundwater district (NGWD), central groundwater district (CGWD) and southern groundwater district (SGWD) in the valley in GIS environment. The study also aimed to demonstrate temporal (seasonal) variation of arsenic. The possible relationship between arsenic and depth of the groundwater was also examined.

2. Materials and Methods

2.1. Study Area

The Kathmandu Valley is in the hilly region of Nepal. It is roughly circular with diameter of 25 km and an average altitude of 1300 m (above sea level) located in the central Nepal Himalaya, lies in between 27°32' and 27°49' North and 85°12' and 85°32' East (**Figure 1**). Its surrounding hills are approximately 2800 m (above sea

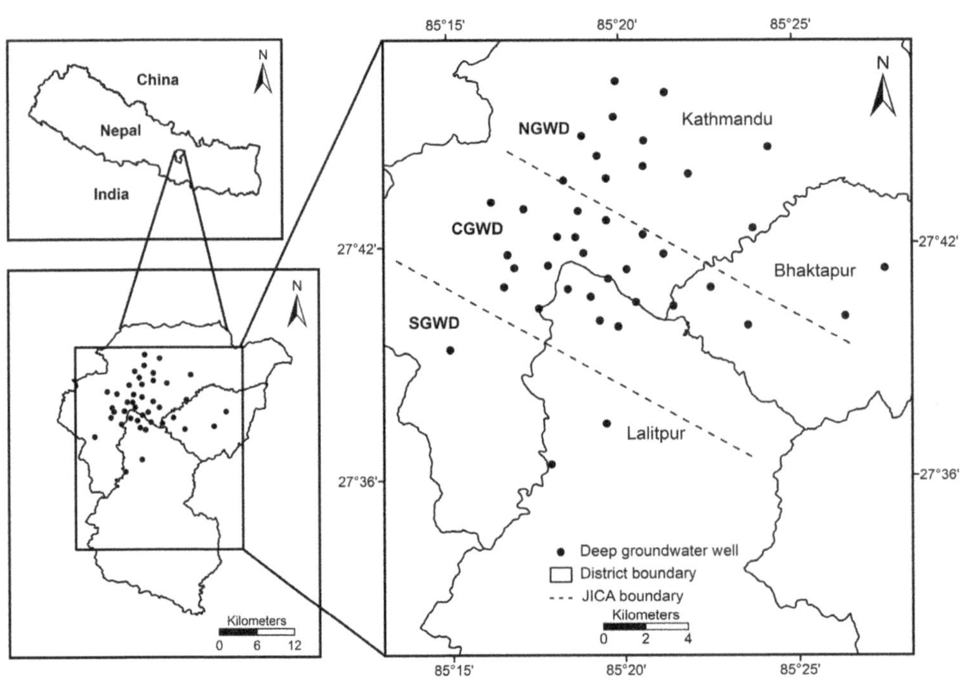

Figure 1. Location map of study area in Kathmandu Valley, Nepal and sampling locations.

level). The area of the Kathmandu Valley is about 650 km^2. The valley comprises of three major cities: Kathmandu, Lalitpur and Bhaktapur. The capital city Kathmandu is the largest city of Nepal. The annual rainfall of the Kathmandu Valley is 1465 mm [20]. The valley receives 80% of annual rainfall during monsoon (June-September) [10].

The Kathmandu Valley is an intermontane basin overlying the bedrock formations up to 500 m of a thick band of Pliocene-quaternary fluvio-lacustrine unconsolidated sediments [21]. The valley sediment is referred as Kathmandu Complex and is composed of Paleozoic and Precambrian-age rocks. The limestone dominates to the south, whereas to the east and west, the valley is bordered by phyllites and siltstones. Granite gneisses thrust into the rocks of the Kathmandu Complex that form the northern border of the valley [22].

Based on hydrochemical and hydrogeological conditions, JICA [23] divided the deep aquifer of the Kathmandu Valley into three groundwater districts: northern, central and southern. The northern groundwater district is considered as main aquifer and has greater abilities to recharge the groundwater. The upper deposits of the district composed of unconsolidated and highly permeable micaceous quartz, sand and gravel of about 60 m thick, interbedded with several impermeable fine layers. In the central groundwater district, the upper deposits are overlaid by impermeable very thick stiff black clay with peat and lignite bands, named as Kalimati Formation. The Kalimati Formation is covered by fluvial deposits of fine to medium sand and silt intercalated with clay and fine gravels. The widespread presence of black clay layer is the barrier in groundwater recharge in the central groundwater district. The urban cores of Kathmandu and Lalitpur are located in this central groundwater district. The southern groundwater district is characterized by a thick impermeable clay and basal gravel of low permeability. But some parts of the eastern area of southern groundwater district have potential for groundwater recharge due to presence of sand and gravel deposits [23]-[25].

2.2. Water Sampling and Analysis

The study was carried out in 41 deep groundwater wells (> 60 m) during pre monsoon and post monsoon seasons in 2013 (**Figure 1**). The study covered the groundwater wells of 84 to 304 m depth. The geo-positions of groundwater sampling locations were determined using global positioning system (GPS). Random sampling technique was used to collect groundwater samples. The locations of all the samples were recorded by handheld Garmin-E GPS and referenced to WGS 84 coordinate system. The high density polyethylene (HDPE) sampling bottles were treated with 5% HNO_3 and then rinsed with double distilled water. Samples from wells were collected by using 500-mL Nalgene (UK) HDPE bottles. A set of samples were collected in sampling bottles after pumping water for five minutes to get the representative samples of groundwater wells. The bottles were labeled with the sample code number. These samples were preserved as per APHA-AWWA-WEF (2005) [26] and then brought to the laboratory for the analysis. The samples were kept at 4˚C prior to analysis. Oxidation reduction potential (ORP), electrical conductivity (EC), pH and temperature were measured in situ at each sampling locations. The ORP and pH were measured by Hanna HI 8314 pH/ORP meter (Italy). The EC was measured by Jenway 4200 conductivity meter (UK).

The analysis of total arsenic, total iron and total manganese were carried out in Nepal standard (NS) certified-CEMAT Water Laboratory by using Varian AA 240 atomic absorption spectrometer (AAS). The analysis of arsenic was carried out using Varian AA 240 atomic absorption spectrometer (Australia) with vapor generation accessory VGA-77 (Australia). The standard solutions produced by Merck, Germany traceable to standard reference material (SRM) of NIST (National Institute of Standards and Technology, Gaithersburg, MD, USA) were used to prepare calibration standards. The samples for the analysis of arsenic, iron and manganese were digested with high purity HNO_3 (Merck) within a week of sample collection as per APHA-AWWA-WEF (2005). Sample digestion with the HNO_3 allows total extraction of the metals from the samples. Three replications of each analysis were performed and mean values were used for calculations. Analytical reagent blanks were prepared for each batch of the digestion set and then analyzed for the same elements as the samples. Analytical precision was in good agreement, generally better than 5% RSD. SPSS version 18 was used for all statistical analyses.

Nepal adopted universal transverse mercator (UTM) projection for the base mapping of the country with some modifications suited to its shape. This is named as modified universal transverse mercator projection. So, all the spatial data layers were maintained in a standard Nepalese coordinate system of modified universal transverse mercator, central meridian 84˚ longitude (*i.e.*, MUTM 84). The software used for mapping and spatial analysis was ArcGIS version 9.3.

3. Results and Discussion

3.1. Physicochemical Parameters and Metals

The summary of values of pH, EC, ORP and the concentrations of metals in the groundwater wells are presented in **Table 1**. The pH were nearly neutral that ranged from 6.4 to 7.9 (mean = 6.8) and 6.3 to 7.9 (mean = 6.7) in pre monsoon and post monsoon, respectively. The EC ranged from 92 to 1715 µS/cm (mean = 567 µS/cm), 93 to 1666 µS/cm (mean = 565 µS/cm) in pre monsoon and post monsoon, respectively. The elevated EC value is mainly due to geological sources since groundwater contamination is less in deep aquifers due to presence of widespread thick lacustrine clay beds that significantly restrict downward percolation [27] and leaching of ions contributing conductivity would be limited.

The mean values for ORP in the groundwater wells were −59 mV in pre monsoon and −49 mV in post monsoon. The negative ORP value was up to −161 mV indicates that the groundwater wells are under reducing condition. A study reported negative ORP value up to −195 mV in deep groundwater [14]. Likewise, another study indicated low ORP value in deep groundwater with mean value of −82 mV in pre monsoon and −56 mV in monsoon [17]. The groundwater wells in the study area are relatively anoxic, as indicated by low ORP values. The dominance of thick lacustrine clay probably restricts the downward diffusion of oxidants such as oxygen in deep groundwater [28].

The mean iron concentration in deep groundwater wells were 3.61 mg/L in pre monsoon and 3.76 mg/L in post monsoon with the highest concentration of 12.99 mg/L in post monsoon. Iron oxides dissolve under strongly acidic and reducing environment. If present in water in excessive amounts, it forms red Fe oxyhydroxide. Reductive dissolution of Fe(III) oxides accounts for the high Fe(II) content in anaerobic water [29] [30]. The chemical composition of the major elements of the sediments, *i.e.*, Fe_2O_3 ranged from 1.48 to 9.55 wt% [31] could be the source of iron in groundwater of the valley. A similar study indicated the Fe_2O_3 contents of the sediments are generally high (ranges < 0.5 to 15 wt%), and are uniformly higher in the fine sediments of the central basin (average 7 wt%) [28]. The overall bulk elemental concentrations in the central basin are greater in comparison with the northern part. The total iron content in the black sticky clay is high (> 7 wt%) [32]. The variation in Fe_2O_3 % and redox state of groundwater might have affected the levels of iron concentration in the groundwater.

Manganese concentration ranged from < 0.02 to 1.55 mg/L (mean = 0.44 mg/L) in pre monsoon, whereas < 0.02 to 1.75 mg/L (mean = 0.42 mg/L) in post monsoon. The chemical composition of the major element of the sediments, *i.e.*, MnO from 0.01 to 0.18 wt% [31] is probable source of manganese in the groundwater of the valley. In the study area, arsenic concentration in the groundwater wells ranged from < 0.003 to 0.137 mg/L (mean = 0.013 mg/L) in pre monsoon and < 0.003 to 0.131 mg/L (mean = 0.011 mg/L) in post monsoon. Arsenic concentration in 17% in pre monsoon and 26% in post monsoon exceeded provisional World Health Organization guideline value for drinking water of 0.010 mg/L [33]. A similar study reported arsenic concentration up to 0.265 mg/L [13]. Iron oxyhydroxides are the common host matter for arsenic, either adsorbed into the surface or co-precipitated [34] and reductive process is responsible for arsenic mobilization by dissolution or desorption in the groundwater [35]-[39], so it is possible that the higher concentration of arsenic in the groundwater is due to the more reducing environment as indicated by lower ORP value. Likewise, under the aerobic and acidic to near neutral conditions typical of many natural environments, arsenic is strongly adsorbed by oxide minerals as the arsenate ion and the concentrations in solution are low [40].

Table 1. Summary of statistical data for physicochemical parameters and metals.

Variable	Unit	Pre monsoon					Post monsoon				
		Min.	Max.	Mean	Med.	SD	Min.	Max.	Mean	Med.	SD
Depth	M	84	304	222.0	247	63.4	84	304	222.0	247	63.4
pH		6.4	7.9	6.8	6.7	0.4	6.3	7.9	6.7	6.6	0.4
EC	µS/cm	92	1715	567	452	381	93	1666	565	456	379
ORP	mV	−161	134	−59	−76	68	−120	135	−49	−56	55
Fe	mg/L	BDL	11.09	3.61	2.88	2.88	0.10	12.99	3.76	2.79	2.87
Mn	mg/L	BDL	1.55	0.44	0.37	0.35	BDL	1.75	0.42	0.31	0.4
As	mg/L	BDL	0.137	0.011	0.003	0.024	BDL	0.131	0.013	0.005	0.023

Min. = Minimum, Max. = Maximum, Med. = Median, SD = Standard deviation, BDL = Below detection limit.

The groundwater wells observed elevated concentrations of iron and manganese. It is observed that the concentrations of iron and manganese are pH dependent and higher aggressiveness of iron and manganese in low pH [41]. Likewise, it is suggested that presence of a reductant (such as organic carbon) is a dominant factor controlling iron and manganese concentrations in groundwater. The oxidation of the reductant would leads to the reduction and solubilisation of iron and manganese [42] and the lacustrine clay in the Kathmandu Valley is rich is organic matter [43] would contribute to the reducing environment. The organic matter may play an important role in the mobilization of arsenic, as reported by many studies from West Bengal (India) and Bangladesh [35] [44] [45]. It has been indicated that the fluvio-lacustrine sediments in the valley are rich in organic matter, and this organic matter may contribute in mobilization of arsenic.

3.2. Correlation between Physicochemical Parameters and Arsenic

The relationships of pH, EC, ORP and metals in the groundwater were examined by Spearman's rank correlation coefficient (**Table 2**). The pH has strong negative correlations with iron and manganese, and weak negative correlation with arsenic ($r = -0.534$, $p < 0.01$; $r = -0.402$, $p < 0.01$; $r = -0.169$, $p < 0.129$) which can be explained by the higher aggressiveness of acidic media towards soil and host rocks that increase the concentrations of the rest of the ions [41]. Though arsenic has negative correlation, *p*-value suggests that there is an insignificant negative correlation between arsenic concentration and pH in the groundwater. The EC shows strong positive correlations (at $p < 0.01$) with iron, manganese and arsenic. Arsenic has strong positive correlations with iron and manganese ($r = 0.384$, $p < 0.01$; $r = 0.447$, $p < 0.01$), which is attributed to common geogenic origin of these metals. The ORP has strong negative correlation with arsenic ($r = -0.492$, $p < 0.01$), which can be explained by reductive arsenic mobilization mechanisms in the groundwater. Likewise, ORP also shows negative correlations with iron and manganese. Reducing environment is responsible for the release of iron as well as manganese through the reduction of Mn(III, IV) (hydr)oxides to soluble Mn(II) and of Fe(III) (hydr)oxides to soluble Fe(II), respectively.

Table 2. Spearman's rank correlation coefficients of physicochemical parameters and metals (n = 82).

Parameter	pH	EC	ORP	Fe	Mn	As
pH	1.000					
EC	−0.332**	1.000				
ORP	0.052	−0.664**	1.000			
Fe	−0.534**	0.591**	−0.570**	1.000		
Mn	−0.402**	0.654**	−0.447**	0.656**	1.000	
As	−0.169	0.463**	−0.492**	0.384**	0.447**	1.000

**Significant value at $p < 0.01$.

3.3. Correlation between Arsenic and Depth of Groundwater

The depth of the deep groundwater wells tested arsenic ranged from 84 to 304 m. The mean and standard deviation (SD) of depth were 222.0 m and 63.4 m respectively (**Table 1**). The study showed weak positive correlations between arsenic and depth of groundwater in pre monsoon and post monsoon ($r = 0.206$, $p = 0.196$; $r = 0.178$, $p = 0.266$), respectively. Though, it showed positive correlations in both seasons, *p*-values suggest that there are insignificant positive correlations between arsenic concentration and depth of groundwater. However, it contradicts with the results shown by the some previous studies [13] [15].

3.4. Temporal (Seasonal) Variation of Arsenic

The temporal (seasonal) variation of the physicochemical parameters and metals were evaluated through season-parameter Spearman's correlation matrix in the groundwater. The measured parameters are not significantly ($p > 0.05$) correlated with the season except for pH ($r = -0.238$, $p < 0.05$), which infers similar distribution of arsenic in both seasons. Arsenic concentrations were insignificantly varied between seasonal groundwater. A study also reported very similar distributions of arsenic for pre monsoon and monsoon [15].

The lack of temporal (seasonal) variation is attributed to less dilution effect of monsoon rainfall in the groundwater. Additionally, contribution of anthropogenic metal contaminations is reluctant in the studied time-series in the groundwater wells. These findings are consistent with the results of previous study which pointed out seasonal variability has no significant effect on deep groundwater quality [17]. The similar studies also indicated no seasonal variability of arsenic in the groundwater [15] [16]. The reports on the temporal variation of arsenic concentration in other parts of the world are inconsistent. Significant variation in arsenic concentration among the seasons was observed in a study [46]. A study spotted no significant monsoonal effects on arsenic distribution [47]. The seasonal variability has little effect in the groundwater [48]. Likewise, limited temporal variability observed in arsenic concentrations in groundwater [49] [50].

3.5. Spatial Distribution of Arsenic

The concentration of arsenic, iron, manganese and ORP value vary significantly in central, northern and southern groundwater districts in the valley (**Figures 2(a)-(d)**). The spatial distribution pattern reveals higher values of arsenic, iron and manganese in central groundwater district. Northern and southern groundwater districts have lower arsenic concentrations in most of the groundwater wells. The concentration of the arsenic in groundwater of the study area increases from northern to southern and showing highest towards central groundwater district. Arsenic concentration was relatively lower in southern groundwater district in comparison with central groundwater district. The spatial distribution of ORP in the groundwater clearly shows most of the groundwater wells have lower ORP value.

The variations of elemental concentration are mainly clay controlled in both the margin and central parts. There is progressive increase in the finer particles and trace elements towards the central part of the sediments from the northern part in the valley [32] which is attributed to decrease in grain size and the concentration of metals in sediments tend to increase in fine grained sediments [32] [51] [52]. The larger particles in sediments have less surface area available for metal hydroxide coatings to form and adsorb arsenic and less adsorbed arsenic contributes a smaller amount of aqueous arsenic in equilibrium with adsorbed arsenic. Therefore, there is in less potential for release of arsenic through reductive mobilization mechanisms [53]. The variation in grain size has role in mobilization of metals in groundwater. Therefore, higher metal concentration is associated with the fine grained sediments in the central groundwater district. Furthermore, higher concentrations of metals in central groundwater district might be due to the fact that the central groundwater district is considered as poorly recharging due to the presence of a thick black clay layer [23].

Groundwater quality depends on the composition of recharging water, the mineralogy and reactivity of the geological formations in aquifers, anthropogenic activities and environmental conditions that may affect the geochemical mobility of certain constituents [54]. Arsenic concentrations in groundwater of the Kathmandu Valley show a wide range and some of groundwater sources investigated were found to be in elevated levels in some parts of the valley could be due to the nature of the sediments there. The high degree of spatial variability in groundwater quality over short distances indicates that groundwater movement has been limited and is poorly mixed [55].

Arsenic concentrations in the sediments of the Kathmandu Valley averaged 8 mg/kg (ranging 3 to 25 mg/kg) similar to the general level seen in modern unconsolidated sediments, typically 5 to 10 mg/kg [32] [36]. The widespread lacustrine clay of the Kathmandu Valley could have greater potentiality for arsenic release [32]. Arsenic mobilization is high in the reducing conditions [36] [56] [57]. The higher concentration of arsenic under reduced groundwater environment may be due to Fe/Mn oxides and direct reduction of As(V) into As(III). After an initial increase, arsenic concentration often decreases again as a function of time below water table due to sulfide precipitation, whereas it increases with increasing sulfate concentrations above water table [58]. Under moderately reduced environment (0 to 100 mV), arsenic solubility seemed to be controlled by the dissolution of Fe oxyhydroxides. But at highly reduced condition, e.g., at −250 mV, arsenic chemistry is dominated by the formation of insoluble sulfides FeAsS, AsS, As_2S_3 [56] attenuating concentration of arsenic in the groundwater.

4. Conclusion

This study has confirmed the presence of higher levels of arsenic in deep groundwater of the Kathmandu Valley. Arsenic showed wide spatial variation ranged from < 0.003 to 0.137 mg/L. In pre monsoon and post monsoon, 17% and 26% of groundwater wells, respectively exceeded arsenic concentration of permissible WHO guideline

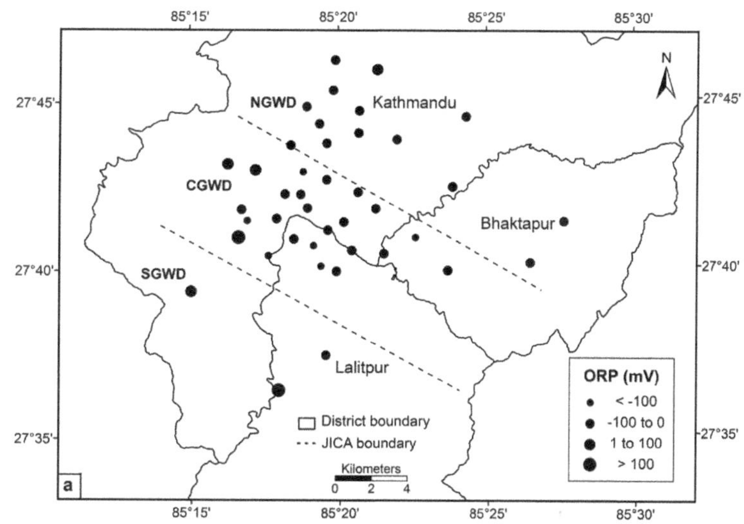
85°15'
85°20'
85°25'
85°30'
27°45'
27°40'
27°35'
N
NGWD
Kathmandu
CGWD
Bhaktapur
SGWD
Lalitpur
District boundary
JICA boundary
Kilometers
0 2 4
ORP (mV)
< -100
-100 to 0
1 to 100
> 100
a

(a)

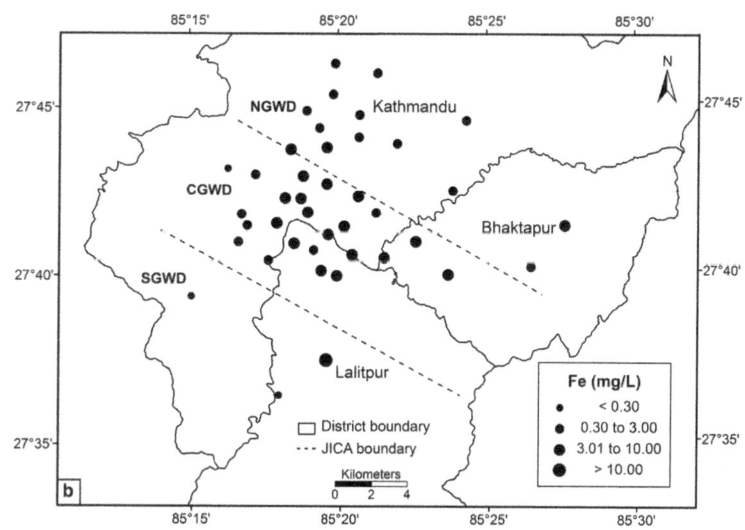
85°15'
85°20'
85°25'
85°30'
27°45'
27°40'
27°35'
N
NGWD
Kathmandu
CGWD
Bhaktapur
SGWD
Lalitpur
District boundary
JICA boundary
Kilometers
0 2 4
Fe (mg/L)
< 0.30
0.30 to 3.00
3.01 to 10.00
> 10.00
b

(b)

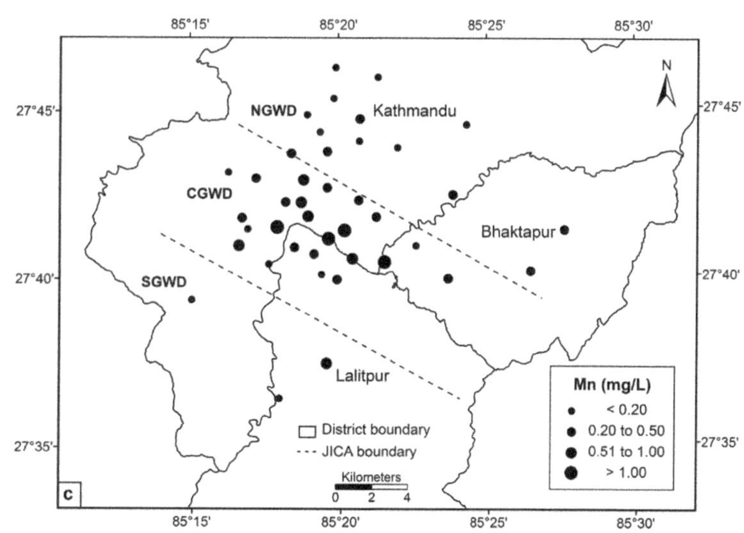
85°15'
85°20'
85°25'
85°30'
27°45'
27°40'
27°35'
N
NGWD
Kathmandu
CGWD
Bhaktapur
SGWD
Lalitpur
District boundary
JICA boundary
Kilometers
0 2 4
Mn (mg/L)
< 0.20
0.20 to 0.50
0.51 to 1.00
> 1.00
c

(c)

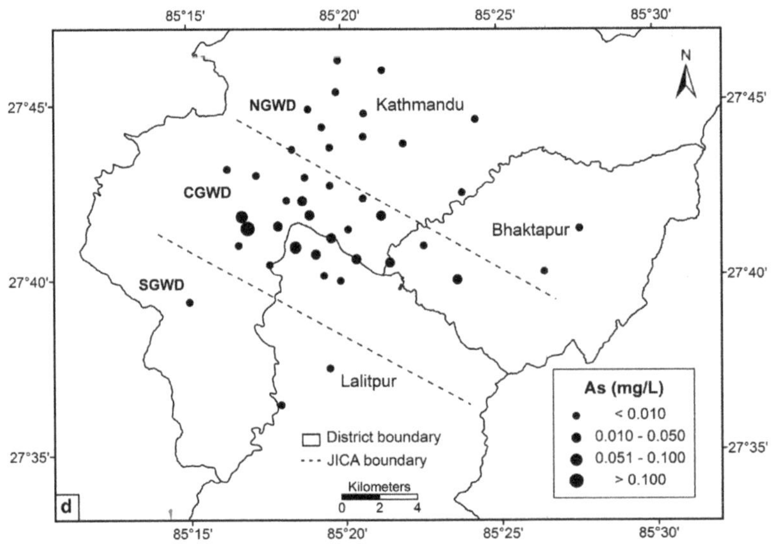

(d)

Figure 2. Spatial distribution: (a) ORP; (b) iron; (c) manganese and (d) arsenic.

value of 0.010 mg/L for drinking water. The arsenic varies spatially with high concentration towards central groundwater district. Strong negative correlation between arsenic and ORP demonstrated reductive arsenic mobilization mechanisms in deep groundwater. Arsenic showed strong correlations with iron, manganese and EC. Iron and manganese are presumably the main factors in regulating release of As in groundwater through reductive mobilization mechanism. The study revealed weak positive correlation between arsenic concentration and depth of deep groundwater. Arsenic is not significantly correlated with season suggesting similar distribution of arsenic in both seasons. The high concentration of arsenic in deep groundwater of some parts of study area particularly in central groundwater district is attributed to the groundwater geochemistry of the study area. Release of arsenic into the groundwater is considered to be due to the natural source under the reductive process.

Acknowledgements

The authors would highly acknowledge Central Department of Environmental Science, Tribhuvan University, Kirtipur, Kathmandu, Nepal for the support of this research work. We would like to thank University Grants Commission, Nepal for a research grant. Thanks also due to CEMAT Water Laboratory, Kathmandu, Nepal for providing laboratory facilities for the analysis of metals using atomic absorption spectrometer (AAS).

References

[1] Jarup, L. (2003) Hazards of Heavy Metal Contamination. *British Medical Bulletin*, **68**, 167-182. http://dx.doi.org/10.1093/bmb/ldg032

[2] Mitchell, E., Frisbie, S. and Sarkar, B. (2011) Exposure to Multiple Metals from Groundwater—A Global Crisis: Geology, Climate Change, Health Effects, Testing and Mitigation. *Metallomics*, **3**, 874-908. http://dx.doi.org/10.1039/c1mt00052g

[3] IARC (2004) IARC Monographs on the Evaluation of Carcinogenic Risks to Humans, Volume 84, Some Drinking-water Disinfectants and Contaminants, Including Arsenic. International Agency for Research on Cancer, Lyons.

[4] Garbarino, J.R., Hayes, H.C., Roth, D.A., Antweiler, R.C., Brinton, T.I. and Taylor, H.E. (1995) Heavy Metals in the Mississippi River. US Geological Survey Circular 1133, Virginia.

[5] Marcovecchio, J.E., Botte, S.E. and Freije, R.H. (2007) Heavy Metals, Major Metals, Trace Elements. In: Nollet, M.L., Ed., *Handbook of Water Analysis*. 2nd Edition, CRC Press, London, 275-311. http://dx.doi.org/10.1201/9781420006315.ch11

[6] Appelo, C.A.J. and Postma, D. (2005) Geochemistry, Groundwater and Pollution. 2nd Edition, A.A. Balkema Publishers, Amsterdam. http://dx.doi.org/10.1201/9781439833544

[7] Mukherjee, A., Sengupta, M.K., Hossain, M.A., Ahamed, S., Das, B., Nayak, B., Lodh, D., Rahman, M.M. and Cha-

kraborti, D. (2006) Arsenic Contamination in Groundwater: A Global Perspective with Emphasis on the Asian Scenario. *Journal of Health, Population and Nutrition*, **24**, 142-163.

[8] Hossain, M.F. (2006) Arsenic Contamination in Bangladesh—An Overview. *Agriculture, Ecosystems and Environment*, **113**, 1-16. http://dx.doi.org/10.1016/j.agee.2005.08.034

[9] CBS (2012) National Population and Housing Census 2011. National Report. Volume 1, NPHC 2011. Nepal Bureau of Statistics, Kathmandu.

[10] Acres International (2004) Optimizing Water Use in Kathmandu Valley (ADB-TA) Project. Final Report. Acres International in Association with Arcadis Euroconsult Land and Water Product Management Group, East Consult (P) Ltd. and Water Asia (P) Ltd.

[11] ICIMOD (2007) Kathmandu Valley Environment Outlook. International Centre for Integrated Mountain Development, Kathmandu.

[12] Jha, M.G., Khadka, M.S., Shrestha, M.P., Regmi, S., Bauld, J. and Jacobson, G. (1997) The Assessment of Groundwater Pollution in Kathmandu, Nepal. Report on Joint Nepal-Australia Project, 1995-96. Australian Geological Survey Organization.

[13] JICA/ENPHO (2005) Arsenic Vulnerability in Groundwater Resources in Kathmandu Valley. Final Report. Japan International Cooperation Agency/Environment and Public Health Organization, Kathmandu.

[14] Khatiwada, N.R., Takizawa, S., Tran, T.V.N. and Inoue, M. (2002) Groundwater Contamination Assessment for Sustainable Water Supply in Kathmandu Valley, Nepal. *Water Science Technology*, **46**, 147-154.

[15] Maharjan, M., Shrestha, B.R., Shrestha, K.B., Shrestha, R.R., Raut-Khadka, R., Shrestha, S.D., Kafle, B. and Ishihara, H. (2006) Arsenic Contamination in Groundwater Resources in Kathmandu Valley. *4th International Symposium on Southeast Asian Water Environment*, Bangkok.

[16] Chapagain, S.K., Shrestha, S., Nakamura, T., Pandey, V.P. and Kazama, F. (2009) Arsenic Occurrence in Groundwater of Kathmandu Valley, Nepal. *Desalination and Water Treatment*, **4**, 248-254. http://dx.doi.org/10.5004/dwt.2009.492

[17] Chapagain, S.K., Pandey, V.P., Shrestha, S., Nakamura, T. and Kazama, F. (2010) Assessment of Deep Groundwater Quality in Kathmandu Valley Using Multivariate Statistical Techniques. *Water Air Soil Pollution*, **210**, 277-288. http://dx.doi.org/10.1007/s11270-009-0249-8

[18] Shrestha, S.M., Rijal, K. and Pokhrel, M.R. (2013) Arsenic Contamination in the Deep and Shallow Groundwater of Kathmandu Valley, Nepal. *Scientific World*, **10**, 25-31. http://dx.doi.org/10.3126/sw.v11i11.8548

[19] Shrestha, S.M., Rijal, K. and Pokhrel, M.R. (2014) Spatial Distribution and Seasonal Variation of Arsenic in Groundwater of Kathmandu Valley, Nepal. *Journal of Institute of Science and Technology, T.U.*, **19**, 7-13.

[20] CBS (2013) Environment Statistics of Nepal. Nepal Bureau of Statistics, Kathmandu.

[21] Yoshida, M. and Igarashi, Y. (1984) Neogene to Quaternary Lacustrine Sediments in the Kathmandu Valley, Nepal. *Journal of Nepal Geological Society*, **4**, 73-100.

[22] Shrestha, S.D., Karmacharya, R. and Rao, G.K. (1996) Estimation of Groundwater Resources in Kathmandu Valley, Nepal. *Journal of Groundwater Hydrology*, **38**, 29-40. http://dx.doi.org/10.5917/jagh1987.38.29

[23] JICA (1990) Groundwater Management Project in the Kathmandu Valley. Final Report to Nepal Water Supply Corporation. Japan International Cooperation Agency, Kathmandu.

[24] Sakai, T., Gajurel, A.P., Tabata, H. and Uprety, B.N. (2001) Small Amplitude Lake-Level Fluctuations Recorded in Aggrading Deltaic of the Upper Pleistocene Thimi and Gokarna Formations, Kathmandu Valley, Nepal. *Journal of Nepal Geological Society*, **25**, 43-52.

[25] Dixit, A. and Upadhya, M. (2005) Augmenting Groundwater in Kathmandu Valley: Challenges and Possibilities. Report to Nepal Water Conservation Foundation, Kathmandu, Nepal.

[26] APHA-AWWA-WEF (2005) Standard Methods for the Examination of Water and Wastewater. 21st Edition, American Public Health Association, American Water Works Association, Water Environment Federation, Washington DC.

[27] Pathak, D.R., Hiratsuka, A., Awata, I. and Chen, L. (2009) Groundwater Vulnerability Assessment in Shallow Aquifer of Kathmandu Valley Using GIS-Based Drastic Model. *Environmental Geology*, **57**, 1569-1578. http://dx.doi.org/10.1007/s00254-008-1432-8

[28] Gurung, J.K. (2007) Geochemical Studies of Sediments and Water, and Implications for Mobilization of Arsenic into Groundwater in Nepal and Japan. Ph.D. Thesis, Shimane University, Japan.

[29] Hem, J.D. (1985) Study and Interpretation of Chemical Characteristics of Natural Water. 3rd Edition, US Geological Survey Water-Supply Paper 2254.

[30] Smedley, P.L. and Kinniburgh, D.G. (2013) Arsenic in Groundwater and the Environment. In: Selinus, O., Ed., *Essential of Medical Geology*, 279-210. http://dx.doi.org/10.1007/978-94-007-4375-5_12

[31] Dill, H.G., Khadka, D.R., Khanal, R., Dohrmann, R., Melcher, F. and Busch, K. (2003) Infilling of the Younger Kathmandu-Banepa Intermontane Lake Basin During the Late Quaternary (Lesser Himalaya, Nepal): A Sedimentological Study. *Journal of Quaternary Science*, **18**, 41-60. http://dx.doi.org/10.1002/jqs.726

[32] Gurung, J.K., Ishiga, H., Khadka, M.S. and Shrestha, N.R. (2007) The Geochemical Study of Fluvio-Lacustrine Aquifers in the Kathmandu Basin (Nepal) and the Implications for Mobilization of Arsenic. *Environmental Geology*, **52**, 503-517. http://dx.doi.org/10.1007/s00254-006-0483-y

[33] WHO (2011) Guidelines for Drinking-Water Quality. 4th Edition, World Health Organization, Geneva.

[34] Pierce, M.L. and Moore, C.B. (1982) Adsorption of Arsenite and Arsenate on Amorphous Iron Hydroxide. *Water Research*, **16**, 1247-1253. http://dx.doi.org/10.1016/0043-1354(82)90143-9

[35] McArthur, J.M., Ravenscroft, P., Safiulla, S. and Thirlwall, M.F. (2001) Arsenic in Groundwater: Testing Pollution Mechanisms for Sedimentary Aquifers in Bangladesh. *Water Research*, **37**, 109-117. http://dx.doi.org/10.1029/2000WR900270

[36] Smedley, P.L. and Kinniburgh, D.G. (2002) A Review of the Source, Behavior and Distribution of Arsenic in Natural Waters. *Applied Geochemistry*, **17**, 517-568. http://dx.doi.org/10.1016/S0883-2927(02)00018-5

[37] Anawar, H.M., Akai, J., Komaki, K., Terao, H., Yoshioka, T., Ishizuka, T., Safiullah, S. and Kato, K. (2003) Geochemical Occurrence of Arsenic in Groundwater of Bangladesh: Sources and Mobilization Processes. *Journal of Geochemical Exploration*, **77**, 109-131. http://dx.doi.org/10.1016/S0375-6742(02)00273-X

[38] Bose, P. and Sharma, A. (2002) Role of Iron in Controlling Speciation and Mobilization of Arsenic in Subsurface Environment. *Water Research*, **36**, 4916-4926. http://dx.doi.org/10.1016/S0043-1354(02)00203-8

[39] Biswas, A., Majumder, S., Neidhardt, H., Halder, D., Bhowmick, S., Mukherjee-Goswami, A., Kundu, A., Saha, D., Berner, Z. and Chatterjee, D. (2011) Groundwater Chemistry and Redox Processes: Depth Dependent Arsenic Release Mechanism. *Applied Geochemistry*, **26**, 516–525. http://dx.doi.org/10.1016/j.apgeochem.2011.01.010

[40] Dzombak, D.A. and Morel, F.M.M. (1990) Surface Complexation Modeling: Hydrous Ferric Oxide, John Wiley & Sons, New York.

[41] Helena, B., Pardo R., Vega, M., Barrado, E., Fernandez, J.M. and Fernandez, L. (2000) Temporal Evolution of Groundwater Composition in an Alluvial Aquifer (Pisuerga River, Spain) by Principal Component Analysis. *Water Research*, **34**, 807-816. http://dx.doi.org/10.1016/S0043-1354(99)00225-0

[42] Daughney, C.J. (2003) Iron and Manganese in New Zealand's Groundwater. *Journal of Hydrology* (*NZ*), **42**, 11-26.

[43] Fujii, R. and Sakai, H. (2001) Palynological Study on the Drilled Sediments from the Kathmandu Basin and Its Paleoclimatic Significances. *Journal of Nepal Geological Society*, **25**, 53-61.

[44] Nickson, R.T., McArthur, J.M., Ravenscroft, P., Burgess, W.G. and Ahmed, K.M. (2000) Mechanism of Arsenic Release to Groundwater, Bangladesh and West Bengal. *Applied Geochemistry*, **15**, 403-413. http://dx.doi.org/10.1016/S0883-2927(99)00086-4

[45] Akai, J., Izumi, K., Fukuhara, H., Masuda H., Nakano, S., Yoshimura, T. Ohfuji, H., Anawar, H. and Akai, K. (2004) Mineralogical and Geomicrobiological Investigations on Groundwater Arsenic Enrichment in Bangladesh. *Applied Geochemistry*, **19**, 215-230. http://dx.doi.org/10.1016/j.apgeochem.2003.09.008

[46] Savarimuthu, X., Hira-Smith, M.M., Yuan, Y., von Ehrenstein, O.S., Das, S., Ghosh, N., Guha Mazumder, D.N. and Smith, A.H. (2006) Seasonal Variation of Arsenic Concentration in Tube Wells in West Bengal, India. *Journal of Health Population and Nutrition*, **24**, 277-281.

[47] Sankar, M.S., Vega, M.A., Defoe, P.P., Kibria, M.G., Ford, S., Telfeyan, K., Neal, A., Mohajerin, T.J., Hettiarachchi, G.M., Barua, S., Hobson, C., Johannesson, K. and Datta, S. (2007) Elevated Arsenic and Manganese in Groundwaters of Murshidabad, West Bengal, India. *Science of the Total Environment*, **488-489**, 570-579. http://dx.doi.org/10.1016/j.scitotenv.2014.02.077

[48] Chen, K., Jio, J.J., Huang, J. and Huang, R. (2007) Multivariate Statistical Evaluation of Trace Elements in Groundwater in a Coastal Area in Shenzhen, China. *Environmental Pollution*, **147**, 771-780. http://dx.doi.org/10.1016/j.envpol.2006.09.002

[49] Cheng, Z., Vangeen, A., Seddique, A.A. and Ahmed, K.M. (2005) Limited Temporal Variability of Arsenic Concentrations in 20 Wells Monitored for 3 Years in Araihazar, Bangladesh. *Environmental Science and Technology*, **39**, 4759-4766. http://dx.doi.org/10.1021/es048065f

[50] Thundiyil, J.G., Yuan, Y., Smith, A.H. and Steinmaus, C. (2007) Seasonal Variation of Arsenic Concentration in Wells in Nevada. *Environmental Research*, **104**, 367-373. http://dx.doi.org/10.1016/j.envres.2007.02.007

[51] Filipek, L.H. and Owen, R.M. (1979) Geochemical Associations and Grain-Size Partitioning of Heavy Metals in Lacustrine Sediments. *Chemical Geology*, **26**, 105-117. http://dx.doi.org/10.1016/0009-2541(79)90033-0

[52] Singh, A.K., Hasnain, S.I. and Banerjee, D.K. (1999) Grain Size and Geochemical Partitioning of Heavy Metals in Se-

diments of the Damodar River—A Tributary of the Lower Ganga, India. *Environmental Geology*, **39**, 90-98. http://dx.doi.org/10.1007/s002540050439

[53] Erickson, M.L. and Barnes, R.J. (2005) Well Characteristics Influencing Arsenic Concentrations in Groundwater. *Water Research*, **39**, 4029-4039. http://dx.doi.org/10.1016/j.watres.2005.07.026

[54] Kouras, A., Katsoyiannis, I. and Voutsa, D. (2007) Distribution of Arsenic in Groundwater in the Area of Chalkidiki, Northern Greece. *Journal of Hazardous Materials*, **147**, 890-899. http://dx.doi.org/10.1016/j.jhazmat.2007.01.124

[55] Smedley, P.L., Nicolli, H.B., MacDonald, D.M.J., Baros, A.J. and Tullio, J.O. (2002) Hydrogeochemistry of Arsenic and Other Inorganic Constituents in Groundwaters from La Pampa, Argentina. *Applied Geochemistry*, **17**, 259-284. http://dx.doi.org/10.1016/S0883-2927(01)00082-8

[56] Carbonell-Barrachina, A.A., Jugsujinda, A., Burlo, F., Delaune, R.D. and Patrick, W.H. (1999) Arsenic Chemistry in Municipal Sewage as affected by Redox Potential and pH. *Water Research*, **34**, 216-224. http://dx.doi.org/10.1016/S0043-1354(99)00127-X

[57] McArthur, J.M., Banerjee, D.M., Hudson-Edwards, K.A., Mishra, R., Purohit, R., Ravenscroft, P., Cronin, A., Howarth, R.J., Chatterjee, A., Talukder, T., Lowry D., Houghton, S. and Chada, D.K. (2004) Natural Organic Matter in Sedimentary Basins and Its Relation to Arsenic in Anoxic Groundwater: The Example of West Bengal and its Worldwide Implications. *Applied Geochemistry*, **19**, 1255-1293. http://dx.doi.org/10.1016/j.apgeochem.2004.02.001

[58] Du Laing, G., Chapagain, S.K., Dewispelaere, M., Meers, E., Kazama, F., Tack, F.M.G., Rinklebe, J. and Verloo, M.G. (2009) Presence and Mobility of Arsenic in Estuarine Wetland Soils of the Scheldt Estuary (Belgium). *Journal of Environmental Monitoring*, **11**, 873-881. http://dx.doi.org/10.1039/b815875d

26

Heavy Metal Concentrations in Pores and Surface Waters during the Emptying of a Small Reservoir

Christian Opp[1], Jens Hahn[2], Nina Zitzer[3], Gabriela Laufenberg[3]

[1]Department of Hydrology and Soil Science, Faculty of Geography, University of Marburg, Marburg, Germany
[2]Department of Geography, Faculty of Mathematic and Natural Science, University of Koblenz-Landau, Koblenz, Germany
[3]Mass Spectrometry, Faculty of Pharmacy, University of Marburg, Marburg, Germany
Email: opp@staff.uni-marburg.de, jenshahn@uni-koblenz.de

Abstract

The impact of reservoir emptying on the concentrations of dissolved heavy metals (As, Cd, Cr, Cu, Fe, Mn, Ni, Pb and Zn) in pore and surface waters was studied in the Aar Reservoir, a small reservoir in central Germany, during and after the emptying process. This study was conducted to observe binding changes within pore waters as well as the input of dissolved heavy metals in waters of the Aar Creek, what becomes possible when the reservoirs water table is removed and lake sediments become exposed. In pore waters, no clear shifting tendencies between dissolved and sorbed heavy metal fractions could be found after the completed sediment exposure. These relatively low dynamics in pore waters can be explained by the fine texture of the lake sediments, which are characterized by a high water holding capacity, what led to high remaining water contents and therefore slowed down the redox changes. A few days after the completed emptying, a general increase of dissolved heavy metal concentrations occurred in running waters of the Aar Creek as a result of pore water drainage. Here, element specific differences in mobilization and transportation abilities were found, what can be reconstructed by the ratio of dissolved heavy metals in pore and surface waters.

Keywords

Heavy Metals, Reservoirs, Sediments, Surface Waters

1. Introduction

Heavy metals with a mass density of >5 g/cm^3 are natural components in rocks, soils and sediments. In most of the cases they occur in relatively low concentrations, why they are also called trace metals [1]. In consequence of human activities (e.g. mining, industrial emissions, car emissions), heavy metals can be distributed or emitted into the environment. When they hereafter entered into sediments and soils, they can be accumulated and increased in their concentrations, because they are non-degradable. Fluvial sediments often contain increased amounts of heavy metals when loaded substrates are entered from the catchment area into the river [2]. Especially the fine-grained sediment fractions and organic components of river and reservoir sediments are able to sorb heavy metals [3] [4].

Since water reservoirs are subject to a continuous sediment input and due to low flow rates at its base mainly fine particles are deposited, large amounts fluvial registered heavy metals can accumulate here [5]. However, the high sorption capacity of fine and organic rich sediments takes place under sufficiently reducing conditions by which the formation of sulfidic compounds counteracts a release of several heavy metals into the pore waters. Dissolved heavy metal contents are also heavily diluted in areas of the water column by the surrounding water body. In addition to the element and substrate specific circumstances especially the prevailing redox conditions determine the binding behavior of heavy metals in sediments [6]. When redox changes occur, particularly before and during the early stabilization of a redox status, an increased release of heavy metals in the dissolved phase becomes possible [7].

The emptying of water reservoirs enhances the oxygen contact of previously water covered sediments. As a result, redox changes inside the pore waters become possible. Simultaneously the hydrostatic uplift is reduced in consequence of the elimination of the water column. This favors a contraction of the exposed lake sediments and the outflow of pore waters, from topographically higher to lower lying areas of the reservoir bottom. Problematic consequences of the targeted emptying of reservoirs are therefore often share shifts between dissolved and fixed heavy metal amounts within the sediments and an enhanced release of elements such as Fe, Mn, N or P in the flowing surface water [8]-[14]. In general, sorption dynamics and heavy metal migration during the emptying of reservoirs were observed very seldom.

Since the Aar Reservoir is characterized by a rural catchment area without mining activities, it should have gotten no excessive heavy metal input in its past. Thus, in the reservoirs sediments rather low to moderate heavy metal contents can be expected. However, during the drainage of large pore water amounts an increasing input of dissolved heavy metals to the receiving running water can be expected, resulting of a reduced dilution by surface waters. The different binding properties of elements might hereby affect their mobilization and transport. To address these issues, dissolved concentrations of As, Cd, Cr, Cu, Fe, Mn, Ni, Pb and Zn in lake sediments, pore waters and surface waters of the Aar Creek were analyzed during and after the emptying process.

2. Study Area

The Aar Reservoir is situated on the eastern edge of the Rhenish Slate Mountains in the upper reaches of the Aar River. It consists of the main reservoir and a pre-reservoir. With an operating space of approximately 4.1 million m^3 and a maximum water depth of 8 m it is a relatively small and shallow reservoir. Due to predominantly upcoming shales and greywackes relatively moderate geogenetic heavy metal contents in the soils of the catchment and in the reservoirs sediments can be expected, what was confirmed by former investigations [15].

3. Methods

An evacuation and emptying of the main basin was carried out from August 2011 for repairing purposes. The sampling of surface waters began on 19th August 2011, even during the emptying process of the reservoir. Pore water sampling started on 30th August 2011, after the bottom of the reservoir had become accessible.

3.1. Sampling

1) Surface water

Water samples were taken by means of PE-syringes at three sampling sites within the reservoir and downstream the dam from a depth of about 10 cm. The freshly collected samples were filtered by syringe filters (Minisart NML, Celluloseacetat-Membran, Fa. Sartorius/Göttingen/Germany) in the field at 0.45 microns, acidified in the ratio 100:1 with nitric acid (HNO_3) and stored in PE-bottles at −18˚C up to their analysis.

2) Pore water

Pore waters were obtained by stationary suction pipettes (Macro Rhizon, Fa. Rizosphere Research Products, Wageningen/The Netherlands), that were installed at three locations of the reservoirs bottom, each in 5 - 10 cm and 15 - 25 cm depth. The used pipettes consisted of non absorbent plastic and a microfiltration membrane, whereby the incoming water has been filtered with a pore size of 0.15 - 0.2 microns during its removal. In order to ensure the representativeness of the samples, the sampling was carried out from each depth level with a minimum of 3 parallel pipettes. The freshly collected samples were combined to a composite sample and separated in two PE bottles. One half of the origin sample was used for field measurements, the second half was acidified in the ratio 100:1 with nitric acid (HNO_3) and frozen at −18°C until the determination of heavy metal contents.

3) Substrate

After completion of the pore water extraction, the sample sites were excavated and surveyed. For the determination of the sediments heavy metal contents substrate samples were taken. All samples were air-dried and homogenized in a porcelain mortar. This was followed by a sieving to <2 mm.

3.2. Field Measurements

At each sampling site pH value and redox potential were determined in the field, immediately after the extraction of pore waters, by means a portable field meter (Multiline P4; Fa. WTW) and adequate electrodes. In parallel, the volumetric soil moisture (HH2, Fa. Delta-T Devices) was recorded in 10 cm depth of the sediments.

3.3. Laboratory Analysis

The proportion of organic matter in the sediments was calculated by determining the loss on ignition for 2 hours at 550°C. The determination of the particle size distribution was performed by wet sieving and pipette analysis. Heavy metal contents of the sediments were determined by aqua regia extraction. The measurement of element concentrations, in aqua regia extracts as well as in water samples, was carried out by ICP-MS (Fa. Thermo Fisher, X-Series 2).

4. Results

The lake sediments were marked by high volumetric water contents and low densities at the beginning of the field campaign. A vegetation cover was absent during the whole sampling period, with the starting evolvement of vascular plants the sampling campaign ended.

Compared within the sampling locations 1 - 3, the sediments textures were relatively homogeneous and contained high silt (45.8 - 65.8 mass %), moderate clay (22.5 - 33.8 mass %) and varying sand (6.0 - 31.3 mass %) contents. The mass of organic matter varied between 4.8 % and 8.6 % and showed decreasing contents with increasing sampling depths. Volumetric soil moisture started with high contents in all sampling locations (77 - 84 vol. %). As an effect of the sediments high water holding capacity, only slight decreases of moisture occurred with enduring sediment exposure, which reached at the end of the sampling campaign 12.8 - 22.4 vol. % of the maximum contents (**Figure 1**). The pore waters pH varied over all sampling locations from pH 6.6 to pH 7.6, while each location and sampling depth showed only very slight pH variations. A similar behavior was found for the redox potential, which locally underlay only small variations with a range of 64 mV to 135 mV in all sampled pore waters. These more or less constant redox states within the whole sampling campaign are a result of high remaining water contents inside the exposed sediments, what hindered an enhanced oxygen exchange between sediments and atmosphere, and therefore an establishment of more oxic conditions.

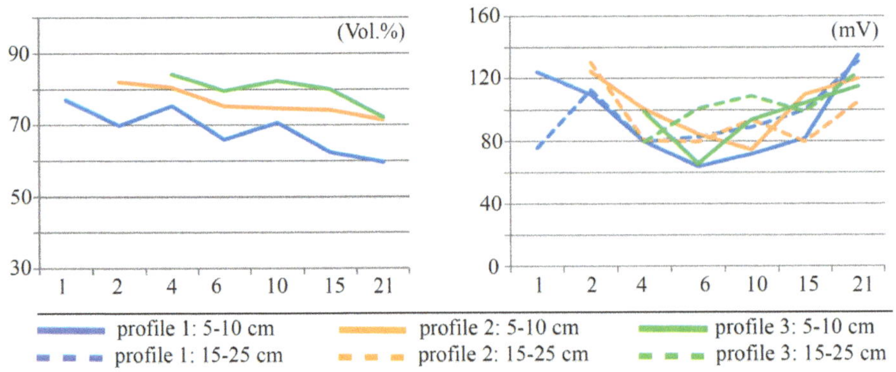

Figure 1. Temporal variations (x-axis: number of days after the complete sediment exposure) of volumetric water content and redox potential (y-axis) in pore waters of sampling profiles 1 - 3.

4.1. Heavy Metal Concentrations in Sediments and Pore Waters

Heavy metal concentrations in the sediments showed no dramatic differences between different sampling locations and sampling depths (**Table 1**). In all, the ranges of As (4.7 - 7.0 mg/kg), Cd (0.2 - 0.3 mg/kg), Cr (23.4 - 39.6 mg/kg), Cu (13.6 - 21.5 mg/kg), Fe (26.7 - 39.2 g/kg), Mn (0.66 - 1.3 g/kg), Ni (23.8 - 37.9 mg/kg), Pb (15.8 - 23.3 mg/kg) and Zn (57.6 - 82.0 mg/kg) attained low to moderate concentrations, compared to the geogenic background values for claystone and greywackes [16].

Dissolved concentrations in the pore waters were also low to moderate (**Table 2**), with ranges of As (3.1 - 34.3 µg/l), Cd (0.01 - 0.17 µg/l), Cr (0.6 - 7.5 µg/l), Cu (0.8 - 17.5 µg/l), Fe (6.2 - 62.5 mg/l), Mn (2.7 - 19.0 mg/l), Ni (0.7 - 7.3 µg/l), Pb (0.2 - 3.1 µg/l) and Zn (11.3 - 256 µg/l). Stronger differences of dissolved metal contents were present within different sampling locations, but no relations of concentration changes and ongoing sediment exposure could be observed. The presence of in all small temporal variations can be seen as a result of a more or less constant redox state with continuous reducing conditions inside the sediments, which is also proven by high remaining concentrations of dissolved Fe and Mn in all sampling depths and locations. Changes in sorption processes and metal mobility should hereby occur only after a longer time of sediment exposure and a stronger decrease of the sediments water content.

4.2. Heavy Metals in Surface Waters

The observation of surface waters of the Aar Creek started at 19th August 2011 downstream the dam (site A) during the emptying process. After the reservoir basin was accessible, additional sampling of creek water started from 30th August 2011 in the reservoirs middle (site B) and root (site C). In waters of the Aar Creek, dissolved heavy metals showed strong variations with ongoing time of sediment exposure. Whereas during the process of reservoir emptying very low concentrations occurred in general, rising concentrations with factors of 10 to 30 could be measured few days after it was completed. This was observed at all sampling locations for all elements, with the exception of dissolved As. The highest concentration peaks were measured at site A, downstream the reservoirs dam (**Figure 2**). A tendency of decreasing concentrations within this export of dissolved metals was found 46 days after completed sediment exposure at the last sampling day (14th October 2011). This is assumed as an effect of a reduced pore water drainage.

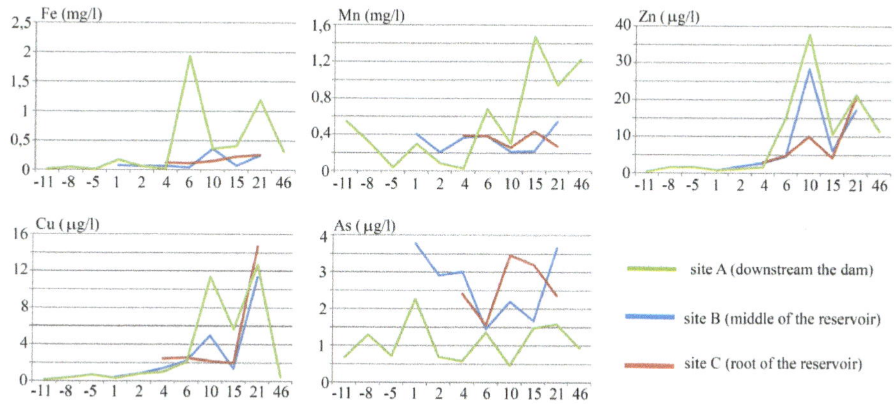

Figure 2. Temporal variations of dissolved Fe, Mn, Zn, Cu and As concentrations in the Aar Creek (number of days before the total exposure of the sediments in negative values, number days after the completed exposure in positive values).

Table 1. Heavy metal concentrations in lake sediments (aqua regia extraction) at the sampling sites profile 1 - 3.

Profile	Depth (cm)	As (mg/kg)	Cd (mg/kg)	Cr (mg/kg)	Cu (mg/kg)	Fe (g/kg)	Mn (g/kg)	Ni (mg/kg)	Pb (mg/kg)	Zn (mg/kg)
1	0 - 5 cm	4.7	0.20	23.4	13.6	26.7	0.85	23.8	16.0	57.6
	6 - 10 cm	4.8	0.19	25.2	14.6	27.9	0.66	25.3	15.8	59.1
2	0 - 5 cm	6.1	0.31	35.7	21.5	38.2	1.31	37.9	20.0	80.4
	6 – 10 cm	7.0	0.30	39.6	21.5	39.2	0.90	21.5	22.6	82.0
3	0 - 5 cm	4.8	0.27	28.9	18.6	31.8	1.05	29.8	20.2	68.8
	6 - 10 cm	5.4	0.27	27.0	16.7	26.9	0,76	27.1	23.3	61.2

Table 2. Median values of dissolved heavy metal concentrations in pore waters in 5 - 10 cm and 15 - 25 cm depth of each sampling profile (n = 6 for each depth).

Profile	Depth (cm)	As (µg/l)	Cd (µg/l)	Cr (µg/l)	Cu (µg/l)	Fe (mg/l)	Mn (mg/l)	Ni (µg/l)	Pb (µg/l)	Zn (µg/l)
1	5 - 10 cm	29.5	0.02	1.6	1.9	15.3	6.6	1.9	0.58	49.1
	15 - 25 cm	28.6	0.02	1.2	7.1	32.8	5.3	1.6	0.70	56.8
2	5 - 10 cm	24.3	0.01	1.5	2.0	18.1	6.8	1.8	0.48	96.8
	15 - 25 cm	28.1	0.02	1.6	4.2	40.7	6.0	2.1	0.91	64.9
3	5 - 10 cm	7.0	0.03	1.07	4.25	12.9	6.11	2.21	1.18	31.1
	15 - 25 cm	19.3	0.02	2.97	3.96	33.1	10.9	2.1	1.08	85.3

In all, the range of metal concentrations in waters of the Aar Creek was for As (0.48 - 3.78 µg/l), Cd (0.003 - 0.076 µg/l), Cr (0.11 - 2.32 µg/l), Cu (0.36 - 14.7 µg/l), Fe (0.02 - 1.94 mg/l), Mn (0.03 - 1.48 mg/l), Ni (0.22 - 5.41 µg/l), Pb (0.072 - 2.95 µg/l) and Zn (0.81 - 38.0 µg/l). The concentration peaks reached at the one hand in all moderate concentrations, which did not exceed water statutory threshold values. At the other hand, these concentrations occurred some days after complete sediment exposure temporarily in a similar concentration scale as in the pore waters. Although the sediments contained low to moderate heavy metal concentrations, the reservoirs emptying led by the drainage of pore waters to a strong increase of the Aar Creeks heavy metal load.

5. Discussion

As a result of reservoir emptying, strong changes in the heavy metal load of the former dammed Aar Creek were observed, whereas pore waters remained more or less constant in their concentrations, pH-values and redox potentials. These relatively low dynamics in pore waters are an effect of the sediments high water holding capacity, high remaining water contents and therefore slowed down redox changes. With ongoing exposure and ongoing drying of the sediments, rising concentration of dissolved Cd, Mn and Zn in the pore waters have to be assumed, resulting from an establishment of more oxic conditions [17] [18]. But, also the amount of pore water drainage and exported dissolved heavy metals should decrease simultaneously to this. Hence, oxidation processes inside the lake sediments should play only a minor role on the export of dissolved heavy metals in downstream situated river zones after the emptying of reservoirs.

As the oxygen supply to dissolved elements rises during and after pore water drainage, the ability of some elements to aquatic transport might be modified. The significance of these processes within metal deposition can be supposed by the comparison of the heavy metal concentrations in creek and pore waters.

As the scale of in running waters occurring concentration rises after sediment exposure is depending on the creeks metal concentrations before emptying, the precision of in **Table 3** presented percentages could be partly limited. Nevertheless, clear differences in the transportability of some elements are visible. Especially the export of dissolved Fe decreases during the drainage of pore waters in the Aar Creek: Although dissolved Fe concentrations rise within the running waters, only 1% of its pore water concentrations are achieved here. As it showed the comparatively highest concentration discrepancy between waters of the Aar Creek and pore waters, a much higher concentration rise should occur, if dissolved Fe was not removed from the water table after pore water drainage. In surface waters, dissolved Mn achieves only 6% of the mean pore water concentrations. As both elements achieve only a low solubility under oxic conditions and the present pH values [19], which are involved in the formation of hydroxides, they are either transferred to the creeks particulate heavy metal fraction or removed from the water column by the formation of precipitates. The same happens to dissolved As, which is under pH 4 - 9 mainly present as negatively charged, highly to Fe-(hydr-)oxides sorbable arsenate $H_2AsO_4^-$ or $HAsO_4^{2-}$ [20]. With the inset of oxic conditions during and after pore water drainage, it should be quickly depleted of its dissolved state.

In opposite, dissolved Cd, Cr, Cu, Ni and Pb underlay only small concentration losses during and after pore water drainage. The relative low differences between concentrations in pore- and surface waters prove a comparably low retention of these elements, in accordance to investigations in brackish sediments [21]. Only dissolved Zn, with 19.3% of the pore water concentrations inside the surface waters, indicates a slightly better retention.

The response of downstream situated aquatic biota to rising concentrations of dissolved heavy metals is hard to assess. This is due to the interaction of several factors as a metals binding state, species and age of a downstream situated organism [22], pH-value [23] or the endurance of an occurring concentration peak, that at least

Table 3. Median values of dissolved heavy metals in pore-(Pw) and surface waters (Sw) and their percental ratio (Sw * [100/Pw]).

Sample	n	As (µg/l)	Cd (µg/l)	Cr (µg/l)	Cu (µg/l)	Fe (mg/l)	Mn (mg/l)	Ni (µg/l)	Pb (µg/l)	Zn (µg/l)
Sw[1]	13	1.57	0.01	0.8	2.25	0.23	0.39	1.28	0.50	10.1
Pw[2]	36	24.1	0.02	1.55	3.72	22.3	6.38	1.9	0.73	52.3
Sw * (100/Pw) (%)		6.5	50.0	51.6	60.5	1.0	6.1	67.4	68.5	19.3

[1] = Surface waters, median values refer only to sampling data that were generated after the sediment exposure was completed (between day 1 to 46 after sediment exposure); [2] = pore waters (median values include all sampling data).

determine the toxicity. The negative impact of enhanced dissolved heavy metal concentration is again relativized by the fact, that it is only one stressor to aquatic biota simultaneous to an intensified export of sediments [24], enhanced concentrations of suspended matter [25], rising concentration of the particulate heavy metal fraction or stronger variations of the creeks discharge.

However, the occurrence of an obvious heavy metal export to the Aar Creek in the absence of metal pollutions inside the lakes sediments, leads to the suggestion that a much higher heavy metal input in running waters might occur in the presence of polluted sediments or under more acidic conditions. This should be the case especially for sediments with high concentrations of Cd, Cr, Cu, Ni and Pb. In consideration to this, the export of dissolved heavy metals and nutrients should always be taken into account during the complete emptying of reservoirs. As some metals seem to be transportable within running waters along distances of several kilometers, they represent a potential risk to downstream situated biological communities or fish farms, when the drainage of pore waters to surface waters occurs.

References

[1] Alloway, B.J., Ed. (1990) Heavy Metals in Soils. Wiley, New York.

[2] Schulz-Zunkel, C. and Krüger, F. (2009) Trace Metal Dynamics in Floodplain Soils of the River Elbe: A Review. *Journal of Environmental Quality*, **38**, 1349-1362. http://dx.doi.org/10.2134/jeq2008.0299

[3] Fenninger, A., Stattegger, K., Manser, P., Plass, N. and Scholger, R. (1988) Sedimentgeologische Untersuchungen an Stauraumsedimenten der Mur: Eine Vorstudie am Beispiel der Staustufe Mellach. *Mitteilungen des Naturwissenschaftlichen Vereins für Steiermark*, **118**, 107-118.

[4] Westrich, B., Kern, U. and Haag, I. (2000) Mobilität von Schadstoffen in den Sedimenten staugeregelter Flüsse-Dynamik und Bilanzierung von Schwebstoffen und Schwermetallen in einer Stauhaltungskette. http://www.fachdokumente.lubw.baden-wuerttemberg.de/servlet/is/5121/

[5] Ulrich, K.U., Paul, L. and Hupfer, M. (2000) Schadstoffgehalte in den Sedimenten von Staugewässern. *Wasser und Boden*, **52/10**, 27-32.

[6] Guo, T., De Laune, R.D. and Patrick, W.H. (1997) The Influence of Sediment Redox Chemistry on Chemically Active Forms of Arsenic, Cadmium, Chromium and Zink in Estuarine Sediment. *Environment International*, **23/3**, 305-316. http://dx.doi.org/10.1016/S0160-4120(97)00033-0

[7] Van den Berg, G.A., Loch, J.P. and Winkels, H.J. (1998) Effect of Fluctuating Hydrological Conditions on the Mobility of Heavy Metals in Soils of a Freshwater Estuary in the Netherlands. *Water, Air and Soil Pollution*, **102**, 377-388. http://dx.doi.org/10.1023/A:1004920700598

[8] Ahearn, D.S. and Dahlgren, R.A. (2005) Sediment and Nutrient Dynamics Following a Low-Head Dam Removal at Murphy Creek, California. *Limnology and Oceanography*, **50**, 1752-1762. http://dx.doi.org/10.4319/lo.2005.50.6.1752

[9] Bushaw-Newton, K., Hart, D., Pizzuto, J., Thomson, J., Egan, J., Ashley, J., Johnson, D., Horwitz, R., Keeley, M., Lawrence, J., Charles, D., Gatenby, C., Kreeger, D., Nightengale, T., Thomas, R. and Velinsky, D. (2002) An Integrative Approach towards Understanding Ecological Responses to Dam Removal: The Manatawny Creec Study. *Journal of the American Water Resources Association*, **38**, 1581-1599. http://dx.doi.org/10.1111/j.1752-1688.2002.tb04366.x

[10] Dagnac, J. (1994) Bilan des depots sedimentaires dans la reservoir de Pareloup (Aveyron, France). *Hydroecologie Applique*, **6**, 59-85. http://dx.doi.org/10.1051/hydro:1994004

[11] Merle, G., Mosnier, D. and Tourenq, J.N. (1994) La vindange de la retenue de Pareloup en 1993: Une etape-clef dans la vie du lac. *Hydroecologie Applique*, **6**, 427-446. http://dx.doi.org/10.1051/hydro:1994018

[12] Orr, C.H., Rogers, K.L. and Stanley, E.H. (2006) Channel Morphology and P Uptake Following Removal of a Small

Dam. *Journal of the North American Benthological Society*, **25**, 556-568. http://dx.doi.org/10.1899/0887-3593(2006)25[556:CMAPUF]2.0.CO;2

[13] Stanley, E.H. and Doyle, M.W. (2002) A Geomorphic Perspective on Nutrient Retention Following Dam Removal. *Bioscience*, **52**, 693-701. http://dx.doi.org/10.1641/0006-3568(2002)052[0693:AGPONR]2.0.CO;2

[14] Ulrich, K.U. (1998) Vergleichende Untersuchungen zu Auswirkungen des Sediments auf die Wasserbeschaffenheit in Trinkwassertalsperren unter Berücksichtigung von Stauspiegelschwankungen. Cuvillier, Göttingen.

[15] Hampel, U. (2013) Schwermetallkonzentrationen in Bach- und Ufersedimenten im Einzugsgebiet der Aartalsperre. Diploma Thesis, Philipps-University of Marburg, Marburg.

[16] LABO (Bund- und Länderarbeitsgemeinschaft Bodenschutz) (2003) Hintergrundwerte für anorganische und organische Stoffe in Böden. 3rd Edition. http//:www.labo-deutschland.de/documents/LABO-HGW-Text_4e3.pdf

[17] Gambrell, R.P., Wiesepape, J.B., Patrick, W.H. and Duff, M.C. (1991) The Effects of pH, Redox, and Salinity on Metal Release from a Contaminated Sediment. *Water, Air and Soil Pollution*, **57-58**, 359-367. http://dx.doi.org/10.1007/BF00282899

[18] Saeki, K., Okazaki, M. and Matsumoto, S. (1993) The Chemical Phase Changes of Heavy Metals with Drying and Oxidation of the Lake Sediments. *Water Research*, **27/7**, 1243-1251. http://dx.doi.org/10.1016/0043-1354(93)90017-C

[19] Alloway, B.J. and Ayres, D.C. (1996) Schadstoffe in der Umwelt. Spektrum, Heidelberg.

[20] Cheng, H., Hu, Y., Luo, J., Xu, B. and Zhao, J. (2009) Geochemical Processes Controlling Fate and Transport of Arsenic in Acid Mine Drainage (AMD) and Natural Systems. *Journal of Hazardous Materials*, **165**, 13-26. http://dx.doi.org/10.1016/j.jhazmat.2008.10.070

[21] Förstner, U., Calmano, W. and Ahlf, G. (1999) Sedimente als Schadstoffsenken und -quellen: Gedächtnis, Schutzgut, Zeitbombe, Endlager. In: Frimmel, F., Ed., Wasser und Gewässer: Ein Handbuch. Spektrum, Heidelberg.

[22] LAWA (Länderarbeitsgemeinschaft Wasser) (1998) Zielvorgaben zum Schutz oberirdischer Binnengewässer, Band II. Kulturbuchverlag, Berlin.

[23] UBA (Umweltbundesamt) (2001) Sachstandsbericht: Abtrag von Kupfer und Zink von Dächern, Dachrinnen und Fallrohren durch Niederschläge. https://www.umweltbundesamt.de/publikationen/abtrag-von-kupfer-zink-von-daechern-dachrinnen

[24] Stanley, E.H. and Doyle, M.W. (2003) Trading off: The Ecological Effects of Dam Removal. *Frontiers in Ecology and the Environment*, **1**, 15-22. http://dx.doi.org/10.1890/1540-9295(2003)001[0015:TOTEEO]2.0.CO;2

[25] Gray L.J. and Ward, J.V. (1982) Effects of Sediment Releases from a Reservoir on Stream Macroinvertebrates. *Hydrobiologia*, **96**, 177-184. http://dx.doi.org/10.1007/BF02185433

27

Hydro-Chemistry of Gokyo Valley, Sagarmatha (Everest) National Park, Nepal

Rajendra Bhandari[1*], Subodh Sharma[1], Juerg Merz[2], Robin Garton[2]

[1]Department of Environmental Science and Engineering, Kathmandu University, Dhulikhel, Nepal
[2]The Glacier Trust, Wiltshire, UK
Email: *rajendra.bhandari@ku.edu.np

Abstract

This research work records the baseline inventory of selected ions in running waters of Gokyo Valley, Nepal and tries to establish relationship of these ions' concentration with elevation gradient. Among these selected hydro-chemical parameters, the average concentration of Ca^{2+}, Mg^{2+}, SO_4^{2-} and total silica were determined to be 3.25 (±2.02), 0.64 (±0.47), 207.2 (±423.01) and 0.36 (±0.17) mg/L respectively. Concentration of these parameters was low compared to earlier studies except SO_4^{2-} that could be justified too since the sampling was performed only once in drier period with low atmospheric input of acid protons required for weathering process. Moreover, Pearson Product-moment Correlation Coefficients (r) for elevation gradient and concentration of Ca^{2+}, Mg^{2+}, SO_4^{2-} and total silica were determined to be +0.350, +0.145, +0.101 and −0.275 respectively. Such weak relationship between these parameters and elevation gradient suggests that the concentration of these ions is related to weathering phenomena, glacier and watershed characteristics and snow cover dynamics in Gokyo Valley.

Keywords

Gokyo Valley, Himalayan Streams, Hydrochemistry, Sagarmatha National Park

1. Introduction

Limnological research in the Himalayas has been carried out since the beginning of the 20th century by Sars, 1903 and Hutchinson, 1932 (As cited in [1]). Present decades have seen more detailed approach with greater focus on morphometric, physiochemical and biological features, primary production and trophic status and par-

*Corresponding author.

ticular emphasis for fish production [1] indicating very low concentration of dissolved minerals at altitude above 4000 m a.s.l. Nevertheless, the studies in the higher Himalayas of Nepal are scanty irrespective to the studies done in high altitude water bodies in Kashmir and Sikkim in the northwest Himalayas [1]. The study carried out by [2] in 24 lakes at altitude between 4500 and 5600 m a.s.l. in the Mount Everest area provides the first data on morphometry, temperature, chemistry and biology. Concerning the overall lake studies performed in Nepal, more than 163 research studies have been reported including master's thesis from 1969 to 2010 [3].

Lakes in Gokyo regions are studied systematically by assigning their names as Lake Cadstre Numbers (LCN) and the results indicate that these lakes are oligotrophic and alkaline and have low ionic content [1]. More recently, various limnological studies have been carried out in Gokyo region specially focusing on lakes [4]. The studies concerning physical-chemical including heavy metals, diatom and macroinvertebrate assemblages have also been carried out [4] [5].

The impact on water chemistry due to anthropogenic pollution from the industrial and agricultural sources has concerned environmentalists and scientists for past decades [6]. However, due to their remoteness high altitude lakes are less often affected by local sources of pollution. But their catchments receive deposition of pollutants due to long distance atmospheric transport [7]. Moreover, the chemistry of running waters draining the high mountain areas of Asia is largely unknown nevertheless; some studies conclude that the degree of pollution of these rivers increases with the distance from the high mountain first order streams and that pollution is increasing through time [6]. Geologically different catchments differ in both the release rates of elements to streams and lakes, and the relative proportion that are released. Yet, for the major ions at least, the relative proportion does not greatly differ in well-watered portion of the temperate, arctic and sub-tropical zones. The most common ion sequences as equivalent weight according to [8] are as follows:

$$Ca^{2+} > Mg^{2+} > Na^{+} > K^{+}$$

$$HCO_3^- > SO_4^{2-} > Cl^-$$

In this study, two cations (Ca^{2+}, Mg^{2+}), sulphate and total silica are taken and their concentrations (mg/L) have been identified and plotted against elevation if there exists any relationship between these ions with elevation gradient.

2. Materials and Methods

2.1. Study Area

Gokyo Valley is situated in Sagarmatha National Park, one of the World Heritage Sites and is a designated Ramsar Site. It is located within 27°52'N to 28°05'N and 86°40'E to 86°43'E. Total population of Khumjung VDC is 1912 with 913 males and 999 females, and has 551 households [9]. For the sampling of water for hydrochemistry, area from Phortse Thanga (3570 m asl) to Gokyo Lake Six (5125 m a.s.l) has been taken as shown in **Figure 1**. Several first order streams that are directly fed to Dudh Koshi River have been sampled. The sampling of water was done in the third week of May, 2013.

2.2. Sampling and Analysis

20 sampling stations were selected systematically and numbered from 1 to 20 as shown in **Figure 2**. Water samples for the analysis of calcium ion, magnesium ion, sulphate ion and total silica were analyzed in Aquatic Ecology Center, Kathmandu University. For the determination of calcium and magnesium ion concentration determination, Atomic Absorption Spectrophotometer (Thermo electron corporation; Model; M5MK2AA system), was used. Similarly for the determination of sulphate ion concentration and total silica concentration, Turbidimetric Spectrophotometric and Molybdosilicate Spectrophotometric (Model: Thermo Spectronic GRNESYS 10 UV) methods were used respectively. The guideline provided by APHA, AWWA and WEF (1998) [10] was strictly followed for the concentration determination of mentioned parameters. Microsoft Office Excel-2007 was used for the processing and analysis of data. This software was also used to construct Trend Graphs and to compute Pearson Product-moment Correlation Coefficient for selected parameters under study.

3. Results and Discussion

One time sampling for two cations (Ca^{2+} and Mg^{2+}), total silica and sulphate was done in 3rd week of May, 2013

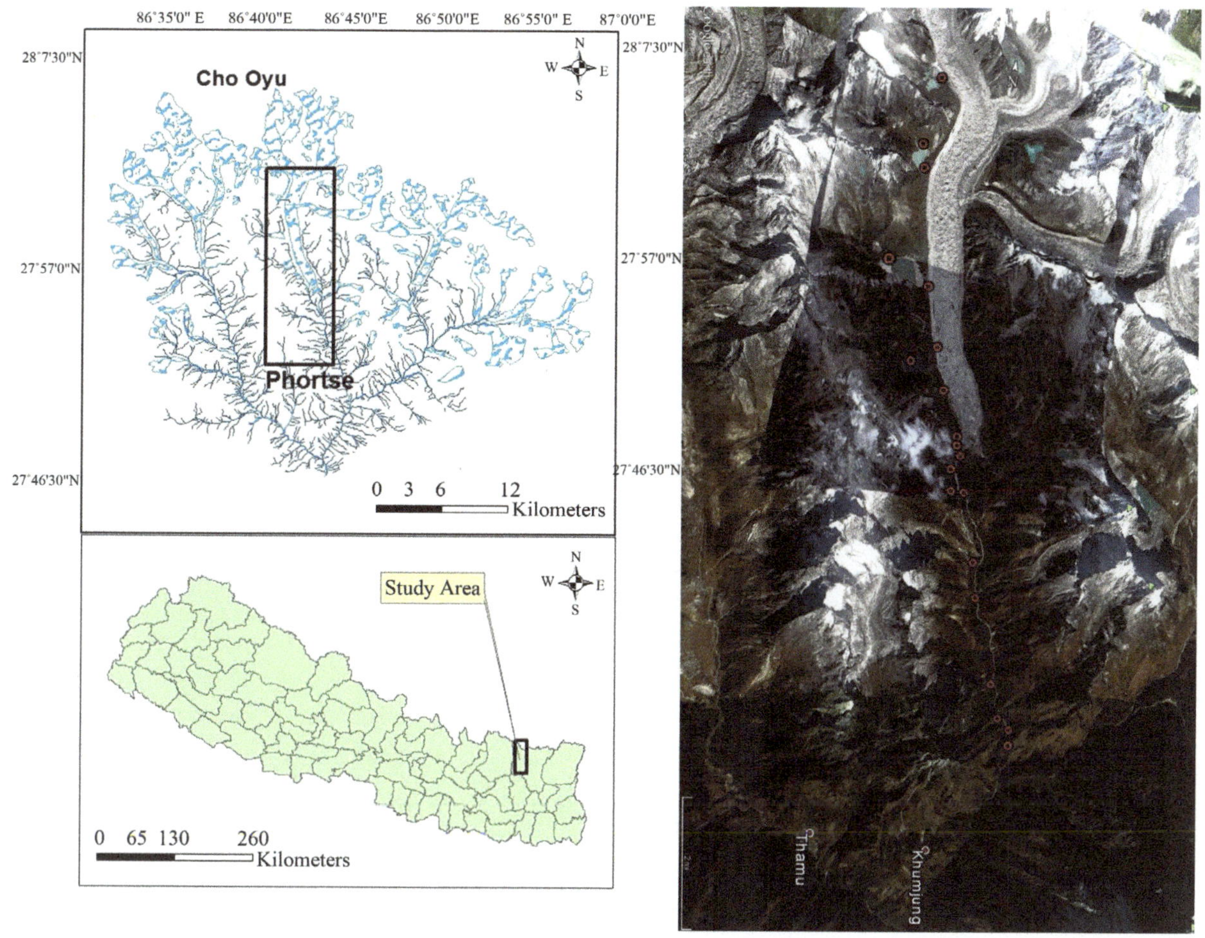

Figure 1. Study area.

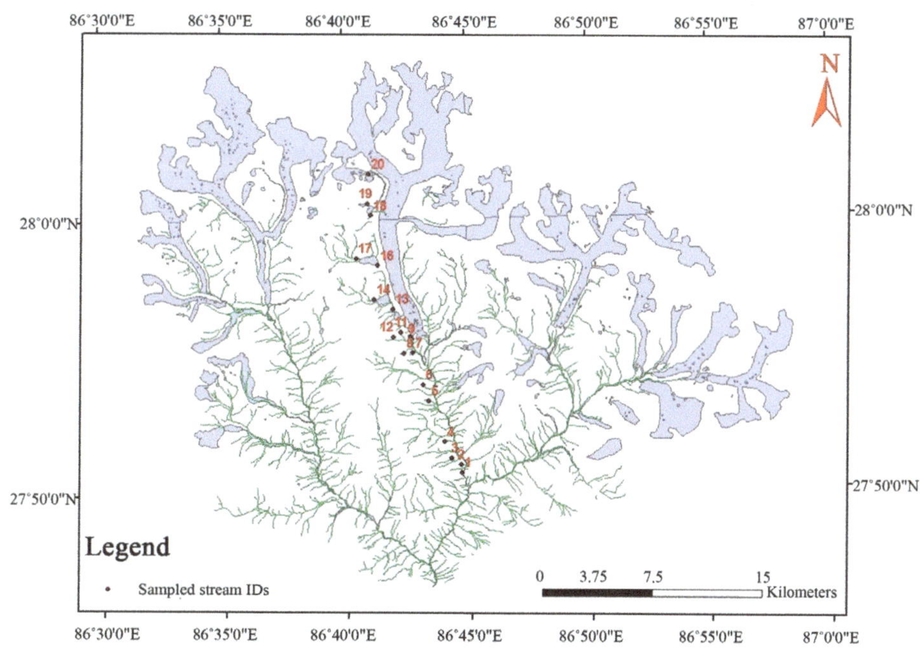

Figure 2. Sampling stations.

so as to examine whether there is any relationship of these parameters with elevation gradient. Sampling points 1 to 8 consist of first order streams which are devoid of lakes. Nevertheless, sampling points 9 to 15 are streams between Gokyo lake series. Sampling points 16, 17, 18, 19 and 20 are strictly located in lakes. Sampling points 16 and 17 are outlet and inlet of Lake 4 respectively. Similarly sampling points 18 and 19 are outlet and inlet of Lake 5 respectively. However, the sampling point 20 lies in Lake 6. The numbering of these streams is presented in **Figure 2**.

The average concentration of calcium ion in the sampling site is found to be 3.25 mg/L with standard deviation of 2.02, statistically as 3.25 (±2.02). Similarly, the average concentration of magnesium ion in mg/L with standard deviation is 0.64 (±0.47). The average concentration of total silica and sulphate in mg/L is 0.36 (±0.17) and 207.20 (±423.01) respectively. Owing to extremely high variability of result in the data sets of sulphate ion due to high concentration in four sites for which geometric mean has been calculated and is 15.16 which is more representative. The statistical results of four parameters that are analysed have been presented in **Table 1**.

Figures 3-6 demonstrate the concentration levels of analyzed four parameters of stream and lake waters in Gokyo region. The concentration of calcium (3.25 ± 2.02) is low compared to the former studies done in Everest region in 1995 and 1998 [6] [11] however, it is very low compared to the study done in 2010 [5]. In similar way,

Table 1. Mean concentration of selected parameters.

Parameters (mg/L)	Mean	Min	Max	Std. Dev	Geomean
Ca^{2+}	3.25	0.56	7.94	2.02	
Mg^{2+}	0.64	0.14	2.14	0.47	
Total silica	0.36	0.13	0.92	0.17	
SO_4^{2-}	207.2	1.46	1291	423.01	15.16

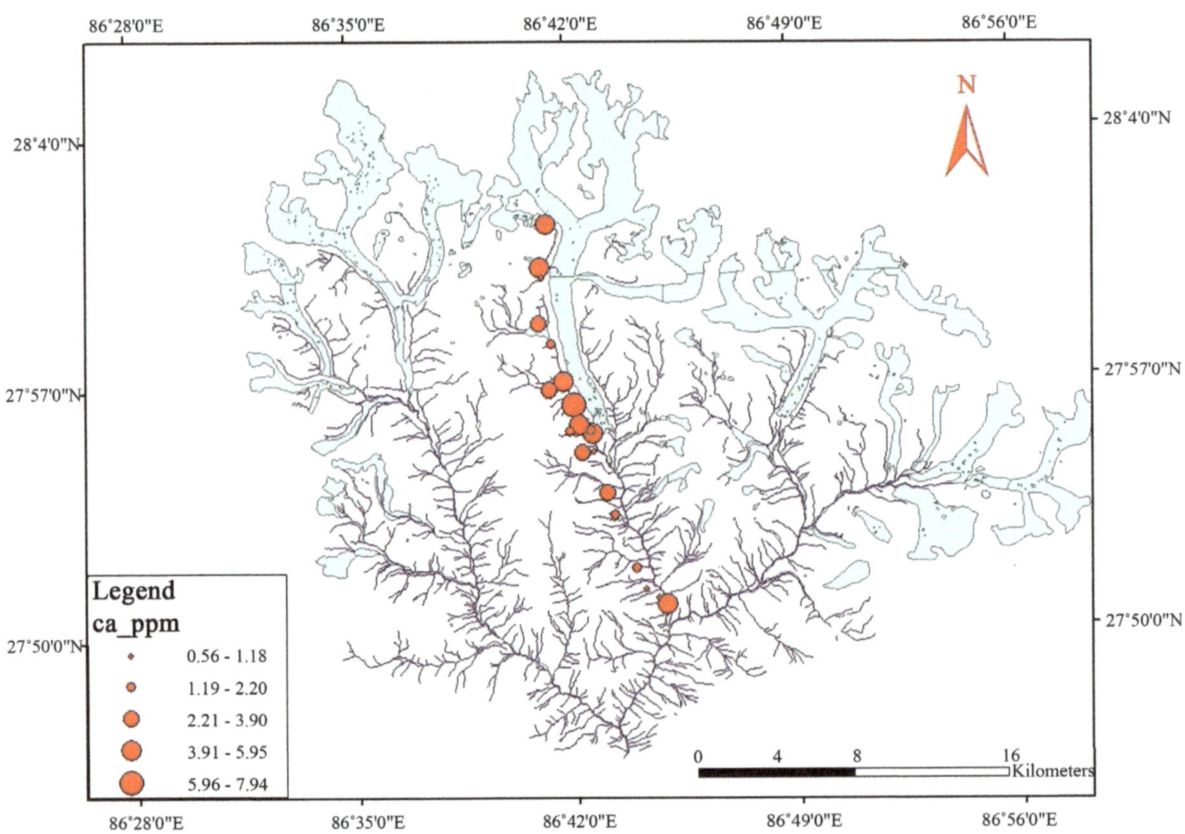

Figure 3. Calcium ion concentration.

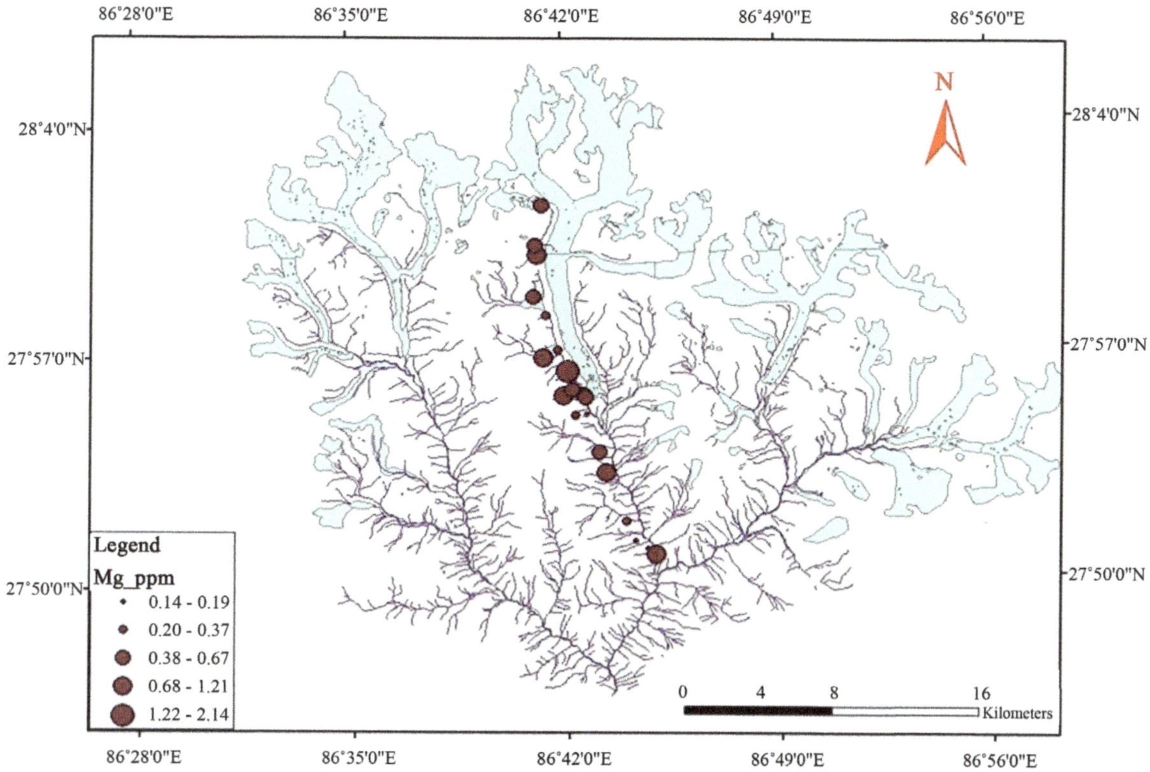

Figure 4. Magnessium ion concentration.

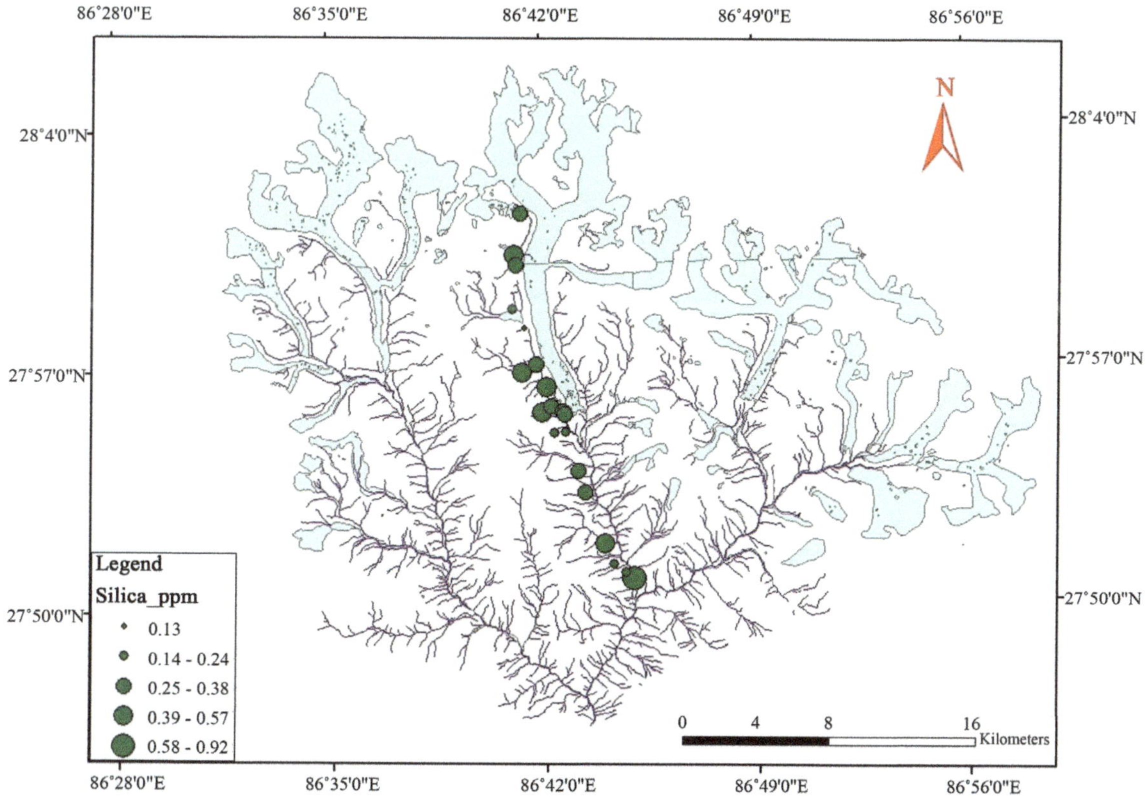

Figure 5. Silica ion concentration.

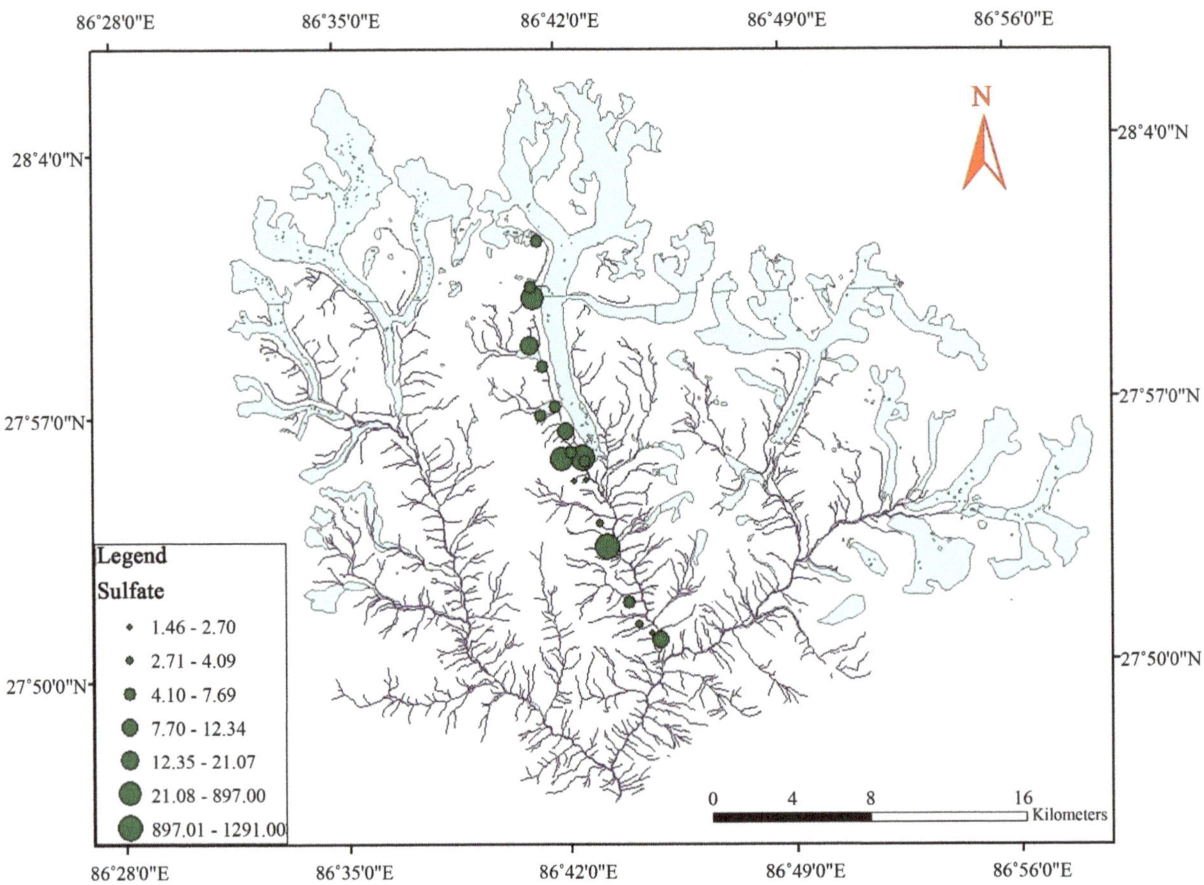

Figure 6. Sulphate ion concentration.

concentration of magnesium (0.64 ± 0.47) and total silica (0.36 ± 0.17) is also low compared to the results of study performed in 1995 and 1998 [6] [11]. However the total silica concentration is approximately similar to the study performed in 2010 [5]. Nevertheless, the sulphate concentration of this study is highly variable and very high compared to the previous studies [6] [11].

This study was done only in one season (3rd week, May 2013) during the dry period with low atmospheric inputs and low stream flows so that the major source of ions and silica to these streams and lakes is from chemical weathering. In the absence of significant atmospheric inputs of acid pollutants, the supply of acid protons required for weathering reactions in alpine glacial environments depends mainly on the dissolution of CO_2 and oxidation of pyrite, the processes which accounts for the dominance of HCO_3^- and SO_4^{2-} amongst anions solution [12]. And the readily weatherable source of Ca^{2+} and Mg^{2+} are alumino-silicates with carbonates that are easily identified in a number of geological formations in these regions [11].

The final objective of this study was to determine whether the ions concentration is related to the elevation gradient or not. For this, scatter chart of elevation withthe concentration of calcium, magnesium and total silica were plotted as shown in **Figures 7-9** respectively.

The Pearson product-moment correlation coefficient was calculated using Microsoft, Excel Workbook for four parameters as mentioned above. Owing to extremely high variability of sulphate concentration the chart is not constructed. The Pearson Product-moment Correlation Coefficients for elevation gradient and concentration of calcium, magnesium, total silica and sulphate are +0.350, +0.145, −0.275 and +0.101 respectively. All the values showed very weak relationship between elevation gradient and ions concentration for Gokyo region. Hence, this study derives the finding that the ions concentrations in Gokyo region are not strictly correlated to eleva-tion gradient. Nevertheless, these are related tothe weathering phenomena, glacier and watershed characteristics and snow cover dynamics [1].

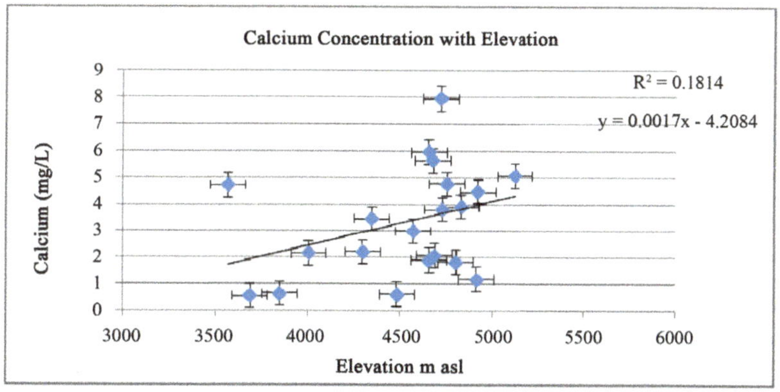

Figure 7. Scatter chart of calcium concentration.

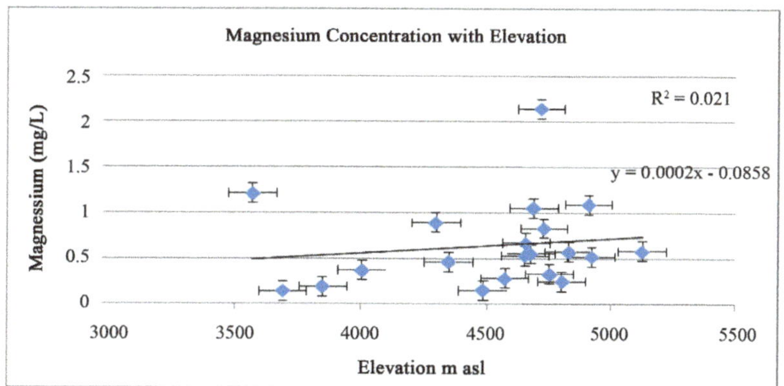

Figure 8. Scatter chart of magnessium concentration.

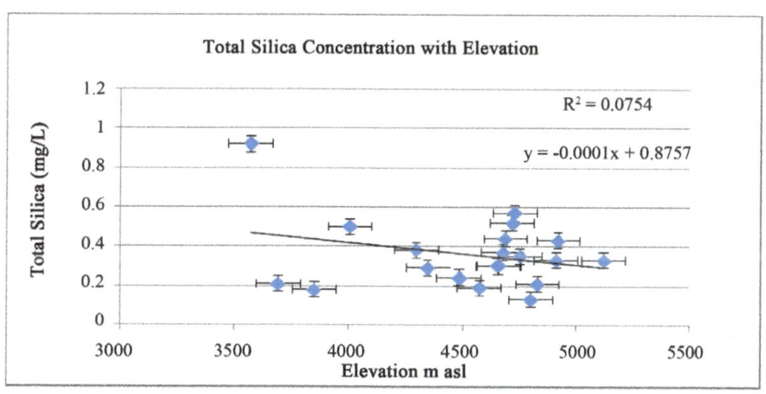

Figure 9. Scatter chart of total silica concentration.

4. Conclusion

The low concentration of calcium and magnesium ions in average compared to past studies was usual in the study site. However, the quantitative presence of total silica has been seen similar in temporal studies. Nevertheless, highly fluctuating concentration of sulphate ions could not be well explained and future studies are suggested. Pearson Product-moment Correlation Coefficient for elevation gradient and concentration of these four parameters showed very weak relationship which proves their disassociation.

Acknowledgements

We are very grateful to The Glacier Trust, UK for their financial assistance under Climate Change Education

Project in Nepal supporting this research work. We are also thankful to the Department of National Park and Wildlife Conservation, Government of Nepal, and villagers in Gokyo Valley for their kind cooperation while conducting this research.

References

[1] Tartari, G.A., Tartari, G. and Mosello, R. (1998) Water Chemistry of High Altitude Lakes in the Khumbu and Imja Kola Valleys (Nepalese Himalayas). *Memorie dell'lstituto Italiano di Idrolnologia*, **57**, 51-76.

[2] Löffler, H. (1969) High Altitude Lakes in Mt Everest Region. *Verhandlungen Internationale Vereinigung Limnologie*, **17**, 373-385.

[3] Bhuju, D.R., Sharma, S., Jha, P.K. and Gaire, N.P. (2012) Scientific Discourse of Lakes in Nepal. *Nepal Journal of Science and Technology*, **13**, 147-158.

[4] Sharma, C.M., Sharma, S., Bajracharya, R.M., Gurung, S., Juttner, I., Kang, S., Zhang, Q. and Li, Q. (2012) First Results on Bathymetry and Limnology of High-Altitude Lakes in the Gokyo Valley, Sagarmatha (Everest) National Park, Nepal. *Limnology*, **13**, 181-192. http://dx.doi.org/10.1007/s10201-011-0366-0

[5] Gurung, S. (2012) High Altitude Aquatic Biodiversity of Gokyo Lake Series in Sagarmatha National Park, Nepal. A Dissertation, Submitted for the Partial Fulfillment of the Requirements for the Doctoral Degree (Ph.D.) in Environmental Science. Central Library, Kathmandu University, Lalitpur.

[6] Reynolds, B., Jenkins, A., Chapman, P.J. and Wilkinson, J. (1998) Stream hydrochemistry of the Khumbu, Annapurna and Langtang Regions of Nepal. Ecovision World Monograph Series, 123-141.

[7] Battarbee, R.W. (2005) Mountain Lakes, Pristine or Polluted? *Limnetica*, **24**, 1-8.

[8] Kallf, J. (2002) Limnology. Prentice Hall. Upper Saddle River.

[9] National Population and Housing Census 2011. Volume 02, NPHC 2011, Government of Nepal National Planning Commission Secretariat Central Bureau of Statistics Kathmandu, Nepal.

[10] APHA, AWWA and WEF (1998) Standard Methods for the Examination of Water and Wastewater. 20th Edition, American Public Health Association, Washington DC.

[11] Reynolds, B., Chapman, B.J., French, M.C., Jenkins, A. and Wheater, H.S. (1995) Major, Minor and Trace Element Chemistry of Surface Waters in the Everest region of Nepal. Proceedings of a Boulder Symposium, IAHS, Publication No. 228.

[12] Tratner, M., Brown, G., Raiswell, R., Sharp, M. and Gurnell, A. (1993) A Conceptual Model of Solute Acquisition by Alpine Glacial Meltwaters. *Journal of Glaciology*, **39**, 573-581.

28

Hydrological Modelling of Sidi Jabeur Watershed (Morocco) Using Spatially Distributed Model ATHYS

Mourad Khattati[1*], Mostapha Serroukh[1], Ismail Rafik[1], Hakim Mesmoudi[1], Brirhet Hassane[2], Yassine Bouslihim[3]

[1]Research Team ERGO, Faculty of Science, Abdelmalek Essaadi University, Tetouan, Morocco
[2]Faculty of Science and Technology, Sidi Mohammed Ben Abdellah University, Fes, Morocco
[3]Faculty of Science and Technology, 1st Hassan University, Settat, Morocco
Email: [*]khattati.mourad@hotmail.fr

Abstract

The objective of this study is to model the hydrology in the Sidi Jabeur basin, located in Bouregreg watershed at the north-central of Morocco, using the spatially distributed model (ATHYS) in order to understand and determine the different watershed hydrological processes. The study requires the collection of a series of data as inputs models namely rainfall data, water quantity, soil occupation, digital terrain model and requires also a calibration in order to evaluate the model in validation phase. The simulation results are obtained from the validation phase aim to replicate the operation of Sidi Jabeur watershed, and present a suitable adjustment perspective of the observed hydrograph. These results show that the objective is achieved and a model distributed like ATHYS plays an effective role in improving the efficiency and presents a high advantage in anticipation of runoff volume.

Keywords

Distributed Model, Hydrology, Sidi Jabeur Watershed, ATHYS

1. Introduction

The water is the most important natural resource especially in the arid or semi-arid zones that face high population growth, scarcity of freshwater, irregularity of rainfall, excessive land use change and increasing vulnerabil-

[*]Corresponding author.

ity to risks such of drought, desertification and pollution. Thus, the availability and the sustainable use of this resource become the core of the local and national strategies and politics in these regions [1].

Managing water resources are mostly required at watershed scale [2] given that is the basic hydrologic unit where can be studied the heterogeneity and complexity of processes and interactions linking land surface, climatic factors and human activities [1]. This adopted approach for assessing water quantity and quality is then expressed as various hydrologic models and tools that try to simulate and predict the watershed response at different spatial and time scales.

The objective of this study is to model the dynamics of Sidi Jabeur watershed functioning with the reconstitution of river flow at outlet (Station Sidi Jabeur). The simulation of water intake in the hydrological station has been done with the use of a spatially distributed model ATHYS. However, the objective is to test the adaptability and ensures a good performance of the model. In this work, the time step of modeling is daily, this choice due to compare modeled flows with occasional discharge measurements on the hydrometric Sidi Jabeur station.

2. Materials and Methods

2.1. Study Area

Sidi Jabeur basin is located at the south of the Bouregreg watershed north-central of Morocco (**Figure 1**). The watershed covers an area of 2910 km^2 with an elevation ranging from 1602 m to 330 m (Sidi Jabeur outlet).

Between 1977-2005, the minimum annual rainfall is 288 mm/year in Ouljet Haboub station, while the maximum is 300 mm/year in Sidi Jabeur station. Furthermore, the annual and monthly average is respectively 294 mm and 28 mm (**Figure 2**).

2.2. Presentation of ATHYS Model

ATHYS (A Telier Hydrologique Spatialisé) is rainfall-runoff transformation model, developed by the Research Institute for Development (IRD) of Montpellier. The basic principles of ATHYS are a hydrological environment for distributed modeling, including a series of models, DEM processing, hydrological and rainfall data and geographical display, spatial data interpolation. The flow generated for a rain event (rainfall-runoff transformation)

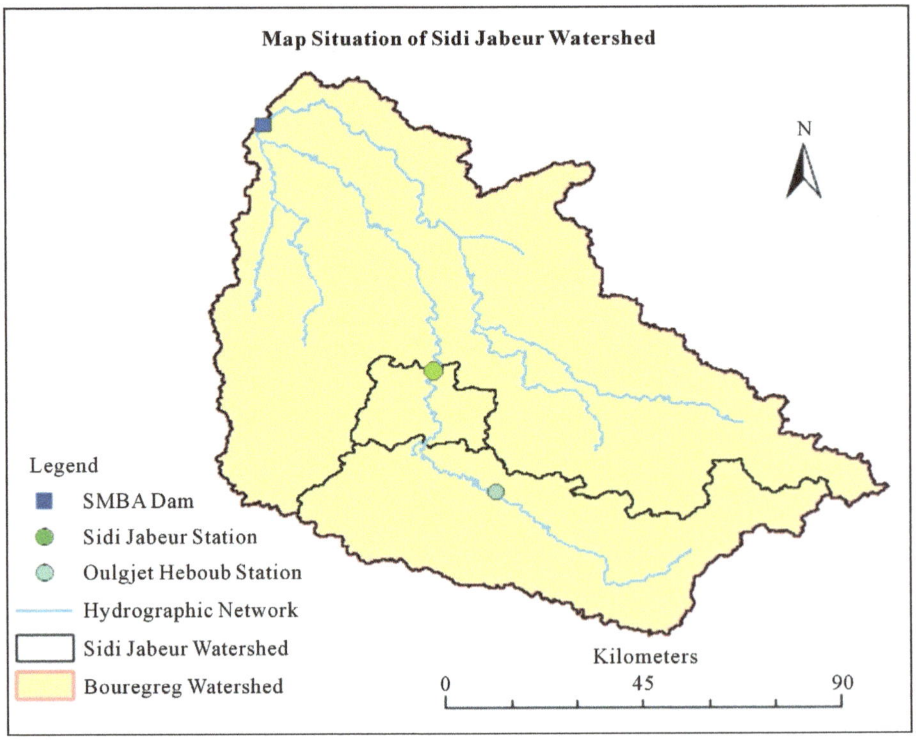

Figure 1. Map situation of Sidi Jabeur watershed.

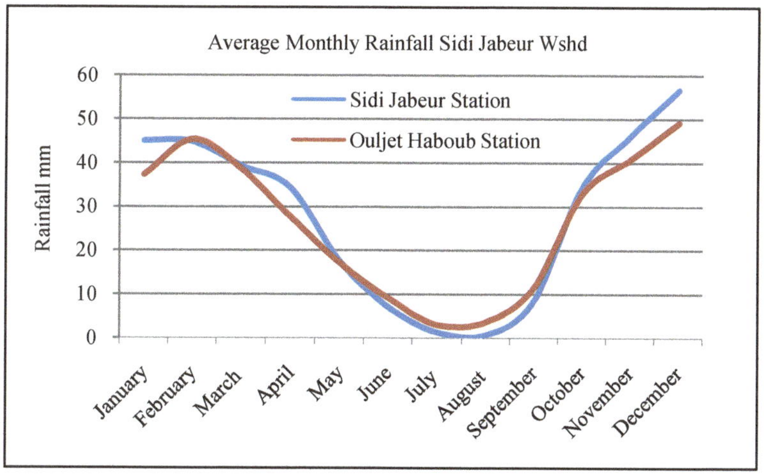

Figure 2. Average monthly rainfall in Sidi Jabeur watershed.

is calculated for each cell, the amount of rain is estimated by the production model which will contribute to run-off. The transfer model calculate the hydrograph produced by each watershed outlet, this calculation is made from the result obtained under the production function.

2.3. Methodology and Data Processing

The spatially distributed model ATHYS are used to simulate the flow rates daily time in Sidi Jabeur station from 1996 to 2004. The calibration and validation period is respectively 1996-1999 and 2000-2004. To estimate the parameters of production and transfer models, a manual procedure by the sensitivity analysis had been used. From input data (digital terrain model, precipitation), the hydrological model ATHYS is used to define the daily runoff flow for each cells of Sidi Jabeur watershed, and also to obtain the drainage network, slope classes… etc. From the effective rainfall, SCS function defines the runoff depth. Moreover, unit hydrograph transfer function calculates the flows for each sub basins outlet, identified as nodes calculation (Ouljet Haboub and Sidi Jabeur). The DEM used is a set of elevation points (three coordinates, x, y and z) located every 90 m. The topography is reconstructed by the assembly of these points (**Figure 3**). The sub-mentioned data help to define the drainage network, slope classes (**Figure 4**), drainage directions (**Figure 5**), sub basins (**Figure 6**).

In this study, three different production classes have been identified from the topography (slopes), as a file re-classified between 447 classes of natural slopes. Low and middle runoff zone correspond to the plains with slopes from 0 to 200 m/km, while the high runoff zone corresponds to the high slopes mainly on reliefs.

Climatologically information is accurately known in each station. This information must be spatially on the catchment area. The method of Thiessen polygons was used to define the influence's area for each station (**Table 1**).

The following table shows the breakdown of the areas of Sidi Jabeur watershed based in the production classes (**Table 2**).

The area of high runoff concerns only 5.15% of the whole watershed area. This area corresponds to the reliefs with bare soil (and/or) with high slopes.

To evaluate our model three statistic coefficients were adopted: Coefficients of Determination (R^2), Nash-Sutcliffe efficiency coefficient (NSE) [3], and RMSE-observations standard deviation ratio (RSR) [4]. The formulas of these coefficients are given in the following equations.

$$\text{NSE} = 1 - \left[\frac{\sum_{i=1}^{n} \left(Y_i^{obs} - Y_i^{sim} \right)^2}{\sum_{i=1}^{n} \left(Y_i^{obs} - Y_i^{\text{mean}} \right)^2} \right] \quad (1)$$

$$\text{RSR} = \left[\frac{\sum_{i=1}^{n} \left(Y_i^{obs} - Y_i^{sim} \right)^2}{\sum_{i=1}^{n} \left(Y_i^{obs} - Y_i^{\text{mean}} \right)^2} \right] \quad (2)$$

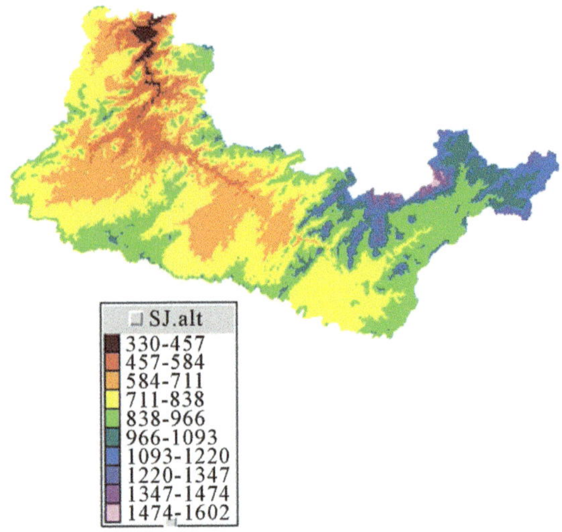

Figure 3. Map of altitudes.

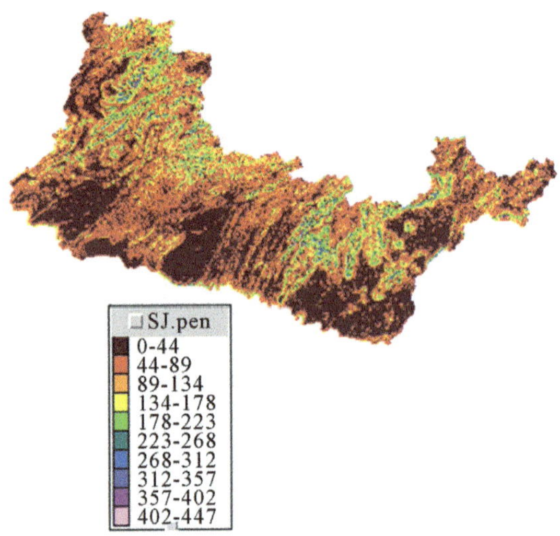

Figure 4. Slope classes.

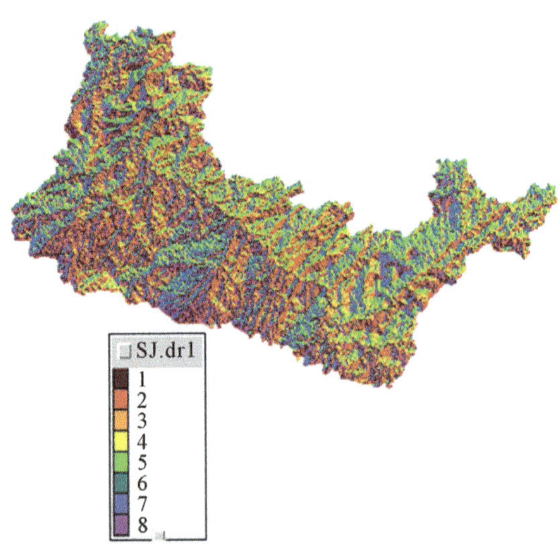

Figure 5. Drainage directions.

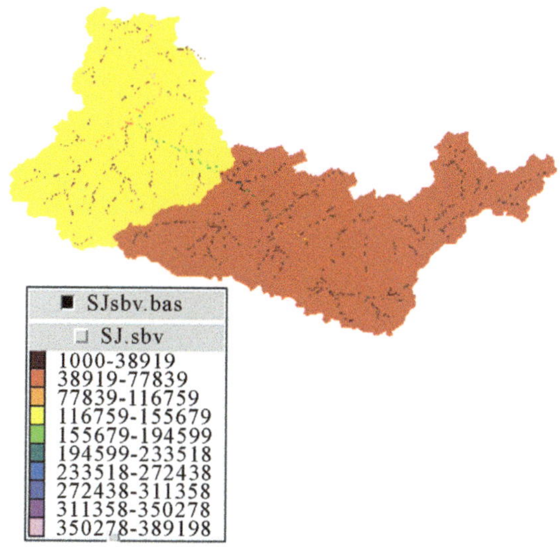

Figure 6. Hydrographic network & sub basins.

Table 1. Weight stations by Theissen method.

Station	Thiessen coefficient	Area (Km^2)	Extent %
Sidi Jabeur	0.367	1067	36.70
Ouljet Haboub	0.633	1843	63.30

Table 2. Production class.

Production class	Area (km^2)	Extent %
1) Low runoff	990	34.02
2) Middle runoff	1770	60.82
3) Rapid runoff	150	5.15
Total	2910	100

3. Results and Discussion

In this study, the functions used by ATHYS model are SCS and unit hydrograph for the production and transfer function. The sensitivity analysis of model has allowed to show the best values of the parameter S (SCS) affecting mainly the peak flow. The decline of the hydrograph explained by the emptying time is generally influenced by the (Ds) parameter. Preferentially, we fix the influenced parameters on the general curve of the hydrograph, namely, the diffusion coefficient (K0), the parameters K1 and Alfa, and the transfer speed of V0, which figures on the transfer function (**Table 3**).

The statistic evaluators showed a good correlation between the monthly observed and simulated river discharge with R^2 of 0.92, NSE of 0.83, and RSR of 0.41 for the calibration period. The validation period revealed good values for R^2 0.90, NSE 0.81 and RSR 0.39, this model performance for both calibration and validation periods is evaluated as "very good performance rating" which is defined by the flowing ranges: 0 to 0.5 for RSR, 0.75 to 1 for NSE (**Figure 7** and **Figure 8**).

4. Conclusion

The approach we adopted was to model the dynamics of Sidi Jabeur watershed functioning with the reconstitution of river flow at outlet (Station Sidi Jabeur) using the spatially distributed model (ATHYS). The model was successfully calibrated in the Sidi Jabeur watershed and produced good simulation results. The evaluation of the

Table 3. Parameters of production and transfer functions.

	Parameter	Prod class 1	Prod class 2	Prod class 3
Production function	STO [mm]	39	43	51
	INF [mm·h^{-1}]	10.3	9.3	9.3
	F [m^{-1}]	41	39.4	38.4
	Ds [j^{-1}]	0.1	0.08	0.08
Transfer Function	V0 [m·s^{-1}]		1.35	
	Alfa [ad]		0.6	
	K0 [ad]		0.2	
	K1 [mn]		0.9	

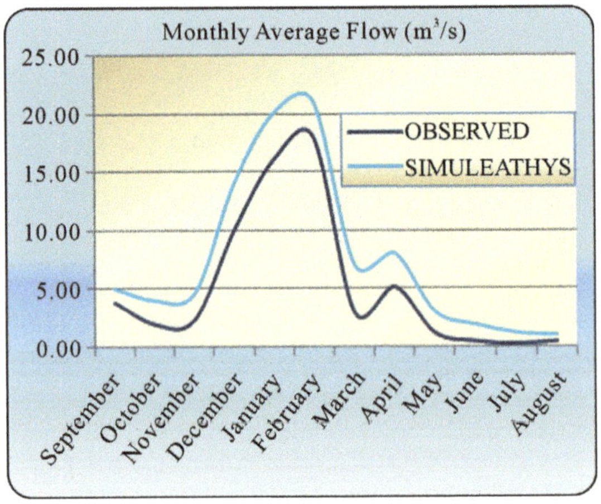

Figure 7. Simulation of mean monthly flows Simulated in Ouljet Haboub station over (1996-2004).

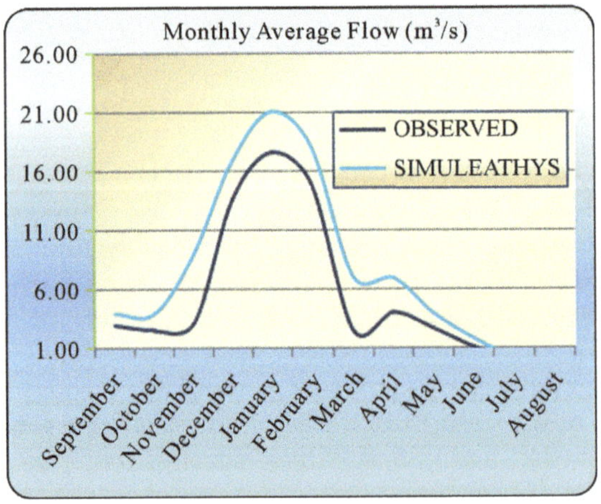

Figure 8. Simulation of mean monthly flows simulated in Sidi Jabeur station over (1996-2004).

model performance was carried out successfully with the recommended statistical coefficients, the comparison of observed and simulated flow stream revealed a Nash-Sutcliffe coefficient and R^2 superior to 0.81 for both calibration and validation periods. This study showed the ability of ATHYS model to be used to simulate the water quantity in semi-arid regions.

Acknowledgements

The authors thank the Ministry of Energy, Mines, Water and Environment (S.Water-Dr.Brirhet) for providing the data on the studied boreholes.

References

[1] Fadil, A., Rhinane, H., Kaoukaya, A., Kharchaf, Y. and Bachir, O. (2011) Hydrologic Modeling of the Bouregreg Watershed (Morocco) Using GIS and SWAT Model. *Journal of Geographic Information System*, **3**, 279. http://dx.doi.org/10.4236/jgis.2011.34024

[2] Sivapalan, M. (2003) Process Complexity at Hillslope Scale, Process Simplicity at the Watershed Scale: Is There a Connection? *Hydrological Processes*, **17**, 1037-1041. http://dx.doi.org/10.1002/hyp.5109

[3] Nash, J.E. and Sutcliffe, J.V. (1970) River Flow Forecasting through Conceptual Models, Discussion of Principles. *Journal of Hydrology*, **10**, 282-290. http://dx.doi.org/10.1016/0022-1694(70)90255-6

[4] Singh, J., Knapp, H.V. and Demissie, M. (2004) Hydrologic Modeling of the Iroquois River Watershed Using HSPF and SWAT. *Journal of the American Water Resources Association*, **41**, 343-360. http://dx.doi.org/10.1111/j.1752-1688.2005.tb03740.x

29

Exploratory Analyses for Water Quality of the Gulf of Mexico Basin, Texas, 2005-2014

Faye Anderson, Najla N. J. Al-Thani

Center for Advances in Water and Air Quality, Lamar University, Beaumont, TX, USA
Email: andersonfaye7@gmail.com

Abstract

With the increased demand for more water sources, the issues of water quality have become more important. This study presents exploratory data analyses on water quality issues related to the Gulf of Mexico Basin in Texas for the last decade: 2005-2014. Dissolved oxygen, water temperature, specific conductance, PH, transparency, sampling depth and *Enterococci* observations for the ten years from 2005 to 2014 were analyzed. Statistical testing showed that the observations followed similar distributions between the years. Thus they were combined for each of the variables. Throughout the ten years transparency, water temperature, specific conductance, and *Enterococci* showed higher variances. Pair-wise correlations were negligible but they had positive spatial autocorrelation. Sampling was constantly conducted in the four months of February, May, August, and November. Water temperature, dissolved oxygen and *Enterococci* concentrations varied significantly during these four months. Univariate spatial regressions showed that *Enterococci* is predicted to increase whereas dissolved oxygen, PH, water temperature, specific conductances were predicted to decrease.

Keywords

Surface Water, Water Quality, Gulf of Mexico Basin, Texas, Multivariate, Spatial, Stormwater Flooding, Community Adaptive Capacity, Resilience

1. Introduction

In the summer of 2015, many warnings were issued in Texas cautioning against water recreational activities because of detection of high levels of *Enterococci*. This was substantiated by the increased risks that face the population especially those engaging in recreational activities in surface waters due to presence of disease-causing microorganisms, which led to the 1999 U.S. Environmental Protection Agency (EPA) Action Plan for Beaches and Recreational Waters [1]. Increased levels of *Enterococci* in water are positively associated with increased cases of gastroenteritis [2]. One of the most important sources of bacterial contamination in the state is storm water flooding which flows into beaches [3]-[5]. This is not an isolated issue as this contamination causes further contamination to groundwater, which is getting increasingly scarce. Increased levels of fecal bacteria is not the only problem facing quality water in Texas [6] [7]. Moreover, these increased detections of *Enterococci* contamination point to issues with community resilience and its adaptive capacity against flooding [8] [9].

This study aimed at assessing surface water quality in the Gulf of Mexico Basin in Texas, in the last decade from 2005 to 2014, by conducting temporal and spatial exploratory analyses for six of the most important parameters: Dissolved oxygen, PH, specific conductance, water temperature, transparency, and levels of *Enterococci*. Dissolved oxygen is one of the most important measurements of fresh water because of its significance to aquatic organisms especially fish. It is measured in milligrams per liter (mg/l). Oxygen is produced during photosynthesis. Excessive plant growth consumes more oxygen than its production leading to decreased levels of dissolved oxygen in the water. Oxygen solubility in water is negatively associated with each water temperature and salinity. Increased levels of bacteria reduce levels of dissolved oxygen in water. Another source of oxygen in water is diffusion from the atmosphere. PH measures acidity of water with a scale from 0 to 14, 7 being neutral and a PH of greater than 7 being alkaline. Just as dissolved oxygen is important for aquatic life, a PH between 5 and 9 is necessary for healthy aquatic organisms' wellbeing. Algal bloom increases PH levels and can be lethal to aquatic organisms. Specific conductance is a measure of water's capability to conduct an electric current and is the ratio of current to voltage between any two points in micro Siemens per centimeter (μS/cm) within water corrected to 25°C. Water temperature is an essential indicator for healthy aquatic life because it can point to levels of dissolved oxygen. Transparency is a measure of water clarity and is important for sunlight to reach aquatic plants. *Enterococci* are a sub-group of fecal streptococci usually found in intestines of mammals and humans. The Texas Commission on Environmental Quality (TCEQ) standard for *Enterococci* in fresh water is 168 most-probable number (MPN)/100 ml and it is measured as an indicator for fecal contamination in water [2] [10]-[12].

2. Methods

Shapefile for Texas surface water quality monitoring stations was downloaded from TCEQ's website (http://www.tceq.state.tx.us/gis/download-tceq-gis-data). Field water quality observations were downloaded from TCEQ Clean Rivers Program (CRP) website (http://www80.tceq.texas.gov/SwqmisWeb/public/crpweb.faces). Variables collected were dissolved oxygen (mg/l), water temperature (°C), specific conductance (μS/cm at 25°C), PH, transparency (Secchi disc) (m), depth (m), and *Enterococci* in the (MPN/100 ml) methodology; for the ten years from 2005 through 2014. In addition, locations of stations were recorded in longitude and latitude (**Figure 1**).

3. Results and Discussion

Data (3880 observations) came from seven monitoring stations: Gulf of Mexico at county line, Gulf of Mexico at Mc Fadden Beach Louisiana, Gulf of Mexico at port Aransas, Gulf of Mexico at Port Isabel, Gulf of Mexico at Sea Rim Park, Gulf of Mexico off Louisiana coast, and Gulf of Mexico Sabine pass (**Figure 1**). Annual observations numbers ranged from 317 (in 2008) to 482 (in 2006). Nevertheless, sampling was not done in every month (**Figure 2**). In the ten years, sampling has always been conducted in the four months of February, May, August, and November, or every three months starting from February. Exploratory data analyses (**Table 1**) showed that Dissolved oxygen averaged 7.07 mg/l, PH averaged 8.02, specific conductance averaged 4,102,676 μS/cm at 25°C, water temperature averaged 22.60°C, transparency averaged 0.83 m, and *Enterococci* averaged 1023.11 MPN/100 ml. Depth of sampling averaged 5.5 m except for transparency (0.30 m). Variation for dissolved oxygen, PH, water temperature, transparency, and sampling depth was not significant, but it was distinctively variant for specific conductance.

In order to test for temporal variation for the six throughout the ten years, Fisher's F-test was used to test the null hypothesis of similar variances and t-test was used to test the null hypothesis of similar means. *P*-values for both tests were greater than the significance level of 0.05. Hence the null hypotheses of similar means and variances could not be rejected for each of the six variables. This means that each variable does not have between-year variation. Therefore, all records for the different years were combined for each variable. Temporal

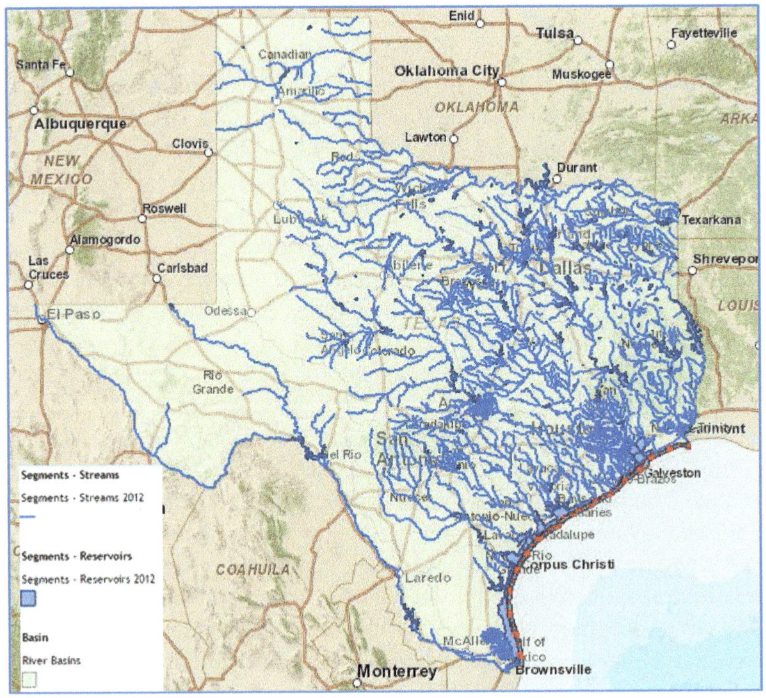

Figure 1. Gulf of Mexico Basin, Texas.

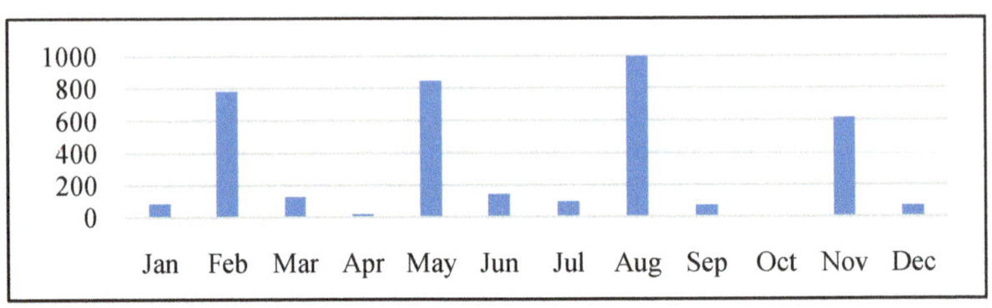

Figure 2. Monthly number of observations, 2005-2014.

Table 1. Descriptive statistics, 2005-2014.

Parameter	Count	Min	Mean	Max	Std. Dev.	Skewness	Kurtosis
Dissolved Oxygen (mg/l)	899	0.60	7.07	15.99	1.72	0.20	1.23
PH	936	6.90	8.02	9.30	0.24	−0.16	4.17
Specific Conductance (μS/cm at 25°C)	918	2200.00	41,026.76	59,700.00	11,353.51	−0.90	1.01
Water Temperature (°C)	936	2.10	22.60	34.60	6.29	−0.40	−0.74
Transparency (m)	191	0.01	0.83	5.00	0.90	1.68	2.79
Depth (m)	3880	0.30	5.28	19.00	4.64	0.56	−0.90
Enterococci (MPN/100 ml)	162	1.00	1023.11	4800	1773.39	1.52	0.52

trends for dissolved oxygen, PH, water temperature, transparency and specific conductance in the past decade are given in **Figure 3**. Over the ten years PH and transparency seemed to be stable, whereas dissolved oxygen, water temperature, specific conductance, and *Enterococci* showed higher variations. Pair-wise correlations were negligible between the six variables as shown in **Table 2**. This does not necessarily mean lack of spatial association. This also signifies the presence of redundant information in georeferenced data [13].

Figure 4 and **Figure 5** give the seasonal trends for dissolved oxygen, transparency, PH, water temperature,

Table 2. Pearson pair-wise correlations.

	Temperature	Transparency	Specific Conductance	Dissolved Oxygen	PH	*Enterococci*
Temperature	1.00	0.03	0.08	0.19	0.07	0.13
Transparency		1.00	0.06	0.08	0.03	0.08
Specific Conductance			1.00	0.18	−0.01	−0.03
Dissolved Oxygen				1.00	−0.02	0.14
PH					1.00	0.03
Enterococci						1.00

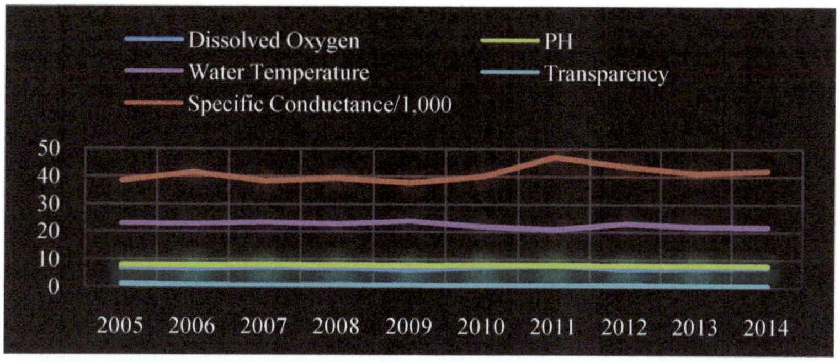

Figure 3. Temporal trends for dissolved oxygen, PH, water temperature, transparency and specific conductance.

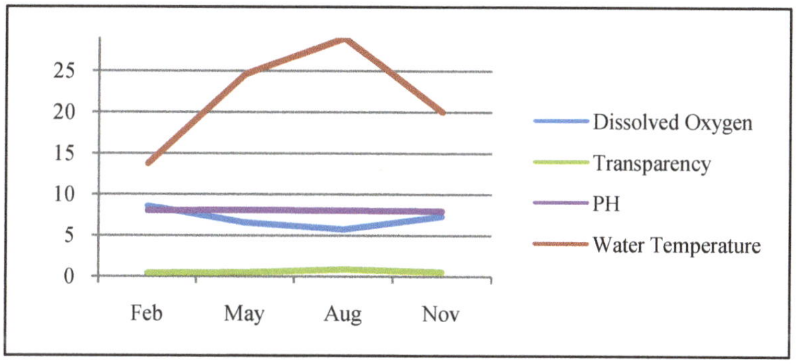

Figure 4. Seasonal trends for dissolved oxygen, PH, water temperature, and transparency.

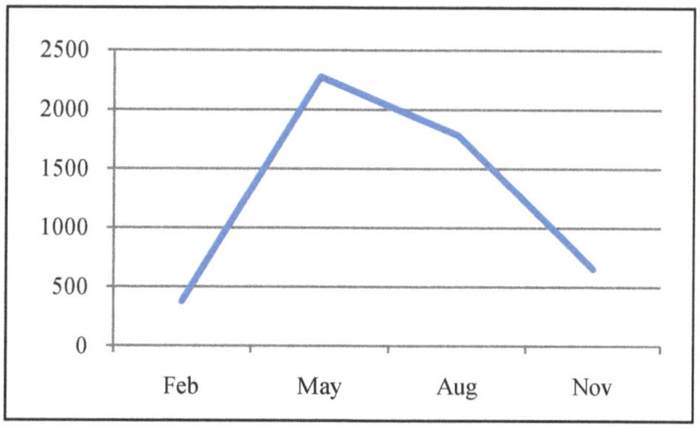

Figure 5. Seasonal trends for *Enterococci* (MPL/100 ml).

Table 3. Results of univariate spatial regressions on longitude and latitude.

Dependent Variable	R^2 (%)
Enterococci	2
Dissolved Oxygen	30
PH	20
Water Temperature	98
Specific Conductance	59

and *Enterococci*. The only observable trend is the decrease of oxygen solubility from May through August accompanied by increase in water temperature. This can be explained by decreased oxygen solubility as water temperature rises [6]. Because the number of transparency observations was way smaller than that for the other variables, it was excluded from subsequent analyses. Skewness of dissolved oxygen, PH, specific conductance, and water temperature were not too far from zero. Nevertheless, Shapiro test for these four variables produced a *p*-value less than the significance level of 0.05. Thus the null hypotheses that they came from normal distribution was rejected and a log transformation was applied. Moran I test for spatial autocorrelation showed that except for transparency, all other variables had positive spatial autocorrelation. That is, their observations cluster together spatially. Univariate regressions on longitude and latitude showed that *Enterococci* is predicted to increase ($R^2 = 0.02$) whereas dissolved oxygen ($R^2 = 0.30$), PH ($R^2 = 0.20$), water temperature ($R^2 = 0.98$), and specific conductance ($R^2 = 0.59$) are predicted to decrease (**Table 3**).

Although this study started with the purpose of quantifying water quality temporal and spatial trends in the Gulf of Mexico Basin in Texas, the results point to major issues that policy makers and researchers need to address. Future studies are recommended to investigate other important parameters like salinity, the investigation of multivariate geostatistical modeling, and the inclusion of more monitoring sites. These can help draw a bigger picture of the factors leading to water contamination and can provide interpretations for issues related to groundwater contamination. More importantly, this study points to lack of sufficient collaboration between risk mitigation planners and public health practitioners to connect the cases of bacterial infections to risks facing the communities especially stormwater flooding. Also, continuous assessment of potential parameters of water quality is recommended. Finally, multi-pollutant framework should be explored.

4. Conclusion

Different analyses were used to assess the temporal and spatial variations of water in the Gulf of Mexico Basin, Texas. Statistical analyses showed marginality of pair-wise associations, presence of positive spatial autocorrelation, minimal temporal variance for the parameters except for specific conductance and *Enterococci*, increased water temperature during the summer and fall, decreased levels of dissolved oxygen during the summer and fall seasons, expected increase in *Enterococci* and forecasted decrease in dissolved oxygen, PH, water temperature, and specific conductance. Despite the need to investigate other important parameters, the results of this study pointed to the need to have more integrated efforts to control storm water flooding in order to control the expected increase in *Enterococci* in surface water and consequently to increase the community's health and adaptive capacity.

References

[1] U.S. Environmental Protection Agency (1999) Action Plan for Beaches and Recreational Waters. EPA/600/R-98/079. Office of Water, U.S. Environmental Protection Agency, Washington DC.

[2] TCEQ (2007) Texas Water Quality Standards. http://www.bayoupreservation.org/html/DavenportOverviewSymposium.pdf

[3] Oshiro, R. and Fujioka, R. (1995) Soil, and Pigeon Droppings: Sources of Indicator Bacteria in the Waters of Hanauma Bay, Oahu, Hawaii. *Water Science & Technology*, **31**, 251-254. http://dx.doi.org/10.1016/0273-1223(95)00275-R

[4] Olapade, O.A., Depas, M.M., Jensen, E.T. and McLellan, S.L. (2006) Microbial Communities and Fecal Indicator

Bacteria Associated with *Cladophora* Mats on Beach Sites along Lake Michigan Shores. *Applied and Environmental Microbiology*, **72**, 1932-1938. http://dx.doi.org/10.1128/aem.72.3.1932-1938.2006

[5] Yoder, J.S., Blackburn, B.G., Craun, G.F., Hill, V., Levy, D.A., Chen, N., Lee, S.H., Calderon, R.L. and Beach, M.J. (2004) Surveillance for Waterborne-Disease Outbreaks Associated with Recreational Water—United States, 2001-2002. *MMWR Surveillance Summaries*, **53**, 1-22.

[6] USGS (2015) Groundwater Quality. http://water.usgs.gov/edu/earthgwquality.html

[7] Anderson, F. (2014) Multivariate Geostatistical Model for Groundwater Constituents in Texas. *International Journal of Geosciences*, **5**, 1609-1617. http://dx.doi.org/10.4236/ijg.2014.513132

[8] Anderson, F. and Al-Thani, N.J. (2015) Sustainability Atlas of Texas Ecoregions. *Journal of Human Resource and Sustainability Studies*, **3**, 203-210. http://dx.doi.org/10.4236/jhrss.2015.34026

[9] Anderson, F. and Al-Thani, N.J. (n.d.) Reviewing Methods of Assessing Community Adaptive Capacity: A Case Study from Jefferson County, Texas. in press.

[10] Ahmed, B. and Sinha, R.K. (2014) Rivers for Life: Proceedings of the International Symposium on River Biodiversity: Ganges-Brahmaputra-Meghna River System. IUCN, Nature.

[11] Fewtrell, L. and Bartram, J. (2001) Water Quality: Guidelines, Standards, and Health: Assessment of Risk and Risk Management for Water-related Infectious Disease. IWA Publishing.

[12] USGS (2015) Water Properties: Dissolved Oxygen. http://water.usgs.gov/edu/dissolvedoxygen.html

[13] Cressie, N. (1991) Statistics for Spatial Data. Wiley, New York.

30

Quantification of River Bank Erosion and Bar Deposition in Chowhali Upazila, Sirajganj District of Bangladesh: A Remote Sensing Study

Md. Shareful Hassan[1], Syed Mahmud-ul-islam[2]

[1]Center for Environmental Change Studies and Management, Dhaka, Bangladesh
[2]Department of Sustainable Development, Environmental Science and Engineering, KTH-Royal Institute of Technology, Stockholm, Sweden
Email: shareful@gmx.com

Abstract

River bank erosion is one of the frequent but the most unpredictable disasters that occur every year in Bangladesh. In this paper, Landsat TM-5 and Landsat-8 imageries from 1989 and 2015 were used to detect changes of present land use, river erosion and bar deposition in Chowhali Upazila, Sirajganj district of Bangladesh. This study reveals that human settlement, forest, seasonal crops and agriculture features decrease, while river coverage increases dramatically. About 1340 hectare areas have been eroded, while 630 hectares are deposited as channel bar in the study area over the last 26 years. Finally, an accuracy assessment is conducted between the test data and each land use feature. The overall classification accuracy was 97% and 98% in 1989 and 2015 respectively. Moreover, 98% accuracy is found in erosion while 97% is found in bar deposition areas.

Keywords

GIS, Remote Sensing, Landsat, River Erosion, Change Detection, Chowhali Bangladesh

1. Introduction

River bank erosion is one of the frequent but the most unpredictable disasters that occur every year in Bangladesh. It is a common geomorphological phenomenon in alluvial floodplain regions. Due to its regularity of every year, river erosion is often called a "silent disaster".

The Ganges, the Brahmaputra and the Meghna river systems comprise one of the biggest combined catchment basins of the world and create the biggest Bengal delta. Annually the river systems carry about two billion tons of sediment to Bengal Delta [1]. Numerous tributaries and connected distributaries of the major river systems are also vulnerable to the riverbank erosion process as well as flood effects. According to Islam and Islam (1985) around 2000-3000 kilometres of riverbank line face erosion annually [2]. Heavy siltation forms channel bars (Char) inside the river course and changes the direction of river flow. The mechanism creates stress on the bank of rivers and erosion occurs. In addition, heavy rainfall and sudden flooding also influence and contribute to river bank erosion.

The erosion process is very active in the Jamuna River, especially during the pre and post monsoon period and during flooding. Every year, large croplands and numerous habitation and infrastructure experience erosion. The impact of natural disasters aggravates normal livelihoods, poverty, public health, and vulnerability. Studies find that an increase of 10 percent maximum discharge of the Jamuna River generates approximately 25 percent increase in riverbank erosion [2]. The Jamuna River is a multi-channel braided river system that frequently develops sand bars and changes in river flow direction [3].

According to research, the Jamuna River embankment soil is made of poor grade sand with high siltation. The soil permeability is found moderately high with low strength properties that increases rapid submerge condition [4]. Earthen embankment does not properly protect the river bank area. In fact, it has been found that poorly earthen embankments of this river area are vulnerable to the impact of the Jamuna river water flow and heavy rainfall [1].

Geographic Information Systems (GIS) and Remote Sensing (RS) are essential tools for detecting changes along coastlines and riverbank erosion. Predictions can be made based on change detection results and trend analysis. Bedini (2007) uses Landsat imageries and GIS tools to detect changes along the coastline between the Semani and Shkumbini rivers and Central Albania and finds important changes [5]. His results detect significant erosion process at the Semani river delta, although the rate of erosion is not uniform. Gogoi (2013) conducts research on the Subansiri River in Assam, India using remote sensing and GIS technology [6]. He uses satellite imageries of 1995 and 2010 and reveals that the width of the Subansiri River has increased and that the erosion process has continued in both banks of the river.

A research work is recently carried out on riverbank erosion in Chandpur district, Bangladesh by using Landsat TM and MSS satellite imageries. According to that GIS based research, the bar deposition process is higher than the riverbank erosion process in the Meghna River in Chandpur. They also estimated that about 3517 square meter areas eroded during 1990 to 2002 [3]. Uddin (2011) conducts an assessment on morphological changes and riverbank erosion of the Jamuna River by using remote sensing [7]. In his research, he uses five different years of Landsat of ETM^+, TM and MSS images and reveals very high siltation rates and erosion results. According to his result, the siltation is 29.82 km^2 and average erosion alongside the Jamuna River is 1235.25 Km^2. Another GIS and remote sensing based investigation are conducted to examine erosion of the Jamuna River bank, where the researchers conclude that the average width of the river is increasing. The rate of increase is about 3 kilometres since last 40 years. They also calculate that average riverbank erosion is as 607 km^2 and average accretion is as 166 km^2 [8].

The main objectives of the study are as follows:

- To assess changes of land use patterns of the study area from 1989 to 2015.
- To estimate riverbank erosion and bar deposition areas in the study area.

2. Study Area

Chowhali Upazila is one of the most vulnerable regions of Bangladesh in terms of riverbank erosion. The Jamuna River along Chowhali Upazila has been suffering from riverbank erosion for long time. The research was conducted in the entire Upazila. This Upazila covers 2,1039 hectares area and located between 24°01' and 24°17'N latitudes and between 89°41'E and 89°59'E longitudes (**Figure 1**). The population of the Upazila is approximately 160,063 [9]. The Upazila is bounded by Shahjadpur and Belkuchi Upazilas on the west, Tangail Sadar and Nagarpur Upazila (Tangail) on the east.

3. Data Collection and Materials

To conduct this research, two multi-date Landsat imageries of 1989 (TM) and 2015 (OLI_TIRS) were used. All

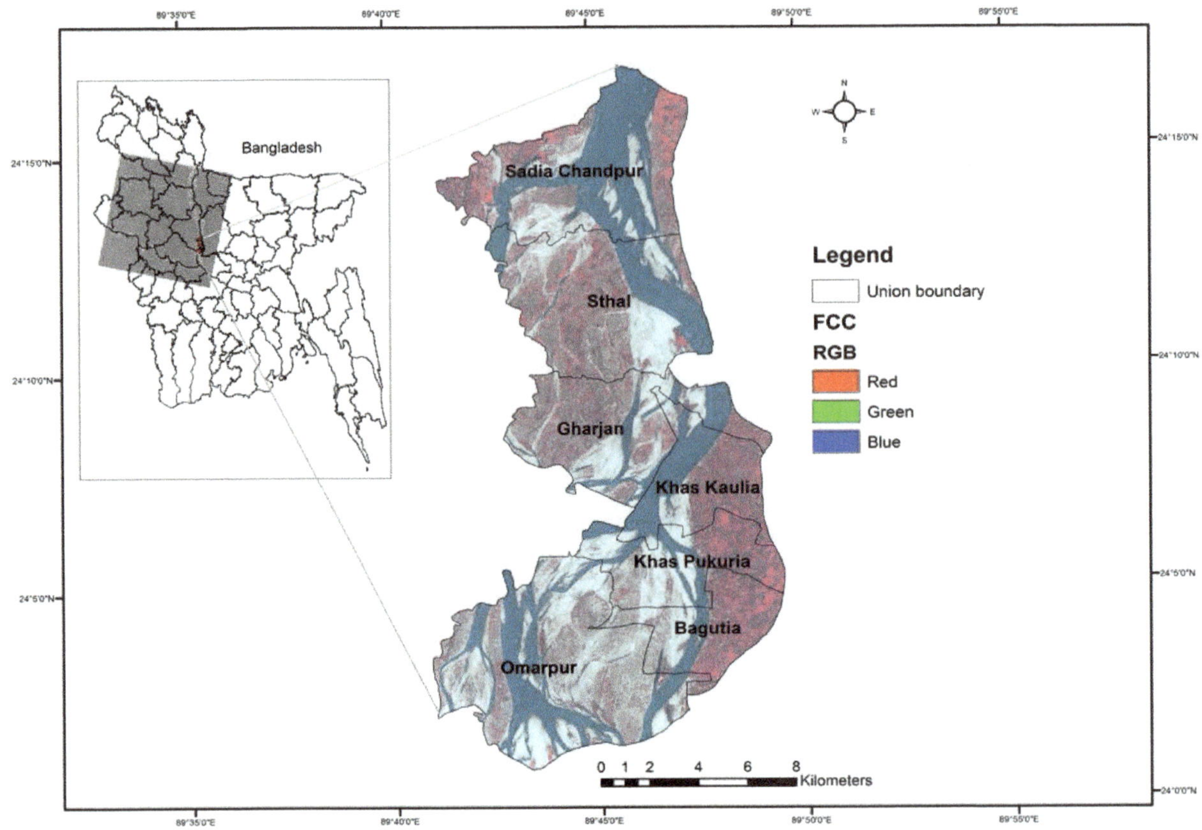

Figure 1. Location of the study area on Landsat False Colour Composite (FCC) image with union boundary shape file.

imageries used in this study were 30 meter resolution with WGS84 datum. A union shape file was collected from the Bangladesh Local Government Engineering Department (LGED) for masking the study area in order to follow digital image processing. ENVI v 4.7 and ArcGIS 10 were intensively used to complete all of the image processing tasks to generate land use, erosion and deposition maps. Moreover, Microsoft Excel was used to calculate areal information of land use, erosion and deposition of the study area. The Landsat imageries information used in the study are summarized in **Table 1**.

These Landsat imageries were freely downloaded from the University of Maryland's Global Land Cover Facility's Earth Science Data Interface (ESDI) website in order to classify landuse, river path, riverbank erosion and bar deposition assessment. In this study, Landsat images were used satisfactorily for the identification of area [10].

4. Methodology

The main methodology adopted in this study is presented in **Figure 2**. A 30 meter spatial resolution of Landsat imageries including blue, green, Red and NIR imageries were selected in order to classify landuse and its area statistics [11].

4.1. Image Processing

Digital image processing is required for satellite data classification. For mapping and extracting the whole features in this study, ENVI and ArcGIS software were used. ENVI software was used for digital image processing and analyzing, such as geometric correction, radiometric correction, atmospheric, noise removal, image clipping, classification and enhancement [10]. All sets of geocoded imageries were masked out by Upazila boundary. Supervised image classification method was used to classify land use features, erosion and deposition in the study. Post classification change detection method was also used in order to assess temporal changes of land use in the study area.

Table 1. Two multi-date image characteristics of Landsat image characteristics.

Landsat	Row/Path	Date of acquisition	Resolution (Meter)	Projection	Channels
Landsat 8-OLI_TIRS	043/138	2015-07-19	30	UTM/WGS 84	Blue, Green, RED, NIR and MIR
TM	044/138	1989-07-11	30	UTM/WGS 84	Blue, Green, RED, NIR and MIR

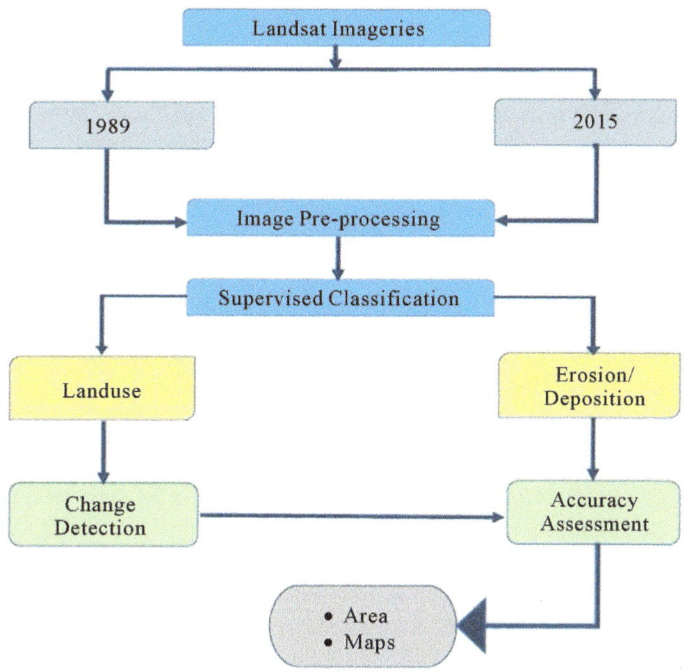

Figure 2. Methodology adopted for carrying out this study.

4.2. Regions of Interest (ROIs)

Regions of Interest (ROIs) are typically used to extract statistics for image classification, masking and other operations [12]. For conducting supervised image classification in this study, Regions of Interest (ROI) were required to classify remotely sensed imageries in a statistical manner. In this study, more than 4000 and 4500 pixels were selected as region of interest and classification accuracy for each feature respectively. Field investigation by GPS, local knowledge on the study area and high resolution google maps were referenced to collect region of interests in order to classify eight land features in this study. The information of each feature is summarized in **Table 2**.

4.3. Supervised Image Classification

Supervised image classification was performed to classify land use features. Supervised classification is a method in which the analyst defines small areas, called regions of interest, on the image which are representative of each desired land cover category [13]. To conduct this image classification, parametric maximum likelihood classifier was executed for all the images. This method assumes a normal distribution of DN values, allowing the function to determine the probability of a pixel belonging to a certain feature class and assign each pixel to the highest probability class [14] [15]. A set of regions of interest was extracted in order classify eight different land features, riverbank erosion and bar deposition in this study.

4.4. Change Detection

Change detection is an important issue for exploring changes of surface phenomena in the earth. Change detection is the process of identifying differences in the state of an object or phenomenon by observing it at different

Table 2. Training sites and pixels with sources used in this research.

Landuse class	Pixels		Sources
	ROIs	Accuracy	
River	4100	4790	GPS
Forest	4400	4678	GPS
Agriculture	4500	4876	GPS
Seasonal crops	4600	4567	GPS and Google Earth
Sand	4900	4786	GPS
Settlement	4300	4876	GPS and Google Earth
Erosion	4798	4656	GPS and community
Deposition	4280	4890	GPS and community

times [16] [17]. In this study, features extracted in 1989 were compared to 2015 classified features. To conduct this exercise, post-classification change detection statistics method was used. This method tabulates the total for each land use cover type and examines the trends between the years [17].

4.5. Accuracy Assessment

Each classified land feature was compared to the reference data in order to create an error matrix. An error matrix is a very effective way of representing accuracy in that the accuracies of each category are plainly described along with both the errors of inclusion (commission errors) and errors of exclusion (omission errors) present in the classification [18]. In this study, all reference data were collected from GPS and high resolution google maps. The whole error matrix, overall accuracy and kappa index were based on the Equations (1)-(3) below:

$$\text{Error matrix, } n = \sum_{i=1}^{K}\sum_{j=1}^{k} nij \quad (1)$$

$$\text{Overall Accuracy} = \frac{\sum_{i=1}^{k} nii}{n} \quad (2)$$

where, *nii* is diagonal elements in error matrix; *n* is total number of samples in error matrix

$$\text{Kappa coefficient, } K = \frac{n\sum_{i=1}^{k} nii - \sum_{i=1}^{k} ni + nj}{N2 - \sum_{i=1}^{k} ni + ni} \quad (3)$$

where, *i* is number of rows, columns in error matric; *N* is total number of observations in error matrix; *nij* is major diagonal element for class *i*; *ni* + is total number of observations in row *i* (right margin); *n* + *j* is total number of observations in column *i* (bottom margin).

5. Results and Discussions

5.1. Land Use Mapping

The land use maps from 1989 and 2015 of the study area were classified based on visual interpretation, local level knowledge and digital image processing methods. United States Geological Survey (USGS) land classification level-2 was also adopted, which is in line with the existing land use categories of the study area. Six different major land use classes were classified in this study (**Figure 3**).

From **Figure 3**, forest was decreased from 1551 to 874 hectares, while the river water coverage was increased from 3453 to 6220 hectares between 1989 to 2015. Due to drastic river erosion by the Jamuna River in the study area, destruction and resettlement in settlement feature was observed. Settlements increased by more than 50% over the study period mainly due to population growth and internal migration. However, settlement patterns

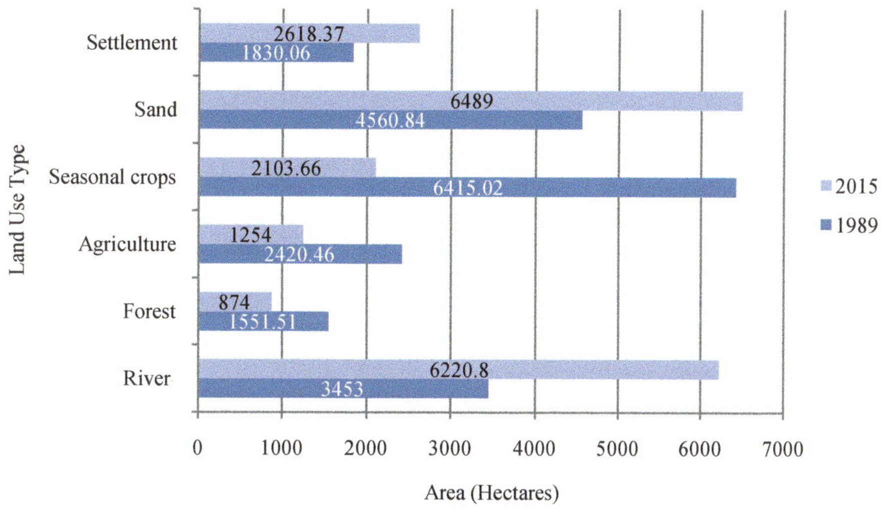

Figure 3. Post classification change detection statistics for assessing different spatial changes of the each land feature over the twenty six years.

were sporadic and non-static due to higher frequency of erosion. Seasonal crop areas have been decreased while sandy areas were expanded due to river channel migration.

5.2. Erosion and Deposition

Riverbank erosion and bar deposition is a common scenario in the study area. In this study, multi-date Landsat images were classified based on supervised method. The two resultant images were superimposed in order to demarcate union wise erosion and deposition areas. Local level consultation and key personnel interviews were conducted to verify classification accuracy. **Figure 4** shows that most of the eroded areas were found in the Sadia Chandpur, Sthal, KashKaulia, KashPukuria and Omarpur unions.

On the other hand, higher bar deposition areas were also found in Sthal, Gharjan, Kash Pukuria, Bagutia and Omarpur unions. The total area of erosion and deposition from the all unions were estimated based on pixels calculation of change detection method. The classified information extracted from Landsat imageries is in **Table 3**. This table reveals that about 1340 hectare areas have been eroded while 630 hectares were deposited in the study area over the twenty six years.

Pahlowan *et al.* (2015) researched the relationship of erosion and accretion in terms of water discharge in the Jamuna River [19]. They found that erosion is always greater than deposition, which is in line with **Table 3**. Moreover, their study shows a strong relationship between water discharge and erosion. It means that if there is more discharge in the river, then the possibility of erosion is higher.

5.3. Accuracy Assessment

Finally, an accuracy assessment was conducted between test data and each land use feature. The overall image classification accuracy was 97% and 98% in 1989 and 2015 respectively. Moreover, 98% accuracy was found in erosion while 97% accuracy was found in bar deposition areas. Poor spectral separability is found in seasonal crops and agriculture due to similar spectral profiles and location. On the other hand, a significant amount of settlements were mixed with forest spectral profiles, as most of the settlements were under forest canopy in the study area. The overall kappa statistics was 0.972 (97%) in landuse mapping while 0.984 (98%) was in erosion and deposition mapping.

6. Conclusion

This paper attempted to detect changes of land use, riverbank erosion and bar deposition in the study area by using multi-date Landsat imageries from 1989 and 2015. From this study, natural vegetation with forest, seasonal crops and agriculture features decrease while human settlement and river coverage are increased dramatically.

Table 3. Erosion and deposition information extracted fromLandsat imageries.

Year	Erosion (hectares)	Deposition (hectares)
1989	1950	1720
2015	3290	2530
Difference	−1340	−630

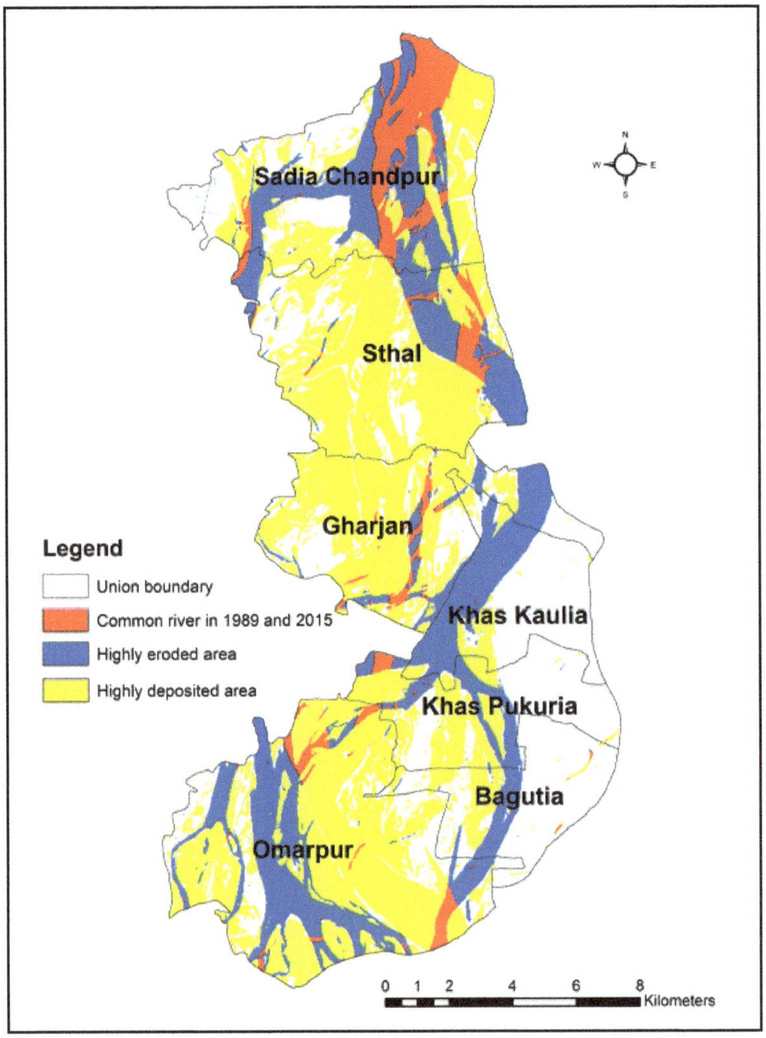

Figure 4. Union wise erosion and bar deposition areas in the study area.

About 1340 hectare areas have been eroded, whereas 630 hectares are deposited as channel bar in the study area over the last twenty-six years. To detect these changes, medium resolution (30 meters) of Landsat imageries are shown to be useful. Further research should be based on high-resolution images with field level ancillary data. Different social aspects should be included in order to enhance future planning related issues of the river management. Moreover, water discharge and rainfall data can be incorporated. Finally, this research will help to formulate any erosion and deposition related local river management as well as policy implications.

References

[1] Hossain, M.Z. and Sakai, T. (2008) Severity of Flood Embankments in Bangladesh and Its Remedial Approach. *Agricultural Engineering International*: *The CIGR E-Journal*, Manuscript LW 08 004, **X**, 1-11.

[2] Islam, M. and Islam, A. (1985) A Brief Account of Bank Erosion, Model Studies and Bank Protective Works in Bangladesh. *REIS Newsletter*, **2**, 11-13.

[3] Nath, B., Naznin, S.N. and Paul, A. (2013) Trends Analysis of River Bank Erosion at Chandpur, Bangladesh: A Remote Sensing and GIS Approach. *International Journal of Geomatics and Geosciences*, **3**, 454-463.

[4] Hossain, B.M., Sakai, T. and Hossain, Z.M. (2010) River Embankment and Bank Failure in Bangladesh: A Study on Geotechnical Characteristics and Stability Analysis. *Proceedings of International Conference on Environmental Aspects of Bangladesh* (*ICEAB*10), Kitakyushu, September 2010, 171-174.

[5] Bedini, E. (2007) Use of GIS and Remote Sensing to Detect Change along the Coastline Segment between Shkumbini and Semani Rivers, Central Albania. *Proceedings of the* 11*th International Congress Bulletin of the Geological Society of Greece*, XXXX, Athens, 24-26 May 2007, 1916-1924

[6] Gogoi, C. and Goswami, D.C. (2013) A Study on Bank Erosion and Bank Line Migration Pattern of the Subansiri River in Assam Using Remote Sensing and GIS Technology. *The International Journal of Engineering and Science* (*IJES*), **2**, 1-6.

[7] Uddin, K., Shrestha, B. and Alam, M.S. (2011) Assessment of Morphological Changes and Vulnerability of River Bank Erosion alongside the River Jamuna Using Remote Sensing. *Journal of Earth Science and Engineering*, **1**, 29-34.

[8] Khan, I., Ahammad, M. and Sarker, S. (2014) A Study on River Bank Erosion of Jamuna River Using GIS and Remote Sensing Technology. *International Journal of Engineering Development and Research*, **2**, 3365-3371.

[9] BBS (Bangladesh Bureau of Statistics) (2012) Community Report Sirajganj Zila June 2012. Population and Housing Census 2011, Statistics and Informatics Division, Ministry of Planning, Dhaka.

[10] Abd, H.A.A. and Alnajjar, H.A. (2013) Maximum Likelihood for Land-Use/Land-Cover Mapping and Change Detection Using Landsat Satellite Images: A Case Study "South Of Johor". *International Journal of Computational Engineering Research*, **3**, 26-33.

[11] Hassan, M.S. and Mahmud-ul-Islam, S. (2014) Identification of Wetland Restoration Areas of Chalan Beel inSirajganj District, Bangladesh Using Integrated GIS and Remote Sensing. *New York Science Journal*, **7**, 94-100.

[12] ENVI (2015) How to Create ROIs in ENVI 5. EXELIS: Visual Information Solutions.
http://www.exelisvis.com/Support/HelpArticles/TabId/185/ArtMID/800/ArticleID/5244/5244.aspx

[13] Rani, M., Kumar, P., Yadav, M. and Hooda, R.S. (2011) Wetland Assessment and Monitoring Using Image Processing Techniques: A Case Study of Ranchi, India. *Journal of Geographic Information System*, **3**, 345-350.
http://dx.doi.org/10.4236/jgis.2011.34032

[14] Lillesand, T.M., Kieffer, W.R. and Chipman, J.W. (2004) Remote Sensing and Image Interpretation. 5th Edition, John Wiley & Sons Inc., New York, 763.

[15] Lee, S. (2011) Detecting Wetland Change through Supervised Classification of Landsat Satellite Imagery within the Tunkwa Watershed of British Columbia, Canada. Bachelor Degree Project in Geomatics, Hogskolan I Gavle, Gävle.

[16] Singh, A. (1989) Digital Change Detection Techniques Using Remotely-Sensed Data. *International Journal of Remote Sensing*, **10**, 989-1000. http://dx.doi.org/10.1080/01431168908903939

[17] Nori, W., Elsiddig, E.N. and Niemeyer, I. (2008) Detection of Land Cover Changes Using Multi-Temporal Satellite Imagery. *The International Archives of the Photogrammetry, Remote Sensing and Spatial Information Sciences*, **37**, 947-952.

[18] Congalton, R.G. (1991) A Review of Assessing the Accuracy of Classifications of Remotely Sensed Data. *Remote Sensing of Environment*, **37**, 35-46. http://dx.doi.org/10.1016/0034-4257(91)90048-B

[19] Pahlowan, E.U. and Hossain, A.T.M.S. (2015) Jamuna River Erosional Hazards, Accretion & Annual Water Discharge—A Remote Sensing & GIS Approach. *Proceedings of the* 36*th International Symposium on Remote Sensing of Environment*, Berlin, 11-15 May 2015, 831-835. http://dx.doi.org/10.5194/isprsarchives-xl-7-w3-831-2015

Permissions

All chapters in this book were first published in GEP, by Scientific Research Publishing; hereby published with permission under the Creative Commons Attribution License or equivalent. Every chapter published in this book has been scrutinized by our experts. Their significance has been extensively debated. The topics covered herein carry significant findings which will fuel the growth of the discipline. They may even be implemented as practical applications or may be referred to as a beginning point for another development.

The contributors of this book come from diverse backgrounds, making this book a truly international effort. This book will bring forth new frontiers with its revolutionizing research information and detailed analysis of the nascent developments around the world.

We would like to thank all the contributing authors for lending their expertise to make the book truly unique. They have played a crucial role in the development of this book. Without their invaluable contributions this book wouldn't have been possible. They have made vital efforts to compile up to date information on the varied aspects of this subject to make this book a valuable addition to the collection of many professionals and students.

This book was conceptualized with the vision of imparting up-to-date information and advanced data in this field. To ensure the same, a matchless editorial board was set up. Every individual on the board went through rigorous rounds of assessment to prove their worth. After which they invested a large part of their time researching and compiling the most relevant data for our readers.

The editorial board has been involved in producing this book since its inception. They have spent rigorous hours researching and exploring the diverse topics which have resulted in the successful publishing of this book. They have passed on their knowledge of decades through this book. To expedite this challenging task, the publisher supported the team at every step. A small team of assistant editors was also appointed to further simplify the editing procedure and attain best results for the readers.

Apart from the editorial board, the designing team has also invested a significant amount of their time in understanding the subject and creating the most relevant covers. They scrutinized every image to scout for the most suitable representation of the subject and create an appropriate cover for the book.

The publishing team has been an ardent support to the editorial, designing and production team. Their endless efforts to recruit the best for this project, has resulted in the accomplishment of this book. They are a veteran in the field of academics and their pool of knowledge is as vast as their experience in printing. Their expertise and guidance has proved useful at every step. Their uncompromising quality standards have made this book an exceptional effort. Their encouragement from time to time has been an inspiration for everyone.

The publisher and the editorial board hope that this book will prove to be a valuable piece of knowledge for researchers, students, practitioners and scholars across the globe.

List of Contributors

A. V. Lukonina
Botanic Garden of "Volgograd State Socio-Pedagogical University", Volgograd, Russia

G. Y. Klinkova
Chair of Park-and-Garden and Landscape Construction of "Volgograd State Socio-Pedagogical University", Volgograd, Russia

N. B. Lopantzeva
Lower Volga and Northern Caspian Region of UNDP/GEF-MNR RF Project "Mainstreaming Biodiversity Conservation into Russia's Energy Sector Policies and Operations", Volgograd, Russia

Peimin Pu
Nanjing Institute of Geography & Limnology, Academia Sinica, Nanjing, China

Jiangping Pu
PLA University of Science and Technology, Nanjing, China

Xuewen Zhou and Hongliang Wang
School of Energy Resources, China University of Geosciences, Beijing, China

Taisiya Bardakhanova, Svetlana Ayusheeva and Zinaida Goryunova
Laboratory of Nature Management Economics, Baikal Institute for Nature Management, Siberian Branch of the Russian Academy of Sciences, Ulan-Ude, Russia

Shivendra Verma, Ramsingh Meena and Anurag Mudgal
Department of Mechanical Engineering, School of Technology, Pandit Deendayal Petroleum University, Gandhinagar, India

Gideon Okpoti Tetteh and Maurice Schönert
BlackBridge AG, Kurfürstendamm 22, Berlin, Germany

Ousmane Laminou Manzo
Department of Agricultural Engineering and Water & Forests, Faculty of Agronomy and Environmental Sciences, Dan Dicko Dankoulodo University of Maradi, Maradi, Niger

Hassidou Saidou
Department of Chemistry, Faculty of Science and Technology, Dan Dicko Dankoulodo University of Maradi, Maradi, Niger

Salamatou Abdourahamane Illiassou and Saoudé Tidjani Idrissa
Department of Biology, Faculty of Science and Technology, Dan Dicko Dankoulodo University of Maradi, Maradi, Niger

Mohammed Abdullah Alghobar and Sidduraiah Suresha
Department of Environmental Science, Yuvaraja's College, University of Mysore, Mysore, India

D. Ciszewski and P. Bijata
AGH-University of Sciences and Technology, Kraków, Poland

Lei Ye, Jianzhong Zhou, Xiaofan Zeng and Muhammad Tayyab
College of Hydropower and Information Engineering, Huazhong University of Science and Technology, Wuhan, China
Hubei Key Laboratory of Digital Valley Science and Technology, Huazhong University of Science and Technology, Wuhan, China

Thangasamy Jeyavel Raja Kumar, Chinadurai Dushiyanthan, Boopalan Thiruneelakandan, Radhakrishnan Suresh, Shanmugam Vasanth Raja and Mailare Senthilkumar
Department of Earth Sciences, Annamalai University, Chidambaram, India

Ch. Ramesh Naidu
GVP College of Engineering (Autonomous), Visakhapatnam, India

M. V. S. S. Giridhar
Jawaharlal Nehru Technological University, Hyderabad, India

Adeola Alex Adesuyi
Environmental Impact Assessment, Environment Department, Shell Petroleum Development Company, Port Harcourt, Nigeria
Environmental Biology Unit, Cell Biology and Genetics Department, University of Lagos, Lagos, Nigeria

Moses Okafor Ngwoke
Environmental Impact Assessment, Environment Department, Shell Petroleum Development Company, Port Harcourt, Nigeria

Modupe Olatunde Akinola, Kelechi Longinus Njoku and Anuoluwapo Omosileola Jolaoso
Environmental Biology Unit, Cell Biology and Genetics Department, University of Lagos, Lagos, Nigeria

Vidula A. Swami and Sushama S. Kulkarni
Rajarambapu Institute of Technology, Sangli, India

Justin Kpagh, Rufus Sha'Ato, Raymond A. Wuana and Terrumun A. Tor-Anyiin
Department of Chemistry, University of Agriculture, Makurdi, Nigeria

Qingfeng Chen, Junjian Ma, Changsheng Zhao and Rongbin Li
Key Laboratory for Applied Technology of Sophisticated Analytical Instruments of Shandong Province, Shandong Provincial Analysis Test Center, Jinan, China

Ola Busari and Jeremiah Mutamba
Trans-Caledon Tunnel Authority, Centurion, South Africa

Daniel Hervé Gouet
Department of Petroleum and Gas Explorations, Institute of Mines and Petroleum Industries, University of Maroua, Maroua, Cameroun

Arsène Meying
Department of Applied Geophysics, Geology and Mining Exploitation College, University of Ngaoundéré, Ngaoundéré, Cameroun

Stéphane Patrick Assembe
Department of Physics, Faculty of Science, University of Yaounde I, Yaoundé, Cameroon

Théophile Ndougsa-Mbarga
Department of Physics, Advanced Teacher's Training College, University of Yaounde I, Yaoundé, Cameroon

Kehinde T. Oyatayo, Godwin A. Songu and Greatest A. Amos
Department of Geography, Kwararafa University, Wukari, Nigeria

Christopher Ndabula
Department of Geography and Regional Planning, Federal University, Dutsin-Ma, Nigeria

Mahmoud Abd El-Kawy Zahran and Yasser Ahmed El-Amier
Botany Department, Faculty of Science, Mansoura University, Mansoura, Egypt

Abdelhamid Ahmed Elnaggar
Soils Department, Faculty of Agriculture, Mansoura University, Mansoura, Egypt

Hoda Abd El-Azim Mohamed and Muhammad Abd El-Hady El-Alfy
Marine Pollution Department, National Institute of Oceanography and Fisheries, Suez, Egypt

Christian Opp
Department of Hydrology and Soil Science, Faculty of Geography, University of Marburg, Marburg, Germany

Bastian Ziebolz
KLT Consult, Hannover/Soest, Germany

Sanit Wongsa
Department of Civil Technology Education, King Mongkut's University of Technology Thonburi (KMUTT), Bangkok, Thailand

Zhiyi Lei
College of Harbour, Coastal and Offshore Engineering, Hohai University, Nanjing, China

Isaiah O. Opadokun, A. E. Falaye and E. K. Ajani
Department of Aquaculture and Fisheries Management, University of Ibadan, Ibadan, Nigeria

Suman Man Shrestha and Kedar Rijal
Central Department of Environmental Science, Tribhuvan University, Kirtipur, Kathmandu, Nepal

Megh Raj Pokhrel
Central Department of Chemistry, Tribhuvan University, Kirtipur, Kathmandu, Nepal

Christian Opp
Department of Hydrology and Soil Science, Faculty of Geography, University of Marburg, Marburg, Germany

Jens Hahn
Department of Geography, Faculty of Mathematic and Natural Science, University of Koblenz-Landau, Koblenz, Germany

Nina Zitzer and Gabriela Laufenberg
Mass Spectrometry, Faculty of Pharmacy, University of Marburg, Marburg, Germany

Rajendra Bhandari and Subodh Sharma
Department of Environmental Science and Engineering, Kathmandu University, Dhulikhel, Nepal

Juerg Merz and Robin Garton
The Glacier Trust, Wiltshire, UK

Mourad Khattati, Mostapha Serroukh, Ismail Rafik and Hakim Mesmoudi
Research Team ERGO, Faculty of Science, Abdelmalek Essaadi University, Tetouan, Morocco

Brirhet Hassane
Faculty of Science and Technology, Sidi Mohammed Ben Abdellah University, Fes, Morocco

Yassine Bouslihim
Faculty of Science and Technology, 1st Hassan University, Settat, Morocco

Faye Anderson and Najla N. J. Al-Thani
Center for Advances in Water and Air Quality, Lamar University, Beaumont, TX, USA

Md. Shareful Hassan
Center for Environmental Change Studies and Management, Dhaka, Bangladesh

Syed Mahmud-ul-islam
Department of Sustainable Development, Environmental Science and Engineering, KTH-Royal Institute of Technology, Stockholm, Sweden